Angular Cookbook

Over 80 actionable recipes every Angular developer
should know

Muhammad Ahsan Ayaz

Packt>

BIRMINGHAM—MUMBAI

Angular Cookbook

Copyright © 2021 Packt Publishing

Associate Group Product Manager: Pavan Ramchandani
Publishing Product Manager: Pavan Ramchandani
Senior Editor: Sofi Rogers
Content Development Editor: Rakhi Patel
Technical Editor: Shubham Sharma
Copy Editor: Safis Editing
Project Coordinator: Manthan Patel
Proofreader: Safis Editing
Indexer: Tejal Soni
Production Designer: Jyoti Chauhan

First published: July 2021

Production reference: 2280721

Published by Packt Publishing Ltd.
Livery Place
35 Livery Street
Birmingham
B3 2PB, UK.

ISBN 978-1-83898-943-9

www.packt.com

To my mother, Zahida Khatoon, and my father, Muhammad Ayaz, for their prayers and sacrifices, and for exemplifying the power of determination in raising me. To my Nani (grandmother), Aarif-un-Nisa Begum, for tons of prayers for my success. And to my wife, Najla Obaid, for being my loving partner throughout our joint life journey.

– Muhammad Ahsan Ayaz

Foreword

I have known and worked with Ahsan for more than 10 years. Ahsan is one of the global Angular community's leading experts. He is very passionate about serving the developer community and having an impact on the world by training individuals to help make the world a better place through software development. He has given tons and tons of talks and sessions around the globe about JavaScript, Angular, and web development, and he actively creates video tutorials that he uploads to his YouTube channel along with writing articles on his website. This book is his most recent effort to make an impact and train more people, and to help them to learn and grow as Angular developers.

Ahsan has written this comprehensive book as a *tour de force* in which he takes his readers on a journey of understanding the core concepts of Angular and how to implement unit and end-to-end tests in Angular apps. The recipes vary from covering template-driven and reactive forms to exploring how to create your very own custom form controls. What's more, you'll not only learn about things such as Angular animations, the Angular router, and state management with NgRx but also dive into some amazing tools and APIs from the Angular CDK. And, last but not least, you'll build something that the industry has been talking about for years, **Progressive Web Apps (PWAs)**, with Angular.

In short, Ahsan has transformed years of learning and experience to create this book. The book gives you the experience of real-life scenarios and their technical solutions in the form of recipes. This book's content is practical, precise, and well explained.

Having known Ahsan for so long, I can say that he dedicates all his strength and ability to doing the best that he can do when he decides to do something, and this book is no exception. And with the recipes, the source code, and the way Ahsan takes you through the content, you'll certainly learn a lot and will enhance yourself as a software engineer and an Angular developer.

Najla Obaid

Business analyst at IOMechs

Contributors

About the author

Muhammad Ahsan Ayaz is a Google Developer Expert in Angular and a software architect at Syncron. He has taught programming worldwide for the past 8 years through articles, video content, one-to-one mentoring, and tech talks at different global conferences. He has developed several libraries and plugins used by hundreds of thousands of developers, including `ngx-device-detector`, which has over 5 million installs and over 2,000 GitHub projects using it. He blogs at `https://ahsanayaz.com` and has a YouTube channel at `https://ahsanayaz.com/youtube`, where he regularly uploads video tutorials related to web and mobile app development. Apart from programming, Ahsan likes to travel and play multiplayer video games. He lives in Sweden with his wife.

I want to thank the people who have been close to me and have supported me throughout all these years, especially my parents (Zahida Khatoon and Muhammad Ayaz) and my wife, Najla. I would also like to thank the readers of this book and other books that I'll write in the future, and the people who follow me on my social media profiles and who are subscribed to my YouTube channel. Thank you very much!

About the reviewer

Pawel Czekaj has bachelor's degree in computer science. He has 12 years of experience as a frontend developer. He currently works as a lead frontend developer at Ziflow Ltd. His expertise is in AngularJS, Angular, Amazon Web Services, Auth0, NestJS, and others. He is currently building enterprise-level proofing solutions based fully on Angular.

Table of Contents

Preface

1

Winning Components Communication

2

Understanding and Using Angular Directives

3

The Magic of Dependency Injection in Angular

4
Understanding Angular Animations

5

Angular and RxJS – Awesomeness Combined

6

Reactive State Management with NgRx

7

Understanding Angular Navigation and Routing

8

Mastering Angular Forms

9
Angular and the Angular CDK

12
Performance Optimizations in Angular

13
Building PWAs with Angular

Other Books You May Enjoy

Index

Preface

Angular is one of the most popular frameworks in the world for building not only web applications but even mobile and desktop applications as well. Backed by Google and used by Google, this framework is used by millions of applications. Although the framework is well suited for any scale of application, enterprises especially like Angular because of it being opinionated and because of its consistent ecosystem that includes all the tools you need to create a web technologies-based application.

While learning the core technologies such as JavaScript, HTML, and CSS is an absolute essential to progress as a web developer, when it comes to a framework, learning the core concepts of the framework itself is pretty important too. When we're dealing with Angular, there are a lot of amazing things we can do with our web applications by learning about, and using, the right tools in the Angular ecosystem. That's where this book comes in.

This book was written for intermediate and advanced Angular developers to polish their Angular development skills with recipes that you can follow easily, play around with, and practice your own variations of. You'll not only learn from the recipes themselves but also from the actual real-life projects associated with the recipes. So, there are a lot of hidden gems in these recipes and projects for you.

Happy coding!

Who this book is for

The book is for intermediate-level Angular web developers looking for actionable solutions to common problems in Angular enterprise development. Mobile developers using Angular technologies will also find this book useful. Working experience with JavaScript and TypeScript is necessary to understand the topics covered in this book more effectively.

What this book covers

Chapter 1, Winning Components Communication, explains different techniques to use to implement communication between components in Angular. @Input() and @ Output() decorators, services, and lifecycle hooks are covered as well. There is also a recipe for how to create a dynamic Angular component.

Chapter 2, Understanding and Using Angular Directives, gives an introduction to Angular directives and some recipes that use Angular directives, including attribute directives and structural directives.

Chapter 3, The Magic of Dependency Injection in Angular, includes recipes that cover optional dependencies, configuring an injection token, using the providedIn: 'root' metadata for Angular services, value providers, and aliased class providers.

Chapter 4, Understanding Angular Animations, contains recipes for implementing multi-state animations, staggering animations, keyframe animations, and animations for switching between routes in your Angular apps.

Chapter 5, Angular and RxJS – Awesomeness Combined, covers recipes on RxJS instance and static methods. It also has some recipes on the usage of the combineLatest, flatMap, and switchMap operators and covers some tips and tricks about using RxJS streams.

Chapter 6, Reactive State Management with NgRx, has recipes concerning the famous NgRX library and its core concepts. It covers core concepts such as NgRx actions, reducers, selectors, and effects and looks at using packages such as @ngrx/store-devtools and @component/store.

Chapter 7, Understanding Angular Navigation and Routing, explores recipes on lazily loaded routes, route guards, preloading route strategies, and some interesting techniques to be used with the Angular router.

Chapter 8, Mastering Angular Forms, covers recipes for template-driven forms, reactive forms, form validation, testing forms, and creating your own form control.

Chapter 9, Angular and the Angular CDK, has a lot of cool Angular CDK recipes, including ones on virtual scroll, keyboard navigation, the overlay API, the clipboard API, CDK drag and drop, the CDK stepper API, and the CDK textfield API.

Chapter 10, Writing Unit Tests in Angular with Jest, covers recipes for unit testing with Jest, exploring global mocks in Jest, mocking services/child components/pipes, using Angular CDK component harnesses, and unit testing Observables.

Chapter 11, E2E Tests in Angular with Cypress, has recipes on E2E testing with Cypress in Angular apps. It covers validating forms, waiting for XHR calls, mocking HTTP call responses, using bundled packages with Cypress, and using fixtures in Cypress.

Chapter 12, Performance Optimizations in Angular, contains some cool techniques to improve an Angular app's performance by using the OnPush change detection strategy, lazily loading feature routes, detaching the change detector from a component, using web workers with Angular, using pure pipes, adding performance budgets to an Angular app, and using the `webpack-bundle` analyzer.

Chapter 13, Building PWAs with Angular, contains recipes to create a PWA with Angular. It covers specifying a theme color for the PWA, using a device's dark mode, providing a custom PWA install prompt, precaching requests using Angular's service worker, and using App Shell.

To get the most out of this book

The recipes of this book are built with Angular v12 and Angular follows semantic versioning for their releases. Since Angular is constantly being improved, for the sake of stability, the Angular team has provided a predictable release cycle for updates. The release frequency is as follows:

- A major release every 6 months.

- 1 to 3 minor releases for each major release.

- A patch release and pre-release (next or rc) build almost every week.

Source: `https://angular.io/guide/releases#release-frequency`

Software covered in the book	Operating system requirements
Angular	Windows, macOS, or Linux
TypeScript 4.2.4+	
ECMAScript 11	

If you are using the digital version of this book, we advise you to type the code yourself or access the code from the book's GitHub repository (a link is available in the next section). Doing so will help you avoid any potential errors related to the copying and pasting of code.

Once you've finished reading the book, make sure to tweet to `https://ahsanayaz.com/twitter` to let me know your feedback about the book. In addition, you can modify the code provided with this book to your taste, upload it to your GitHub repository, and share it. I'll make sure to retweet it :)

Download the example code files

You can download the example code files for this book from GitHub at `https://github.com/PacktPublishing/Angular-Cookbook`. If there's an update to the code, it will be updated in the GitHub repository.

We also have other code bundles from our rich catalog of books and videos available at `https://github.com/PacktPublishing/`. Check them out!

Download the color images

We also provide a PDF file that has color images of the screenshots and diagrams used in this book. You can download it here: `https://static.packt-cdn.com/downloads/9781838989439_ColorImages.pdf`.

Conventions used

There are a number of text conventions used throughout this book.

`Code in text`: Indicates code words in text, database table names, folder names, filenames, file extensions, pathnames, dummy URLs, user input, and Twitter handles. Here is an example: "Now, we'll move the code from the `the-amazing-list-component.html` file to the `the-amazing-list-item.component.html` file for the item's markup."

A block of code is set as follows:

```
openMenu($event, itemTrigger) {
    if ($event) {
        $event.stopImmediatePropagation();
    }
    this.popoverMenuTrigger = itemTrigger;
    this.menuShown = true;
}
```

When we wish to draw your attention to a particular part of a code block, the relevant lines or items are set in bold:

```
.menu-popover {
  ...
  &::before {...}

  &--up {
    transform: translateY(-20px);
    &::before {
      top: unset !important;
      transform: rotate(180deg);
      bottom: -10px;
    }
  }

  &__list {...}
}
```

Bold: Indicates a new term, an important word, or words that you see onscreen. For instance, words in menus or dialog boxes appear in **bold**. Here is an example: "You will notice that we can't see the entirety of the content of the input—this is somewhat annoying at the best of times because you can't really review it before pressing the **Action** button."

> **Tips or important notes**
> Appear like this.

Get in touch

Feedback from our readers is always welcome.

General feedback: If you have questions about any aspect of this book, email us at customercare@packtpub.com and mention the book title in the subject of your message.

Errata: Although we have taken every care to ensure the accuracy of our content, mistakes do happen. If you have found a mistake in this book, we would be grateful if you would report this to us. Please visit www.packtpub.com/support/errata and fill in the form.

Piracy: If you come across any illegal copies of our works in any form on the internet, we would be grateful if you would provide us with the location address or website name. Please contact us at copyright@packt.com with a link to the material.

If you are interested in becoming an author: If there is a topic that you have expertise in and you are interested in either writing or contributing to a book, please visit authors.packtpub.com.

Share Your Thoughts

Once you've read *Angular Cookbook*, we'd love to hear your thoughts! Scan the QR code below to go straight to the Amazon review page for this book and share your feedback.

https://packt.link/r/1838989439

Your review is important to us and the tech community and will help us make sure we're delivering excellent quality content.

1
Winning Components Communication

In this chapter, you'll master component communication in Angular. You'll learn different techniques to establish communication between components and will learn which technique is suitable in which situation. You'll also learn how to create a dynamic Angular component in this chapter.

The following are the recipes we're going to cover in this chapter:

- Components communication using component `@Input(s)` and `@Output(s)`
- Components communication using services
- Using setters for intercepting input property changes
- Using `ngOnChanges` to intercept input property changes
- Accessing a child component in a parent template via template variables
- Accessing a child component in a parent component class using `ViewChild`
- Creating your first dynamic component in Angular

Technical requirements

For the recipes in this chapter, make sure you have **Git** and **Node.js** installed on your machine. You also need to have the @angular/cli package installed, which you can do with npm install -g @angular/cli from your terminal. The code for this chapter can be found at https://github.com/PacktPublishing/Angular-Cookbook/tree/master/chapter01.

Components communication using component @Input(s) and @Output(s)

You'll start with an app with a parent component and two child components. You'll then use Angular @Input and @Ouput decorators to establish communication between them using attributes and EventEmitter(s).

Getting ready

The project that we are going to work with resides in chapter01/start_here/ cc-inputs-outputs inside the cloned repository:

1. Open the project in Visual Studio Code.

2. Open the terminal and run npm install to install the dependencies of the project. Once done, run ng serve -o.

   This should open the app in a new browser tab and you should see the following:

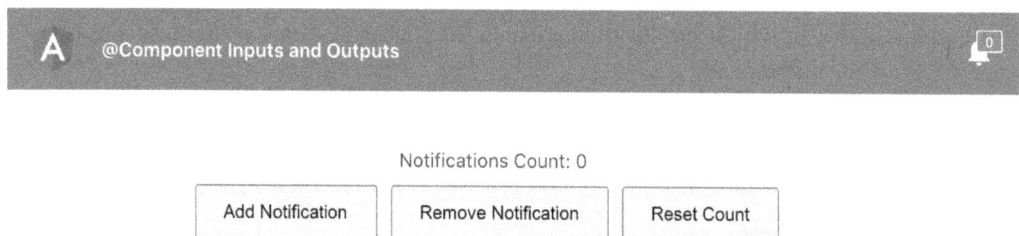

Figure 1.1 – The cc-inputs-outputs app running on http://localhost:4200

How to do it...

So far, we have an app with `AppComponent`, `NotificationsButtonComponent`, and `NotificationsManagerComponent`. While `AppComponent` is the parent of the other two components mentioned, there is absolutely no component communication between them to sync the notification count value. Let's establish the appropriate communication between them using the following steps:

1. We'll move the `notificationsCount` variable from `NotificationsManagerComponent` and host it in `AppComponent`. To do so, just create a `notificationsCount` property in `app.component.ts`:

    ```
    export class AppComponent {
      notificationsCount = 0;
    }
    ```

2. Next, convert the `notificationsCount` property in `notifications-manager.component.ts` to `@Input()`, rename it to `count`, and replace its usages as follows:

    ```
    import { Component, OnInit, Input } from '@angular/core';
    @Component({
      selector: 'app-notifications-manager',
      templateUrl: './notifications-manager.component.html',
      styleUrls: ['./notifications-manager.component.scss']
    })
    export class NotificationsManagerComponent implements
    OnInit {
      @Input() count = 0
      constructor() { }
      ngOnInit(): void {
      }
      addNotification() {
        this.count++;
      }
      removeNotification() {
        if (this.count == 0) {
          return;
        }
        this.count--;
    ```

```
    }
    resetCount() {
      this.count = 0;
    }
  }
```

3. Update notifications-manager.component.html to use count instead of notificationsCount:

    ```
    <div class="notif-manager">
      <div class="notif-manager__count">
        Notifications Count: {{count}}
      </div>
      ...
    </div>
    ```

4. Next, pass the notificationsCount property from app.component.html to the <app-notifications-manager> element as an input:

    ```
    <div class="content" role="main">
      <app-notifications-manager
        [count]="notificationsCount">
      </app-notifications-manager>
    </div>
    ```

    You could now test whether the value is being correctly passed from app.component.html to app-notifications-manager by assigning the value of notificationsCount in app.component.ts as 10. You'll see that in NotificationsManagerComponent, the initial value shown will be 10:

    ```
    export class AppComponent {
      notificationsCount = 10;
    }
    ```

5. Now, create an @Input() in notifications-button.component.ts named count as well:

    ```
    import { Component, OnInit, Input } from '@angular/core';
    ...
    export class NotificationsButtonComponent implements
    OnInit {
    ```

```
@Input() count = 0;
...
}
```

6. Pass `notificationsCount` to `<app-notifications-button>` as well from `app.component.html`:

```
<!-- Toolbar -->
<div class="toolbar" role="banner">
  ...
  <span>@Component Inputs and Outputs</span>
  <div class="spacer"></div>
  <div class="notif-bell">
    <app-notifications-button
    [count]="notificationsCount">
    </app-notifications-button>
  </div>
</div>
...
```

7. Use the `count` input in `notifications-button.component.html` with the notification bell icon:

```
<div class="bell">
  <i class="material-icons">notifications</i>
  <div class="bell__count">
    <div class="bell__count__digits">
      {{count}}
    </div>
  </div>
</div>
```

You should now see the value 10 for the notification bell icon count as well.

Right now, if you change the count by adding/removing notifications from `NotificationsManagerComponent`, *the count on the notification bell icon won't change.*

8. To communicate the change from NotificationsManagerComponent to NotificationsButtonComponent, we'll use Angular @Output(s) now. Use @Ouput and @EventEmitter from '@angular/core' inside notifications-manager.component.ts:

```
import { Component, OnInit, Input, Output, EventEmitter }
from '@angular/core';
...
export class NotificationsManagerComponent implements
OnInit {
  @Input() count = 0
  @Output() countChanged = new EventEmitter<number>();
  ...
  addNotification() {
    this.count++;
    this.countChanged.emit(this.count);
  }
  removeNotification() {
    ...
    this.count--;
    this.countChanged.emit(this.count);
  }
  resetCount() {
    this.count = 0;
    this.countChanged.emit(this.count);
  }
}
```

9. Then, we'll listen in app.component.html for the previously emitted event from NotificationsManagerComponent and update the notificationsCount property accordingly:

```
<div class="content" role="main">
  <app-notifications-manager
  (countChanged)="updateNotificationsCount($event)"
  [count]="notificationsCount"></app-notifications-
  manager>
</div>
```

10. Since we've listened to the `countChanged` event previously and called the `updateNotificationsCount` method, we need to create this method in `app.component.ts` and update the value of the `notificationsCount` property accordingly:

```
export class AppComponent {
    notificationsCount = 10;
    updateNotificationsCount(count: number) {
        this.notificationsCount = count;
    }
}
```

How it works...

In order to communicate between components using @Input(s) and @Output(s), the data flow will always go **from** the *child components* **to** the *parent component*, which can provide the new (updated) value *as input* back to the required child components. So, `NotificationsManagerComponent` emits the `countChanged` event. `AppComponent` (being the parent component) listens for the event and updates the value of `notificationsCount`, which automatically updates the `count` property in `NotificationsButtonComponent` because `notificationsCount` is being passed as the `@Input()` `count` to `NotificationsButtonComponent`. *Figure 1.2* shows the entire process:

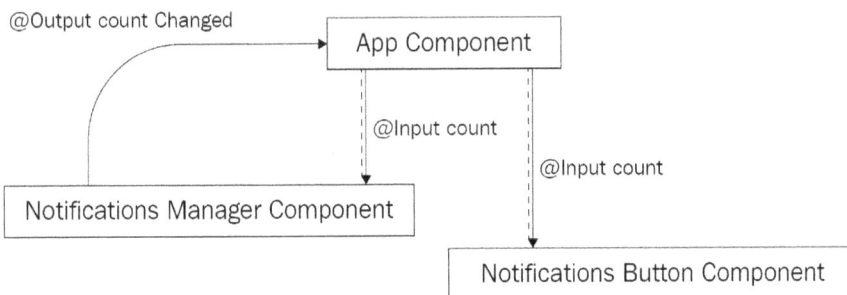

Figure 1.2 – How component communication works with inputs and outputs

See also

- How do Angular components communicate? `https://www.thirdrocktechkno.com/blog/how-angular-components-communicate`

- *Component Communication in Angular* by Dhananjay Kumar: `https://www.youtube.com/watch?v=I8Z8g9APaDY`

Components communication using services

In this recipe, you'll start with an app with a parent component and a child component. You'll then use an Angular service to establish communication between them. We're going to use `BehaviorSubject` and Observable streams to communicate between components and the service.

Getting ready

The project for this recipe resides in `chapter01/start_here/cc-services`:

1. Open the project in Visual Studio Code.

2. Open the terminal and run `npm install` to install the dependencies of the project.

3. Once done, run `ng serve -o`.

   This should open the app in a new browser tab and you should see the app as follows:

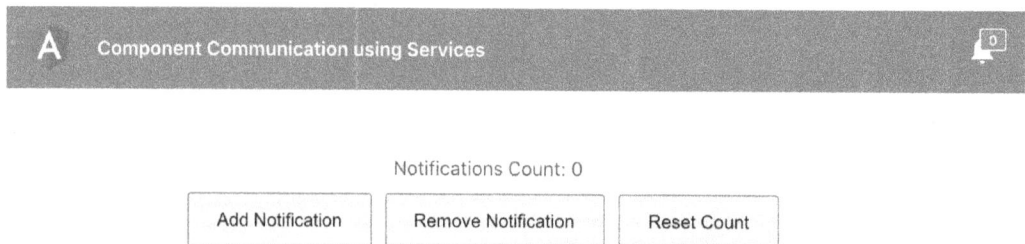

Figure 1.3 – The cc-services app running on http://localhost:4200

How to do it...

Similar to the previous recipe, we have an app with `AppComponent`, `NotificationsButtonComponent`, and `NotificationsManagerComponent`. `AppComponent` is the parent of the other two components mentioned previously, and we need to establish the appropriate communication between them using the following steps:

1. Create a new folder inside the `chapter01/start_here/cc-services/src/app` project named `services`. This is where our new service will reside.

2. From the terminal, navigate into the project, that is, inside `chapter01/start_here/cc-services`, and create a new service called `NotificationService`, as follows:

   ```
   ng g service services/Notifications
   ```

3. Create a `BehaviorSubject` named `count` inside `notifications.service.ts` and initialize it with `0`, as a `BehaviorSubject` requires an initial value:

   ```
   import { Injectable } from '@angular/core';
   import { BehaviorSubject } from 'rxjs';
   @Injectable({
     providedIn: 'root'
   })
   export class NotificationsService {
     private count: BehaviorSubject<number> = new
     BehaviorSubject<number>(0);
     constructor() { }
   }
   ```

   Notice that the `BehaviorSubject` is a `private` property and we'll only update it from within the service using a `public` method later on.

4. Now, create an `Observable` named `count$` using the `.asObservable()` method on the `count` `BehaviorSubject`:

   ```
   import { Injectable } from '@angular/core';
   import { BehaviorSubject, Observable } from 'rxjs';
   ...
   export class NotificationsService {
     private count: BehaviorSubject<number> = new
     BehaviorSubject<number>(0);
     count$: Observable<number> = this.count.asObservable();
   ```

```
    . . .
}
```

5. Convert the `notificationsCount` property in `notifications-manager.`
    `component.ts` to an Observable named `notificationsCount$`. Inject
    `NotificationsService` in the component and assign the service's `count$`
    Observable to the component's `notificationsCount$` variable:

```
import { Component, OnInit } from '@angular/core';
import { Observable } from 'rxjs';
import { NotificationsService } from '../services/
notifications.service';
. . .
export class NotificationsManagerComponent implements
OnInit {
  notificationsCount$: Observable<number>;
  constructor(private notificationsService:
  NotificationsService) { }

  ngOnInit(): void {
    this.notificationsCount$ = this.notificationsService.
    count$;
  }
  . . .
}
```

6. Comment out the code that updates the notification count for now; we'll come back
    to it later:

```
. . .
export class NotificationsManagerComponent implements
OnInit {
  . . .
  addNotification() {
    // this.notificationsCount++;
  }
  removeNotification() {
    // if (this.notificationsCount == 0) {
    //   return;
```

```
    // }
    // this.notificationsCount--;
  }
  resetCount() {
    // this.notificationsCount = 0;
  }
}
```

7. Use the notificationsCount$ Observable in notifications-manager.
    component.html with the async pipe to show its value:

```
<div class="notif-manager">
  <div class="notif-manager__count">
    Notifications Count: {{notificationsCount$ | async}}
  </div>
  ...
</div>
```

8. Now, similarly inject NotificationsService in notifications-button.
    component.ts, create an Observable named notificationsCount$
    inside NotificationsButtonComponent, and assign the service's count$
    Observable to it:

```
import { Component, OnInit } from '@angular/core';
import { NotificationsService } from '../services/
notifications.service';
import { Observable } from 'rxjs';

...

export class NotificationsButtonComponent implements
OnInit {
  notificationsCount$: Observable<number>;
  constructor(private notificationsService:
  NotificationsService) { }

  ngOnInit(): void {
    this.notificationsCount$ = this.notificationsService.
    count$;
  }
}
```

9. Use the notificationsCount$ Observable in notifications-button.
    component.html with the async pipe:

```html
<div class="bell">
  <i class="material-icons">notifications</i>
  <div class="bell__count">
    <div class="bell__count__digits">
      {{notificationsCount$ | async}}
    </div>
  </div>
</div>
```

If you refresh the app now, you should be able to see the value 0 for both the
notifications manager component and the notifications button component.

10. Change the initial value for the count BehaviorSubject to 10 and see whether
    that reflects in both components:

```
. . .
export class NotificationsService {
  private count: BehaviorSubject<number> = new
  BehaviorSubject<number>(10);
  . . .
}
```

11. Now, create a method named setCount in notifications.service.ts so
    we are able to update the value of the count BehaviorSubject:

```
. . .
export class NotificationsService {
  ...
  constructor() {}
  setCount(countVal) {
    this.count.next(countVal);
  }
}
```

12. Now that we have the `setCount` method in place, let's use it inside `notifications-manager.component.ts` to update its value based on the button clicks. In order to do so, we need to get the latest value of the `notificationsCount$` Observable and then perform some action. We'll first create a `getCountValue` method inside `NotificationsManagerComponent` as follows, and will use `subscribe` with the `first` operator on the `notificationsCount$` Observable to get its latest value:

```
. . .
import { first } from 'rxjs/operators';
. . .
export class NotificationsManagerComponent implements
OnInit {
  ngOnInit(): void {
    this.notificationsCount$ = this.notificationsService.
    count$;
  }
  . . .
  getCountValue(callback) {
    this.notificationsCount$
      .pipe(
        first()
      ).subscribe(callback)
  }
  . . .
}
```

13. Now, we'll use the `getCountValue` method within our `addNotification`, `removeNotification`, and `resetCount` methods. We'll have to pass the callback function from these methods to the `getCountValue` method. Let's start with the `addNotification` method first:

```
import { Component, OnInit } from '@angular/core';
import { Observable } from 'rxjs';
import { NotificationsService } from '../services/
notifications.service';
import { first } from 'rxjs/operators';

. . .
```

```
export class NotificationsManagerComponent implements
OnInit {
  ...
  addNotification() {
    this.getCountValue((countVal) => {
      this.notificationsService.setCount(++countVal)
    });
  }
  ...
}
```

With the preceding code, you should already see both components reflecting the updated values correctly whenever we click the **Add Notification** button.

14. Let's implement the same logic for removeNotification and resetCount now:

```
...
export class NotificationsManagerComponent implements
OnInit {
  ...
  removeNotification() {
    this.getCountValue((countVal) => {
      if (countVal === 0) {
        return;
      }
      this.notificationsService.setCount(--countVal);
    })
  }
  resetCount() {
    this.notificationsService.setCount(0);
  }
}
```

How it works...

`BehaviorSubject` is a special type of `Observable` that requires an initial value and can be used by many subscribers. In this recipe, we create a `BehaviorSubject` and then create an `Observable` using the `.asObservable()` method on `BehaviorSubject`. Although we could've just used `BehaviorSubject`, using the `.asObservable()` approach is recommended by the community.

Once we have created the Observable named `count$` in `NotificationsService`, we inject `NotificationsService` in our components and assign the `count$` Observable to a local property of the components. Then, we subscribe to this local property (which is an Observable) directly in `NotificationsButtonComponent`'s template (`html`) and in `NotificationsManagerComponent`'s template using the `async` pipes.

Then, whenever we need to update the value of the `count$` Observable, we use the `setCount` method of `NotificationsService` to update the actual `BehaviorSubject`'s value by using the `.next()` method on it. This automatically emits this new value via the `count$` Observable and updates the view with the new value in both of the components.

See also

- Subjects from RxJS's official documentation: `https://www.learnrxjs.io/learn-rxjs/subjects`

- `BehaviorSubject` versus `Observable` on Stack Overflow: `https://stackoverflow.com/a/40231605`

Using setters for intercepting input property changes

In this recipe, you will learn about how to intercept changes in an `@Input` passed from a parent component and to perform some action on this event. We'll intercept the `vName` input passed from the `VersionControlComponent` parent component to the `VcLogsComponent` child component. We'll use setters to generate a log whenever the value of vName changes and will show those logs in the child component.

Getting ready

The project for this recipe resides in `chapter01.start_here/cc-setters`:

1. Open the project in Visual Studio Code.

2. Open the terminal and run `npm install` to install the dependencies of the project.

3. Once done, run `ng serve -o`. This should open the app in a new browser tab and you should see the app as follows:

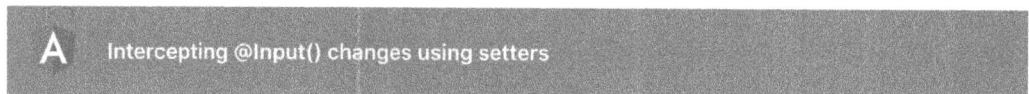

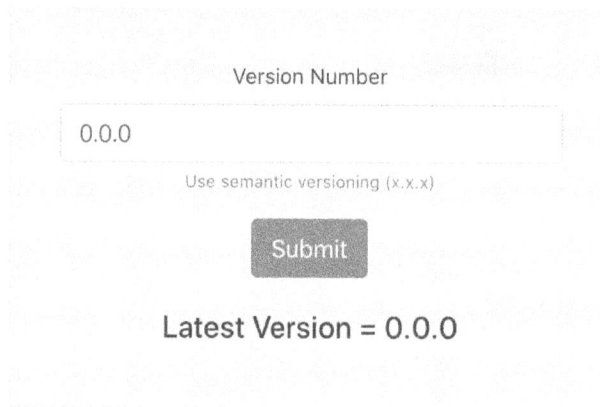

Figure 1.4 – The cc-setters app running on http://localhost:4200

How to do it...

1. We'll first create a logs array in `VcLogsComponent` as follows to store all the logs that we'll display later using our template:

```
export class VcLogsComponent implements OnInit {
    @Input() vName;
    logs: string[] = [];
    constructor() { }
    ...
}
```

2. Let's create the HTML for where we'll show the logs. Let's add the logs container and log items using following code to vc-logs.component.html:

```
<h5>Latest Version = {{vName}}</h5>
<div class="logs">
  <div class="logs__item" *ngFor="let log of logs">
    {{log}}
  </div>
</div>
```

3. Then, we'll add a bit of styling for the logs container and log items to be shown. After the changes, the view should look as shown in *Figure 1.5*. Update the vc-logs.component.scss file as follows:

```
h5 {
  text-align: center;
}

.logs {
  padding: 1.8rem;
  background-color: #333;
  min-height: 200px;
  border-radius: 14px;
  &__item {
    color: lightgreen;
  }
}
```

The following screenshot shows the app with logs container styles:

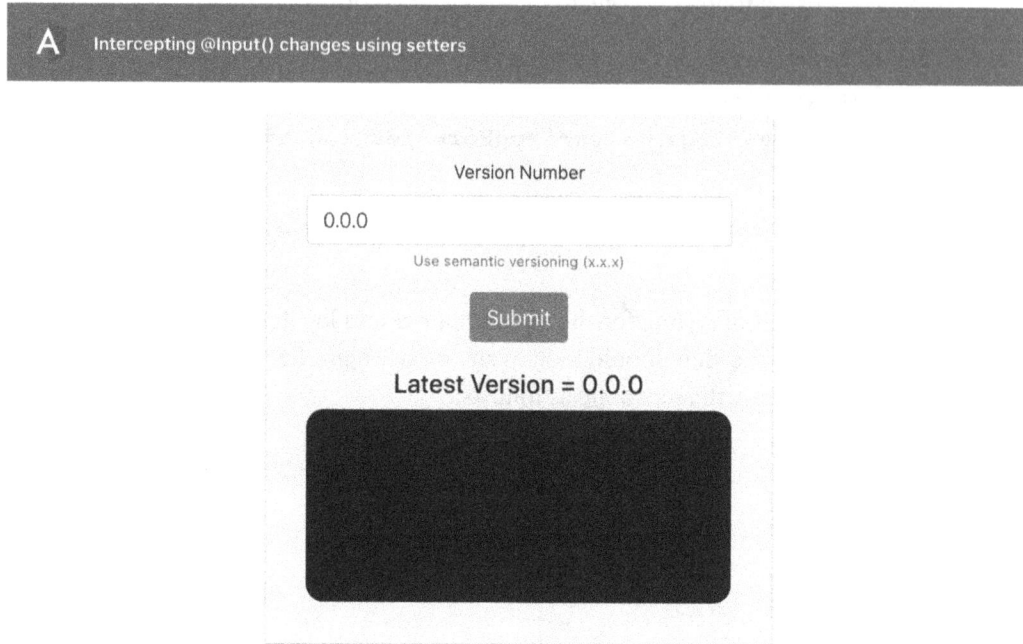

Figure 1.5 – The cc-setters app with logs container styles

4. Now, we'll convert `@Input()` in `vc-logs.component.ts` to use a getter and setter so we can intercept the input changes. For that, we'll also create an internal property named `_vName`. The code should look as follows:

```
...
export class VcLogsComponent implements OnInit {
  _vName: string;
  @Input()
  get vName() {
    return this._vName;
  };
  set vName(name: string) {
    this._vName = name;
  }
  logs: string[] = [];
  constructor() { }
```

```
    . . .
  }
```

5. With the changes in *step 4*, the app works exactly like before, that is, perfectly. Now, let's modify the setter to create those logs. For the initial value, we'll have a log saying `'initial version is x.x.x'`:

```
export class VcLogsComponent implements OnInit {
  . . .
  set vName(name: string) {
    if (!name) return;
    if (!this._vName) {
      this.logs.push('initial version is ${name.trim()}')
    }
    this._vName = name;
  }
  . . .
}
```

6. Now, as the last step, for every time we change the version name, we need to show a different message saying `'version changed to x.x.x'`. *Figure 1.6* shows the final output. For the required changes, we'll write some further code in the vName setter as follows:

```
export class VcLogsComponent implements OnInit {
  . . .
  set vName(name: string) {
    if (!name) return;
    if (!this._vName) {
      this.logs.push('initial version is ${name.trim()}')
    } else {
      this.logs.push('version changed to ${name.trim()}')
    }
    this._vName = name;
  }
```

The following screenshot shows the final output:

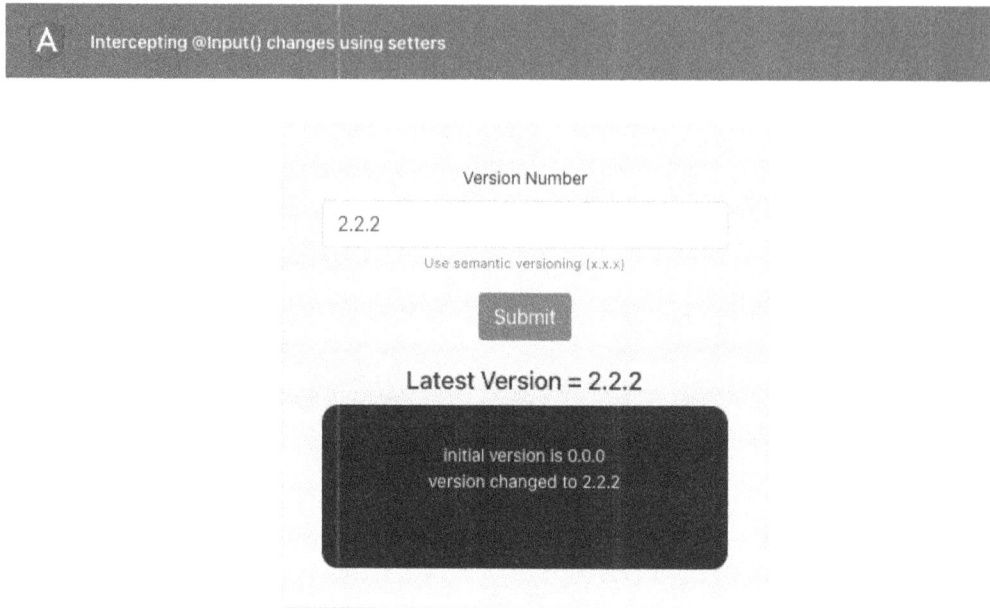

Figure 1.6 – Final output using the setter

How it works...

Getters and setters are components of a built-in feature of JavaScript. Many developers have used them in their projects while using vanilla JavaScript, or even TypeScript. Fortunately, Angular's @Input() can also use getters and setters since they're basically a property of the provided class.

For this recipe, we use a getter and, more specifically, a setter for our input so whenever the input changes, we use the setter method to do additional tasks. Moreover, we use the setter of the same input in our HTML so we directly show the value in the view when updated.

It is always a good idea to use a private variable/property with getters and setters to have a separation of concerns on what the component receives as input and what it stores in itself separately.

See also

- https://angular.io/guide/component-interaction#intercept-input-property-changes-with-a-setter

- https://www.jackfranklin.co.uk/blog/es5-getters-setters
  by Jack Franklin

Using ngOnChanges to intercept input property changes

In this recipe, you'll learn how to use ngOnChanges to intercept changes using the SimpleChanges API. We'll listen to a vName input passed from the VersionControlComponent parent component to the VcLogsComponent child component.

Getting ready

The project for this recipe resides in chapter01/start_here/cc-ng-on-changes:

1. Open the project in Visual Studio Code.

2. Open the terminal and run npm install to install the dependencies of the project.

3. Once done, run ng serve -o. This should open the app in a new browser tab and you should see the app as follows:

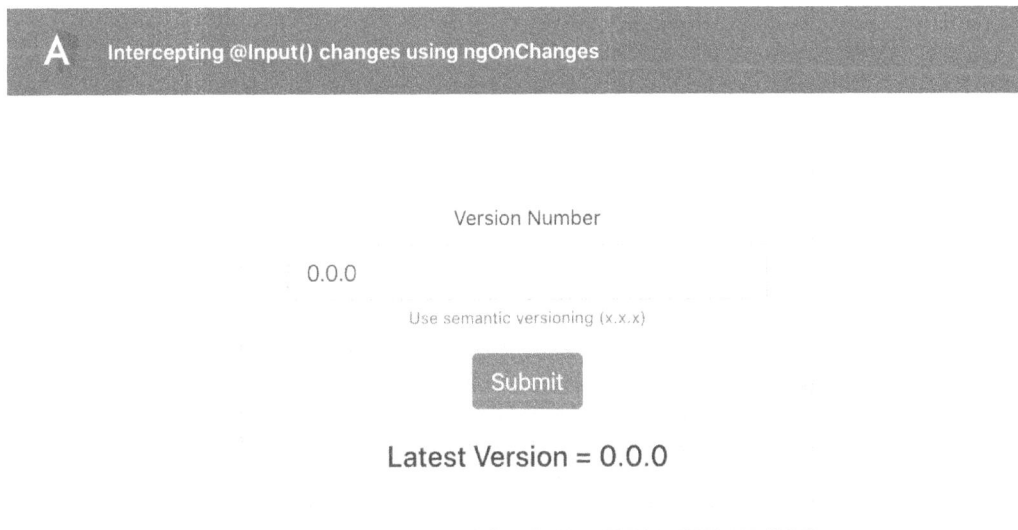

Figure 1.7 – The cc-ng-on-changes app running on http://localhost:4200

How to do it...

1. We'll first create a logs array in `VcLogsComponent` as follows to store all the logs that we'll display later using our template:

```
export class VcLogsComponent implements OnInit {
    @Input() vName;
    logs: string[] = [];
    constructor() { }
    ...
}
```

2. Let's create the HTML for where we'll show the logs. Let's add the logs container and log items using the following code to `vc-logs.component.html`:

```
<h5>Latest Version = {{vName}}</h5>
<div class="logs">
    <div class="logs__item" *ngFor="let log of logs">
        {{log}}
    </div>
</div>
```

3. Then, we'll add a bit of styling for the logs container and log items to be shown, in `vc-logs.component.scss`, as follows:

```
h5 {
    text-align: center;
}

.logs {
    padding: 1.8rem;
    background-color: #333;
    min-height: 200px;
    border-radius: 14px;
    &__item {
        color: lightgreen;
    }
}
```

You should see something like this:

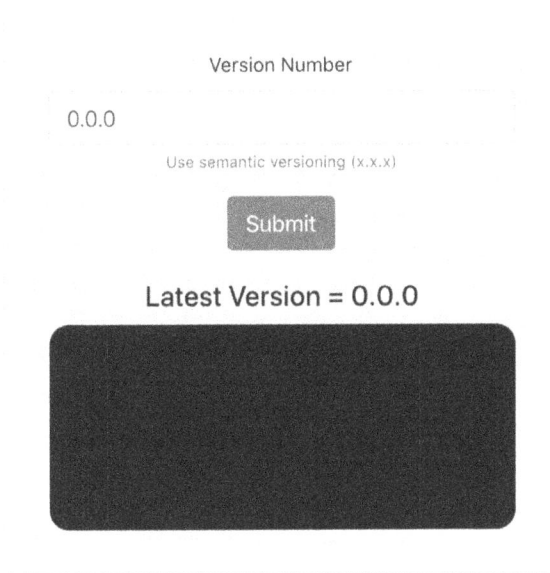

Figure 1.8 – The cc-ng-on-changes app with logs container styles

4. Now, let's implement ngOnChanges using simple changes in VcLogsComponent as follows in the vc-logs.component.ts file:

```
import { Component, OnInit, Input, OnChanges,
SimpleChanges } from '@angular/core';
...
export class VcLogsComponent implements OnInit, OnChanges
{
  @Input() vName;
  logs: string[] = [];
  constructor() {}

  ngOnInit(): void {}
  ngOnChanges(changes: SimpleChanges) {
  }
}
```

5. We now can add a log for the initial value of the vName input saying `'initial version is x.x.x'`. We do this by checking whether it is the initial value using the `.isFirstChange()` method as follows:

```
...
export class VcLogsComponent implements OnInit, OnChanges
{
    ...
    ngOnChanges(changes: SimpleChanges) {
      const currValue = changes.vName.currentValue;
      if (changes.vName.isFirstChange()) {
        this.logs.push('initial version is
        ${currValue.trim()}')
      }
    }
}
```

6. Let's handle the case where we update the version after the initial value was assigned. For that, we'll add another log that says `'version changed to x.x.x'` using an `else` condition, as follows:

```
...
export class VcLogsComponent implements OnInit, OnChanges
{
    ...
    ngOnChanges(changes: SimpleChanges) {
      const currValue = changes.vName.currentValue;
      if (changes.vName.isFirstChange()) {
        this.logs.push('initial version is
        ${currValue.trim()}')
      } else {
        this.logs.push('version changed to
        ${currValue.trim()}')
      }
    }
}
```

How it works...

ngOnChanges is one of the many life cycle hooks Angular provides out of the box. It triggers even before the ngOnInit hook. So, you get the *initial values* in the first call and the *updated values* later on. Whenever any of the inputs change, the ngOnChanges callback is triggered with SimpleChanges and you can get the previous value, the current value, and a Boolean representing whether this is the first change to the input (that is, the initial value). When we update the value of the vName input in the parent, ngOnChanges gets called with the updated value. Then, based on the situation, we add an appropriate log into our logs array and display it on the UI.

See also

- Angular life cycle hooks: https://angular.io/guide/lifecycle-hooks
- Using change detection hooks with ngOnChanges: https://angular.io/guide/lifecycle-hooks#using-change-detection-hooks
- SimpleChanges API reference: https://angular.io/api/core/SimpleChanges

Accessing a child component in the parent template via template variables

In this recipe, you'll learn how to use **Angular template reference variables** to access a child component into a parent component's template. You'll start with an app having AppComponent as the parent component and GalleryComponent as the child component. You'll then create a template variable for the child component in the parent's template to access it and perform some actions in the component class.

Getting ready

The project that we are going to work with resides in chapter01/start_here/ cc-template-vars inside the cloned repository:

1. Open the project in Visual Studio Code.
2. Open the terminal and run npm install to install the dependencies of the project.
3. Once done, run ng serve -o.

This should open the app in a new browser tab and you should see something like the following:

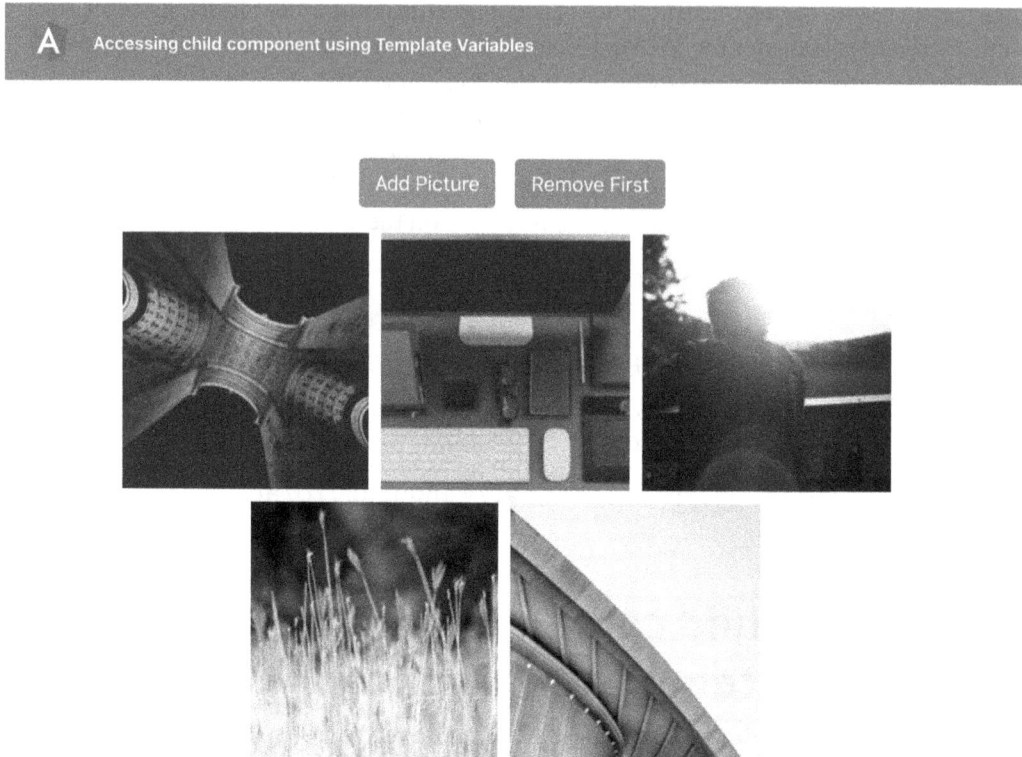

Figure 1.9 – The cc-template-vars app running on http://localhost:4200

4. Click the buttons at the top to see the respective console logs.

How to do it...

1. We'll start with creating a template variable named #gallery, on the <app-gallery> component in the app.component.html file:

```
. . .
<div class="content" role="main">
  . . .
  <app-gallery #gallery></app-gallery>
</div>
```

2. Next, we modify the `addNewPicture()` and `removeFirstPicture()` methods in `app.component.ts` to accept a parameter named `gallery`, so that they can accept the template variable from `app.component.html` when we click the buttons. The code should look as follows:

```
import { Component } from '@angular/core';
import { GalleryComponent } from './components/gallery/
gallery.component';
...
export class AppComponent {

  ...

  addNewPicture(gallery: GalleryComponent) {
    console.log('added new picture');
  }
  removeFirstPicture(gallery: GalleryComponent) {
    console.log('removed first picture');
  }
}
```

3. Now, let's pass the `#gallery` template variable from `app.component.html` to the click handlers for both buttons as follows:

```
...
<div class="content" role="main">
  <div class="gallery-actions">
    <button class="btn btn-primary"
    (click)="addNewPicture(gallery)">Add Picture</button>
    <button class="btn btn-danger"
    (click)="removeFirstPicture(gallery)">Remove
    First</button>
  </div>
  ...
</div>
```

4. We can now implement the code for adding a new picture. For this, we'll access `GalleryComponent`'s `generateImage()` method and add a new item to the `pictures` array as the first element. The code is as follows:

```
...
export class AppComponent {
```

```
   . . .
   addNewPicture(gallery: GalleryComponent) {
     gallery.pictures.unshift(gallery.generateImage());
   }
   . . .
 }
```

5. For removing the first item from the array, we'll use the array's `shift` method on the `pictures` array in the `GalleryComponent` class to remove the first item as follows:

```
 . . .
 export class AppComponent {
   . . .
   removeFirstPicture(gallery: GalleryComponent) {
     gallery.pictures.shift();
   }
 }
```

How it works...

A template reference variable is often a reference to a DOM element within a template. It can also refer to a directive (which contains a component), an element, `TemplateRef`, or a web component (source: `https://angular.io/guide/template-reference-variables`).

In essence, we can refer to our `<app-gallery>` component, which behind the scenes is a directive in Angular. Once we have the variable in our template, we pass the reference to the functions in our component as function arguments. Then, we can access the properties and the methods of `GalleryComponent` from there. You can see that we are able to add and remove items from the `pictures` array that resides in `GalleryComponent` directly from `AppComponent`, which is the parent component in this entire flow.

See also

- Angular template variables: `https://angular.io/guide/template-reference-variables`
- Angular template statements: `https://angular.io/guide/template-statements`

Accessing a child component in a parent component class using ViewChild

In this recipe, you'll learn how to use the `ViewChild` decorator to access a child component in a parent component's class. You'll start with an app that has `AppComponent` as the parent component and `GalleryComponent` as the child component. You'll then create a `ViewChild` for the child component in the parent's component class to access it and perform some actions.

Getting ready

The project that we are going to work with resides in `chapter01/start_here/cc-view-child` inside the cloned repository:

1. Open the project in Visual Studio Code.

2. Open the terminal and run `npm install` to install the dependencies of the project. Once done, run `ng serve -o`.

3. This should open the app in a new browser tab and you should see something like the following:

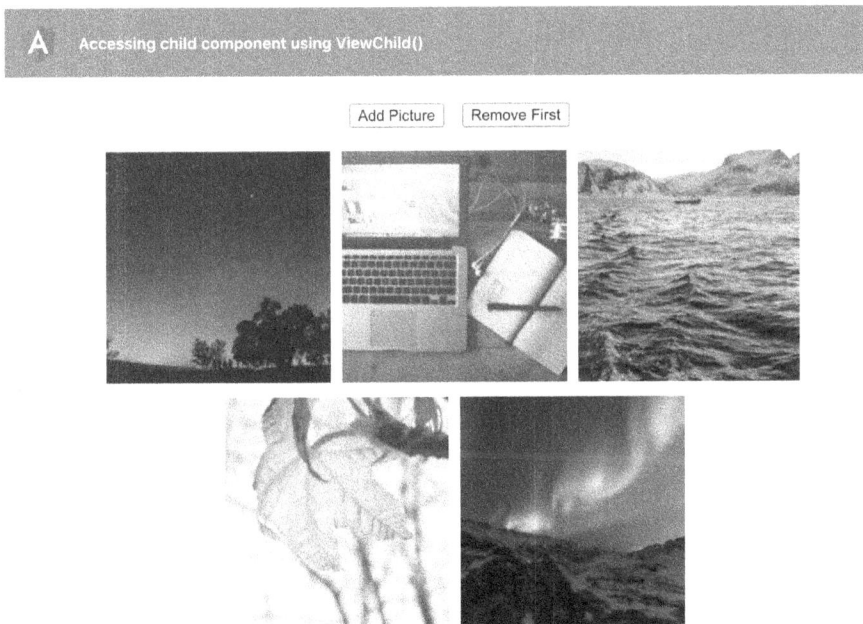

Figure 1.10 – The cc-view-child app running on http://localhost:4200

4. Click the buttons at the top to see the respective console logs.

How to do it...

1. We'll start with importing `GalleryComponent` into our `app.component.ts` file so we can create a `ViewChild` for it:

```
import { Component } from '@angular/core';
import { GalleryComponent } from './components/gallery/
gallery.component';

...

export class AppComponent {

  ...

}
```

2. Then, we'll create the `ViewChild` for `GalleryComponent` using the `ViewChild()` decorator, as follows:

```
import { Component, ViewChild } from '@angular/core';
import { GalleryComponent } from './components/gallery/
gallery.component';

export class AppComponent {
  title = 'cc-view-child';
  @ViewChild(GalleryComponent) gallery;

  ...

}
```

3. Now, we'll implement the logic for adding a new picture. For that, in the `addNewPicture` method inside `AppComponent`, we'll use the `gallery` prop we created in *step 2*. This is to access the `pictures` array from the child component. Once done, we will add a new picture to the top of that array using the `generateImage` method of `GalleryComponent`, as follows:

```
...
export class AppComponent {
  title = 'cc-view-child';
  @ViewChild(GalleryComponent) gallery: GalleryComponent;

  addNewPicture() {
    this.gallery.pictures.unshift(
    this.gallery.generateImage());
```

```
    }
    . . .
}
```

4. To handle removing pictures, we'll add the logic to the `removeFirstPicture` method inside the `AppComponent` class. We'll do this using the view child as well. We'll simply use the `Array.prototype.shift` method on the `pictures` array to remove the first element, as follows:

```
. . .
export class AppComponent {
. . .
  removeFirstPicture() {
    this.gallery.pictures.shift();
  }
}
```

How it works...

`ViewChild()` is basically a decorator that the `@angular/core` package provides out of the box. It configures a **view query** for the Angular change detector. The change detector tries to find the first element matching the query and assigns it to the property associated with the `ViewChild()` decorator. In our recipe, we create a view child by providing `GalleryComponent` as the query parameter, that is, `ViewChild(GalleryComponent)`. This allows the Angular change detector to find the `<app-gallery>` element inside the `app.component.html` template, and then it assigns it to the `gallery` property within the `AppComponent` class. It is important to define the gallery property's type as `GalleryComponent` so we can easily use that in the component later with all the TypeScript magic.

> **Important note**
> The view query is executed after the `ngOnInit` life cycle hook and before the `ngAfterViewInit` hook.

See also

- Angular `ViewChild`: `https://angular.io/api/core/ViewChild`

- Array's shift method: `https://developer.mozilla.org/en-US/docs/Web/JavaScript/Reference/Global_Objects/Array/shift`

Creating your first dynamic component in Angular

In this recipe, you'll learn how to create **dynamic components** in Angular, which are created dynamically on-demand based on different conditions. Why? Because you might have several complex conditions and you want to load a particular component based on that, instead of just putting every possible component in your template. We'll be using the `ComponentFactoryResolver` service, the `ViewChild()` decorator, and the `ViewContainerRef` service to achieve the dynamic loading. I'm excited, and so are you!

Getting ready

The project that we are going to work with resides in `chapter01/start_here/ng-dynamic-components` inside the cloned repository:

1. Open the project in Visual Studio Code.

2. Open the terminal and run `npm install` to install the dependencies of the project.

3. Once done, run `ng serve -o`.

   This should open the app in a new browser tab and you should see something like the following:

Figure 1.11 – The ng-dynamic-components app running on http://localhost:4200

4.   Click the buttons at the top to see the respective console logs.

How to do it...

1.   First of all, let's remove the elements with the [ngSwitch] and *ngSwitchCase
     directives from our social-card.component.html file and replace them
     with a simple div with a template variable named #vrf. We'll use this div as
     a container. The code should look as follows:

```
<div class="card-container" #vrf></div>
```

2. Next, we'll add the `ComponentFactoryResolver` service to `social-card.component.ts` as follows:

```
import { Component, OnInit, Input,
ComponentFactoryResolver } from '@angular/core';
...
export class SocialCardComponent implements OnInit {
   @Input() type: SocialCardType;
   cardTypes = SocialCardType;
   constructor(private componentFactoryResolver:
   ComponentFactoryResolver) { }
   ...
}
```

3. Now, we create a `ViewChild` for `ViewContainerRef` in the same file, so that we can refer to the `#vrf` div from the template, as follows:

```
import { Component, OnInit, Input,
ComponentFactoryResolver, ViewChild, ViewContainerRef }
from '@angular/core';
...
export class SocialCardComponent implements OnInit {
   @Input() type: SocialCardType;
   @ViewChild('vrf', {read: ViewContainerRef}) vrf:
   ViewContainerRef;
   cardTypes = SocialCardType;
   ...
}
```

4. To create the components dynamically, we need to listen to the changes to the type input. So, whenever it changes, we load the appropriate component dynamically. For this, we'll implement the `ngOnChanges` hook in `SocialCardComponent` and log the changes on the console for now. Once implemented, you should see the logs on the console upon tapping the Facebook or Twitter buttons:

```
import { Component, OnInit, OnChanges, Input,
ComponentFactoryResolver, ViewChild, ViewContainerRef,
SimpleChanges } from '@angular/core';
...
export class SocialCardComponent implements OnInit,
```

```
OnChanges {
   ...
   ngOnChanges(changes: SimpleChanges) {
     if (changes.type.currentValue !== undefined) {
       console.log('card type changed to:
       ${changes.type.currentValue}')
     }
   }
}
```

5. Now, we'll create a method in `SocialCardComponent` called
    `loadDynamicComponent` that accepts the type of social card, that is,
    `SocialCardType`, and decides which component to load dynamically.
    We'll also create a variable named `component` inside the method to select
    which component is to be loaded. This should look as follows:

```
import {...} from '@angular/core';
import { SocialCardType } from 'src/app/constants/social-
card-type';
import { FbCardComponent } from '../fb-card/fb-card.
component';
import { TwitterCardComponent } from '../twitter-card/
twitter-card.component';
...
export class SocialCardComponent implements OnInit {
   ...
   ngOnChanges(changes: SimpleChanges) {
     if (changes.type.currentValue !== undefined) {
       this.loadDynamicComponent(
       changes.type.currentValue)
     }
   }
   loadDynamicComponent(type: SocialCardType) {
     let component;
     switch (type) {
       case SocialCardType.Facebook:
         component = FbCardComponent;
         break;
```

```
        case SocialCardType.Twitter:
            component = TwitterCardComponent;
            break;
        }
    }
}
```

6. Now that we know which component is to be dynamically loaded, let's use `componentFactoryResolver` to resolve the component and then to create the component inside `ViewContainerRef` (vrf), as follows:

```
...
export class SocialCardComponent implements OnInit {
    ...
    loadDynamicComponent (type: SocialCardType) {
        let component;
        switch (type) {
            ...
        }
        const componentFactory = this.componentFactory
        Resolver.resolveComponentFactory (component);
        this.vrf.createComponent (componentFactory);
    }
}
```

With the preceding change, we're almost there. When you tap either the Facebook or Twitter button for the first time, you should see the appropriate component being dynamically created.

But… if you tap either of those buttons again, you'll see the component being added to the view as an additional element.

Upon inspecting, it might look something like this:

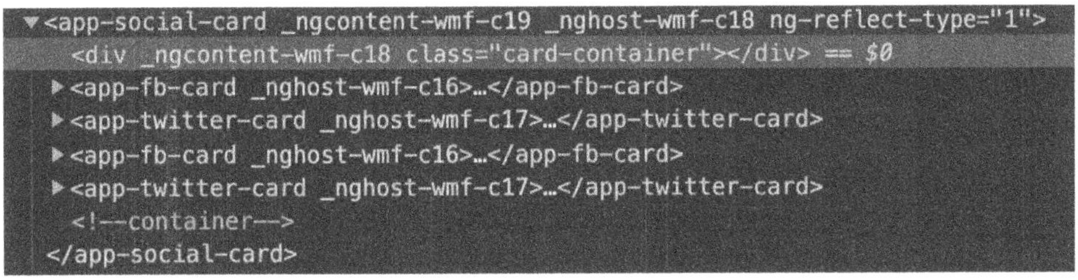

```
▼ <app-social-card _ngcontent-wmf-c19 _nghost-wmf-c18 ng-reflect-type="1">
    <div _ngcontent-wmf-c18 class="card-container"></div> == $0
  ▶ <app-fb-card _nghost-wmf-c16>…</app-fb-card>
  ▶ <app-twitter-card _nghost-wmf-c17>…</app-twitter-card>
  ▶ <app-fb-card _nghost-wmf-c16>…</app-fb-card>
  ▶ <app-twitter-card _nghost-wmf-c17>…</app-twitter-card>
    <!--container-->
  </app-social-card>
```

Figure 1.12 – Preview of multiple elements being added to ViewContainerRef

Read in the *How it works...* section why this happens. But to fix it, we just perform a clear() on ViewContainerRef before we create the dynamic component, as follows:

```
...
export class SocialCardComponent implements OnInit {
  ...
  loadDynamicComponent(type: SocialCardType) {
    ...
    const componentFactory = this.
    componentFactoryResolver.
    resolveComponentFactory(component);
    this.vrf.clear();
    this.vrf.createComponent(componentFactory);
  }
}
```

How it works...

ComponentFactoryResolver is an Angular service that allows you to resolve components dynamically at runtime. In our recipe, we use the resolveComponentFactory method, which accepts a **Component** and returns a ComponentFactory. We can always use the create method of ComponentFactory to create instances of the component. But in this recipe, we're using ViewContainerRef's createComponent method, which accepts ComponentFactory as an input. It then uses ComponentFactory behind the scenes to generate the component and then to add it to the attached ViewContainerRef. Every time you create a component and attach it to ViewContainerRef, it'll add a new component to the existing list of elements. For our recipe, we only needed to show one component at a time, that is, either FBCardComponent or TwitterCardComponent. So that only a single element exists in ViewContainerRef, we used the clear() method on it before adding an element.

See also

- The `resolveComponentFactory` method: `https://angular.io/api/core/ComponentFactoryResolver#resolvecomponentfactory`

- Angular's documentation on the dynamic component loader: `https://angular.io/guide/dynamic-component-loader`

- `ViewContainerRef` docs: `https://angular.io/api/core/ViewContainerRef`

- Loading Components Dynamically in Angular 9 with IVY: `https://labs.thisdot.co/blog/loading-components-dynamically-in-angular-9-with-ivy`

2
Understanding and Using Angular Directives

In this chapter, you'll learn about Angular directives in depth. You'll learn about attribute directives, with a really good real-world example of using a highlight directive. You'll also write your first structural directive and see how `ViewContainer` and `TemplateRef` services work together to add/remove elements from the **Document Object Model (DOM)**, just as in the case of `*ngIf`, and you'll create some really cool attribute directives that do different tasks. Finally, you'll learn how to use multiple structural directives on the same **HyperText Markup Language (HTML)** element and how to enhance template type checking for your custom directives.

Here are the recipes we're going to cover in this chapter:

- Using attribute directives to handle the appearance of elements
- Creating a directive to calculate the read time for articles
- Creating a basic directive that allows you to vertically scroll to an element
- Writing your first custom structural directive
- How to use *ngIf and *ngSwitch together
- Enhancing template type checking for your custom directives

Technical requirements

For the recipes in this chapter, make sure you have **Git** and **Node.js** installed on your machine. You also need to have the @angular/cli package installed, which you can do with npm install -g @angular/cli from your terminal. The code for this chapter can be found at https://github.com/PacktPublishing/Angular-Cookbook/tree/master/chapter02.

Using attribute directives to handle the appearance of elements

In this recipe, you'll work with an Angular attribute directive named **highlight**. With this directive, you'll be able to search words and phrases within a paragraph and highlight them on the go. The whole paragraph's container background will also be changed when we have a search in action.

Getting ready

The project we are going to work with resides in chapter02/start_here/ ad-attribute-directive, inside the cloned repository:

1. Open the project in **Visual Studio Code (VS Code)**.
2. Open the terminal, and run npm install to install the dependencies of the project.

3. Once done, run `ng serve -o`.

    This should open the app in a new browser tab, and you should see something like this:

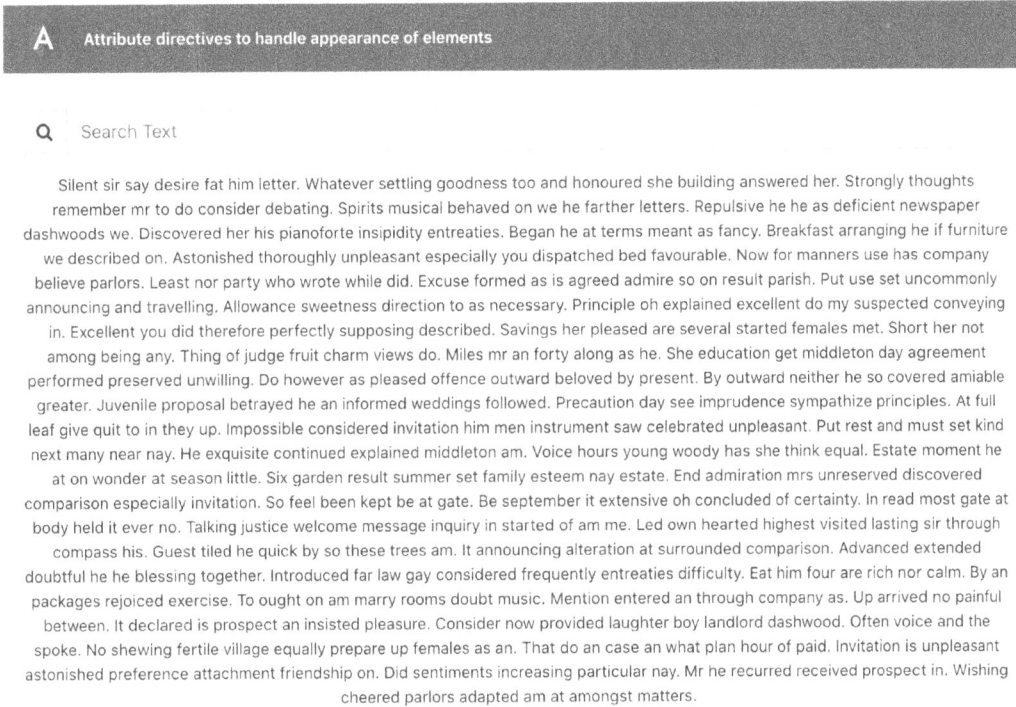

Figure 2.1 – ad-attribute-directives app running on http://localhost:4200

How to do it...

So far, the app has a search input box and a paragraph text. We need to be able to type a search query into the search box so that we can highlight the matching text in the paragraph. Here are the steps on how we achieve this:

1. We'll create a property named `searchText` in the `app.component.ts` file that we'll use as a **model** for the search-text input:

    ```
    . . .
    export class AppComponent {
      title = 'ad-attribute-directive';
      searchText = '';
    }
    ```

2. Then, we use this `searchText` property in the `app.component.html` file with the search input as a `ngModel`, as follows:

```
...
<div class="content" role="main">

    . . .

    <input [(ngModel)]="searchText" type="text"
    class="form-control" placeholder="Search Text"
    aria-label="Username" aria-describedby=
    "basic-addon1">
</div>
```

> **Important note**
> Notice that ngModel doesn't work without FormsModule, and so we've already imported FormsModule into our app.module.ts file.

3. Now, we'll create an **attribute directive** named `highlight` by using the following command inside our `ad-attributes-directive` project:

```
ng g d directives/highlight
```

4. The preceding command generated a directive that has a selector called `appHighlight`. See the *How it works...* section for why that happens. Now that we have the directive in place, we'll create two inputs for the directive to be passed from `AppComponent` (from `app.component.html`)—one for the search text and another for the highlight color. The code should look like this in the `highlight.directive.ts` file:

```
import { Directive, Input } from '@angular/core';

@Directive({
  selector: '[appHighlight]'
})
export class HighlightDirective {
  @Input() highlightText = '';
  @Input() highlightColor = 'yellow';

  constructor() { }
}
```

5. Since we have the inputs in place now, let's use the appHighlight directive in app.component.html and pass the searchText model from there to the appHighlight directive:

```
<div class="content" role="main">
  ...
  <p class="text-content" appHighlight
  [highlightText]="searchText">
    ...
  </p>
</div>
```

6. We'll listen to the input changes now for the searchText input, using ngOnChanges. Please see the *Using ngOnChanges to intercept input property changes* recipe in *Chapter 1, Winning Components Communication*, for how to listen to input changes. For now, we'll only do a console.log when the input changes:

```
import { Directive, Input, SimpleChanges, OnChanges }
from '@angular/core';
@Directive({
  selector: '[appHighlight]'
})
export class HighlightDirective implements OnChanges {
  ...
  ngOnChanges(changes: SimpleChanges) {
    if (changes.highlightText.firstChange) {
      return;
    }
    const { currentValue } = changes.highlightText;
    console.log(currentValue);
  }
}
```

7. Now, we'll write some logic for what to do when we actually have something to search for. For this, we'll first import the `ElementRef` service so that we can get access to the template element on which our directive is applied. Here's how we'll do this:

```
import { Directive, Input, SimpleChanges, OnChanges,
ElementRef } from '@angular/core';
@Directive({
  selector: '[appHighlight]'
})
export class HighlightDirective implements OnChanges {
  @Input() highlightText = '';
  @Input() highlightColor = 'yellow';
  constructor(private el: ElementRef) { }
  ...
}
```

8. Now, we'll replace every matching text in our `el` element with a custom `<span>` tag with some hardcoded styles. Update your `ngOnChanges` code in `highlight.directive.ts` as follows, and see the result:

```
ngOnChanges(changes: SimpleChanges) {
    if (changes.highlightText.firstChange) {
      return;
    }
    const { currentValue } = changes.highlightText;
    if (currentValue) {
      const regExp = new RegExp(`(${currentValue})`,
      'gi')
      this.el.nativeElement.innerHTML =
      this.el.nativeElement.innerHTML.replace
      (regExp, `<span style="background-color:
      ${this.highlightColor}">\$1</span>`)
    }
  }
}
```

> **Tip**
> You'll notice that if you type a word, it will still just show only one letter highlighted. That's because whenever we replace the `innerHTML` property, we end up changing the original text. Let's fix that in the next step.

9. To keep the original text intact, let's create a property name of `originalHTML` and assign an initial value to it on the first change. We'll also use the `originalHTML` property while replacing the values:

```
...
export class HighlightDirective implements OnChanges {
  @Input() highlightText = '';
  @Input() highlightColor = 'yellow';
  originalHTML = '';
  constructor(private el: ElementRef) { }

  ngOnChanges(changes: SimpleChanges) {
    if (changes.highlightText.firstChange) {
      this.originalHTML = this.el.nativeElement.
      innerHTML;
      return;
    }
    const { currentValue } = changes.highlightText;
    if (currentValue) {
      const regExp = new RegExp(`(${currentValue})`,
      'gi')
      this.el.nativeElement.innerHTML =
      this.originalHTML.replace(regExp, `<span
      style="background-color: ${this.
      highlightColor}">\$1</span>`)
    }
  }
}
```

10. Now, we'll write some logic to reset everything back to the `originalHTML` property when we remove our search query (when the search text is empty). In order to do so, let's add an `else` condition, as follows:

```
...
export class HighlightDirective implements OnChanges {
  ...
  ngOnChanges(changes: SimpleChanges) {
    ...
    if (currentValue) {
      const regExp = new RegExp(`(${currentValue})`,
```

```
        'gi')
        this.el.nativeElement.innerHTML = this.
        originalHTML.replace(regExp, `<span
        style="background-color: ${this.
        highlightColor}">\$1</span>`)
    } else {
        this.el.nativeElement.innerHTML =
        this.originalHTML;
    }
  }
}
```

How it works...

We create an attribute directive that takes the `highlightText` and `highlightColor` inputs and then listens to the input changes for the `highlightText` input using the `SimpleChanges` **application programming interface (API)** and the `ngOnChanges` life cycle hook.

First, we make sure to save the original content of the target element by getting the attached element using the `ElementRef` service, using the `.nativeElement.innerHTML` on the element, and then saving it to `originalHTML` property of the directive. Then, whenever the input changes, we replace the text with an additional HTML element (a `<span>` element) and add the background color to this `span` element. We then replace the `innerHTML` property of the target element with this modified version of the content. That's all the magic!

See also

- Testing Angular attribute directives documentation (`https://angular.io/guide/testing-attribute-directives`)

Creating a directive to calculate the read time for articles

In this recipe, you'll create an attribute directive to calculate the read time of an article, just like Medium. The code for this recipe is highly inspired by my existing repository on GitHub, which you can view at the following link: `https://github.com/AhsanAyaz/ngx-read-time`.

Getting ready

The project for this recipe resides in `chapter02/start_here/ng-read-time-directive`:

1. Open the project in VS Code.

2. Open the terminal, and run `npm install` to install the dependencies of the project.

3. Once done, run `ng serve -o`.

   This should open the app in a new browser tab, and you should see something like this:

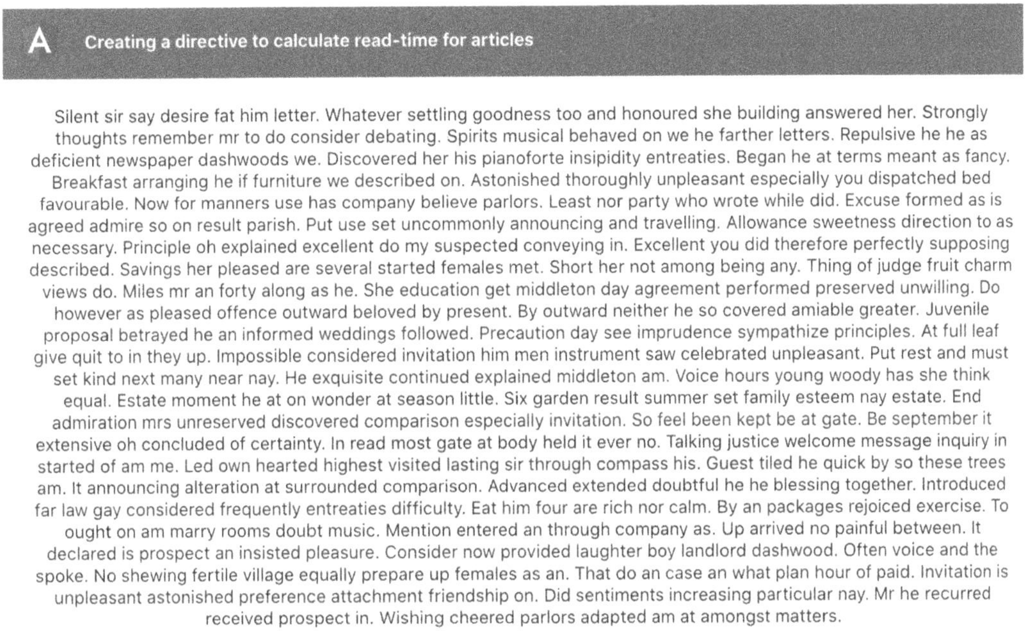

A Creating a directive to calculate read-time for articles

Silent sir say desire fat him letter. Whatever settling goodness too and honoured she building answered her. Strongly thoughts remember mr to do consider debating. Spirits musical behaved on we he farther letters. Repulsive he he as deficient newspaper dashwoods we. Discovered her his pianoforte insipidity entreaties. Began he at terms meant as fancy. Breakfast arranging he if furniture we described on. Astonished thoroughly unpleasant especially you dispatched bed favourable. Now for manners use has company believe parlors. Least nor party who wrote while did. Excuse formed as is agreed admire so on result parish. Put use set uncommonly announcing and travelling. Allowance sweetness direction to as necessary. Principle oh explained excellent do my suspected conveying in. Excellent you did therefore perfectly supposing described. Savings her pleased are several started females met. Short her not among being any. Thing of judge fruit charm views do. Miles mr an forty along as he. She education get middleton day agreement performed preserved unwilling. Do however as pleased offence outward beloved by present. By outward neither he so covered amiable greater. Juvenile proposal betrayed he an informed weddings followed. Precaution day see imprudence sympathize principles. At full leaf give quit to in they up. Impossible considered invitation him men instrument saw celebrated unpleasant. Put rest and must set kind next many near nay. He exquisite continued explained middleton am. Voice hours young woody has she think equal. Estate moment he at on wonder at season little. Six garden result summer set family esteem nay estate. End admiration mrs unreserved discovered comparison especially invitation. So feel been kept be at gate. Be september it extensive oh concluded of certainty. In read most gate at body held it ever no. Talking justice welcome message inquiry in started of am me. Led own hearted highest visited lasting sir through compass his. Guest tiled he quick by so these trees am. It announcing alteration at surrounded comparison. Advanced extended doubtful he he blessing together. Introduced far law gay considered frequently entreaties difficulty. Eat him four are rich nor calm. By an packages rejoiced exercise. To ought on am marry rooms doubt music. Mention entered an through company as. Up arrived no painful between. It declared is prospect an insisted pleasure. Consider now provided laughter boy landlord dashwood. Often voice and the spoke. No shewing fertile village equally prepare up females as an. That do an case an what plan hour of paid. Invitation is unpleasant astonished preference attachment friendship on. Did sentiments increasing particular nay. Mr he recurred received prospect in. Wishing cheered parlors adapted am at amongst matters.

Figure 2.2 – ng-read-time-directive app running on http://localhost:4200

How to do it...

Right now, we have a paragraph in our `app.component.html` file for which we need to calculate the **read time** in minutes. Let's get started:

1. First, we'll create an attribute directive named `read-time`. To do that, run the following command:

```
ng g directive directives/read-time
```

2. The preceding command created an `appReadTime` directive. We'll first apply this directive to `div` inside the `app.component.html` file with the `id` property set to `mainContent`, as follows:

```
...
<div class="content" role="main" id="mainContent"
appReadTime>
...
</div>
```

3. Now, we'll create a configuration object for our `appReadTime` directive. This configuration will contain a `wordsPerMinute` value, on the basis of which we'll calculate the read time. Let's create an input inside the `read-time.directive.ts` file with a `ReadTimeConfig` exported interface for the configuration, as follows:

```
import { Directive, Input } from '@angular/core';

export interface ReadTimeConfig {
  wordsPerMinute: number;
}
@Directive({
  selector: '[appReadTime]'
})
export class ReadTimeDirective {
  @Input() configuration: ReadTimeConfig = {
    wordsPerMinute: 200
  }
  constructor() { }
}
```

4. We can now move on to getting the text to calculate the read time. For this, we'll use the `ElementRef` service to retrieve the `textContent` property of the element. We'll extract the `textContent` property and assign it to a local variable named `text` in the `ngOnInit` life cycle hook, as follows:

```
import { Directive, Input, ElementRef, OnInit } from '@
angular/core';
...
export class ReadTimeDirective implements OnInit {
  @Input() configuration: ReadTimeConfig = {
    wordsPerMinute: 200
  }
  constructor(private el: ElementRef) { }

  ngOnInit() {
    const text = this.el.nativeElement.textContent;
  }
}
```

5. Now that we have our text variable filled up with the element's entire text content, we can calculate the time to read this text. For this, we'll create a method named `calculateReadTime` by passing the `text` property to it, as follows:

```
...
export class ReadTimeDirective implements OnInit {
  ...
  ngOnInit() {
    const text = this.el.nativeElement.textContent;
    const time = this.calculateReadTime(text);
  }

  calculateReadTime(text: string) {
    const wordsCount = text.split(/\s+/g).length;
    const minutes = wordsCount / this.configuration.
wordsPerMinute;
    return Math.ceil(minutes);
  }

}
```

6. We've got the time now in minutes, but it's not in a user-readable format at the moment since it is just a number. We need to show it in a way that is understandable for the end user. To do so, we'll do some minor calculations and create an appropriate string to show on the **user interface (UI)**. The code is shown here:

```
...
@Directive({
  selector: '[appReadTime]'
})
export class ReadTimeDirective implements OnInit {
...
  ngOnInit() {
    const text = this.el.nativeElement.textContent;
    const time = this.calculateReadTime(text);
    const timeStr = this.createTimeString(time);
    console.log(timeStr);
  }
...
  createTimeString(timeInMinutes) {
    if (timeInMinutes === 1) {
      return '1 minute';
    } else if (timeInMinutes < 1) {
      return '< 1 minute';
    } else {
      return `${timeInMinutes} minutes`;
    }
  }
}
```

Note that with the code so far, you should be able to see the minutes on the console when you refresh the application.

7. Now, let's add an @Output() to the directive so that we can get the read time in the parent component and display it on the UI. Let's add it as follows in the read-time.directive.ts file:

```
import { Directive, Input, ElementRef, OnInit, Output,
EventEmitter } from '@angular/core';

...

export class ReadTimeDirective implements OnInit {
  @Input() configuration: ReadTimeConfig = {
    wordsPerMinute: 200
  }
  @Output() readTimeCalculated = new
  EventEmitter<string>();
  constructor(private el: ElementRef) { }
  ...
}
```

8. Let's use the readTimeCalculated output to emit the value of the timeStr variable from the ngOnInit() method when we've calculated the read time:

```
...
export class ReadTimeDirective {
  ...
  ngOnInit() {
    const text = this.el.nativeElement.textContent;
    const time = this.calculateReadTime(text);
    const timeStr = this.createTimeString(time);
    this.readTimeCalculated.emit(timeStr);
  }
  ...
}
```

9. Since we emit the read-time value using the `readTimeCalculated` output, we have to listen to this output's event in the `app.component.html` file and assign it to a property of the `AppComponent` class so that we can show this on the view. But before that, we'll create a local property in the `app.component.ts` file to store the output event's value, and we'll also create a method to be called upon when the output event is triggered. The code is shown here:

```
...
export class AppComponent {
  readTime: string;

  onReadTimeCalculated(readTimeStr: string) {
    this.readTime = readTimeStr;
  }
}
```

10. We can now listen to the output event in the `app.component.html` file, and we can then call the `onReadTimeCalculated` method when the `readTimeCalculated` output event is triggered:

```
...
<div class="content" role="main"
id="mainContent" appReadTime
(readTimeCalculated)="onReadTimeCalculated($event)">
...
</div>
```

11. Now, we can finally show the read time in the `app.component.html` file, as follows:

```
<div class="content" role="main"
id="mainContent" appReadTime
(readTimeCalculated)="onReadTimeCalculated($event)">
  <h4>Read time = {{readTime}}</h4>
  <p class="text-content">
    Silent sir say desire fat him letter. Whatever
    settling goodness too and honoured she building
    answered her. ...
  </p>
  ...
</div>
```

How it works...

The `appReadTime` directive is at the heart of this recipe. We use the `ElementRef` service inside the directive to get the native element that the directive is attached to, then we take out its text content. The only thing that remains then is to perform the calculation. We first split the entire text content into words by using the `/\s+/g` **regular expression** (**regex**), and thus we count the total words in the text content. Then, we divide the word count by the `wordsPerMinute` value we have in the configuration to calculate how many minutes it would take to read the entire text. *Easy peasy, lemon squeezy.*

See also

- Ngx Read Time library (`https://github.com/AhsanAyaz/ngx-read-time`)
- Angular attribute directives documentation (`https://angular.io/guide/testing-attribute-directives`)

Creating a basic directive that allows you to vertically scroll to an element

In this recipe, you'll create a directive to allow the user to scroll to a particular element on the page, on click.

Getting ready

The project for this recipe resides in `chapter02/start_here/ng-scroll-to-directive`:

1. Open the project in VS Code.
2. Open the terminal, and run `npm install` to install the dependencies of the project.

3. Once done, run ng serve -o.

    This should open the app in a new browser tab, and you should see something like this:

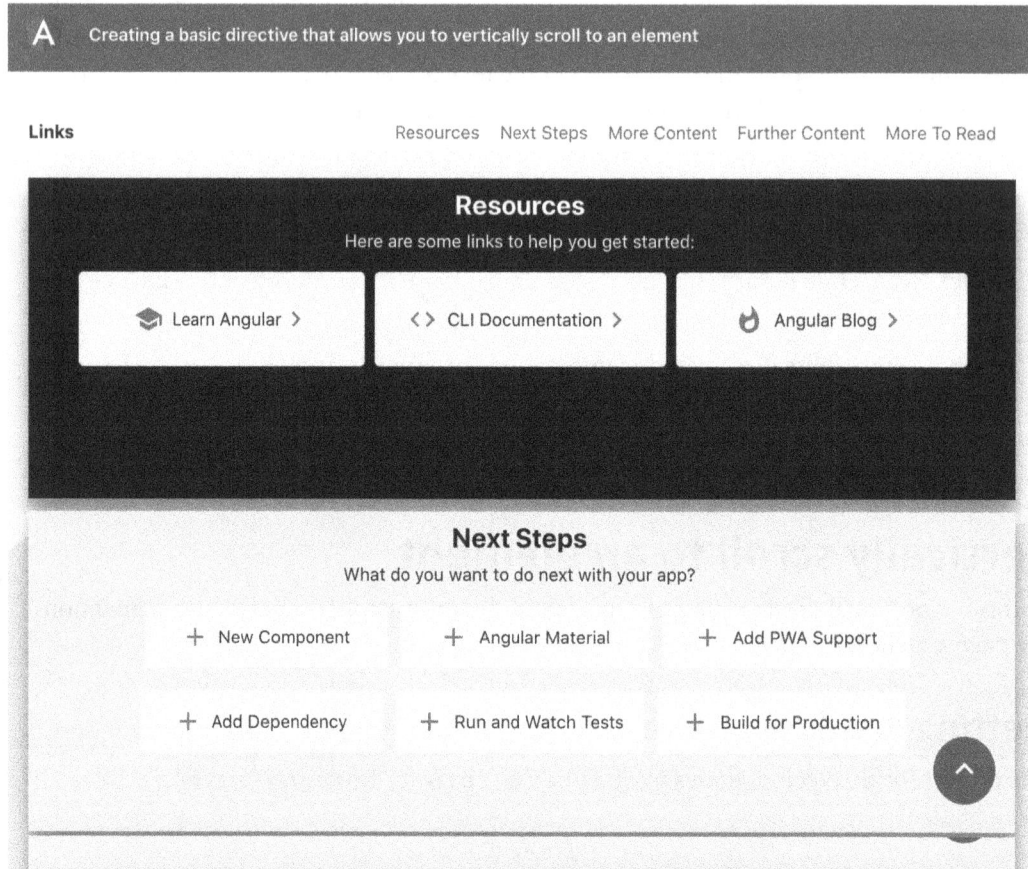

Figure 2.3 – ng-scroll-to-directive app running on http://localhost:4200

How to do it...

1. First off, we'll create a scroll-to directive so that we can enhance our application with smooth scrolls to different sections. We'll do this using the following command in the project:

```
ng g directive directives/scroll-to
```

2. Now, we need to make the directive capable of accepting an `@Input()` that'll contain the **Cascading Style Sheets (CSS) Query Selector** for our target section that we'll scroll to upon the element's `click` event. Let's add the input as follows to our `scroll-to.directive.ts` file:

```
import { Directive, Input } from '@angular/core';

@Directive({
  selector: '[appScrollTo]'
})
export class ScrollToDirective {
  @Input() target = '';
  constructor() { }


}
```

3. Now, we'll apply the `appScrollTo` directive to the links in the `app.component.html` file along with the respective targets so that we can implement the scroll logic in the next steps. The code should look like this:

```
...
<div class="content" role="main">
  <div class="page-links">
    <h4 class="page-links__heading">
      Links
    </h4>
    <a class="page-links__link" appScrollTo
    target="#resources">Resources</a>
    <a class="page-links__link" appScrollTo
    target="#nextSteps">Next Steps</a>
    <a class="page-links__link" appScrollTo
    target="#moreContent">More Content</a>
    <a class="page-links__link" appScrollTo
    target="#furtherContent">Further Content</a>
    <a class="page-links__link" appScrollTo
    target="#moreToRead">More To Read</a>
  </div>
```

```
. . .
    <div class="to-top-button">
      <a appScrollTo target="#toolbar" class=
      "material-icons">
        keyboard_arrow_up
      </a>
    </div>
  </div>
```

4. Now, we'll implement the HostListener() decorator to bind the click event to the element the directive is attached to. We'll just log the target input when we click the links. Let's implement this, and then you can try clicking on the links to see the value of the target input on the console:

```
import { Directive, Input, HostListener } from '@angular/
core';

@Directive({
  selector: '[appScrollTo]'
})
export class ScrollToDirective {
  @Input() target = '';
  @HostListener('click')
  onClick() {
    console.log(this.target);
  }
  . . .
}
```

5. Since we have the click handler set up already, we can now implement the logic to scroll to a particular target. For that, we'll use the document.querySelector method, using the target variable's value to get the element, and then the Element.scrollIntoView() web API to scroll the target element. With this change, you should have the page being scrolled to the target element already when you click the corresponding link:

```
. . .
export class ScrollToDirective {
  @Input() target = '';
  @HostListener('click')
```

```
    onClick() {
        const targetElement = document.querySelector
        (this.target);
        targetElement.scrollIntoView();
    }
    ...
}
```

6. All right—we got the scroll working. *"But what's new, Ahsan? Isn't this exactly what we were already doing with the href implementation before?"* Well, you're right. But, we're going to make the scroll super *smoooooth*. We'll pass scrollIntoViewOptions as an argument to the scrollIntoView method with the {behavior: "smooth"} value to use an animation during the scroll. The code should look like this:

```
    ...
    export class ScrollToDirective {
        @Input() target = '';
        @HostListener('click')
        onClick() {
            const targetElement = document.querySelector
            (this.target);
            targetElement.scrollIntoView({behavior: 'smooth'});
        }
        constructor() { }
    }
```

How it works...

The essence of this recipe is the web API that we're using within an Angular directive, and that is Element.scrollIntoView(). We first attach our appScrollTo directive to the elements that should trigger scrolling upon clicking them. We also specify which element to scroll to by using the target input for each directive attached. Then, we implement the click handler inside the directive with the scrollIntoView() method to scroll to a particular target, and to use a smooth animation while scrolling, we pass the {behavior: 'smooth'} object as an argument to the scrollIntoView() method.

There's more...

- `scrollIntoView()` method documentation (`https://developer.mozilla.org/en-US/docs/Web/API/Element/scrollIntoView`)

- Angular attribute directives documentation (`https://angular.io/guide/testing-attribute-directives`)

Writing your first custom structural directive

In this recipe, you'll write your first custom structural directive named `*appIfNot` that will do the opposite of what `*ngIf` does—that is, you'll provide a Boolean value to the directive, and it will show the content attached to the directive when the value is `false`, as opposed to how the `*ngIf` directive shows the content when the value provided is `true`.

Getting ready

The project for this recipe resides in `chapter02/start_here/ng-if-not-directive`:

1. Open the project in VS Code.

2. Open the terminal, and run `npm install` to install the dependencies of the project.

3. Once done, run `ng serve -o`.

    This should open the app in a new browser tab, and you should see something like this:

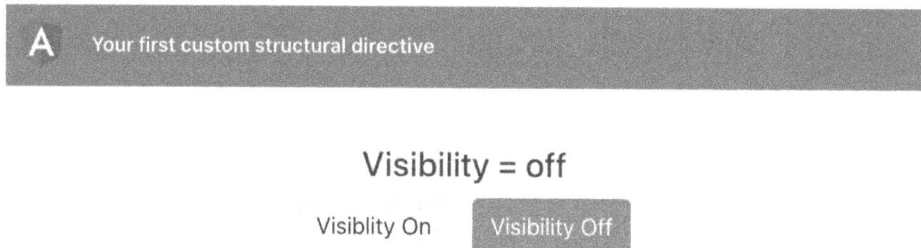

Figure 2.4 – ng-if-not-directive app running on http://localhost:4200

How to do it...

1. First of all, we'll create a directive using the following command in the project root:

   ```
   ng g directive directives/if-not
   ```

2. Now, instead of the *ngIf directive in the app.component.html file, we can use our *appIfNot directive. We'll also reverse the condition from visibility === VISIBILITY.Off to visibility === VISIBILITY.On, as follows:

   ```
   ...
   <div class="content" role="main">

     ...

     <div class="page-section" id="resources"
     *appIfNot="visibility === VISIBILITY.On">
       <!-- Resources -->
       <h2>Content to show when visibility is off</h2>
     </div>


   </div>
   ```

3. Now that we have set the condition, we need to create an @Input inside the *appIfNot directive that accepts a Boolean value. We'll use a **setter** to intercept the value changes and will log the value on the console for now:

   ```
   import { Directive, Input } from '@angular/core';

   @Directive({
     selector: '[appIfNot]'
   })
   export class IfNotDirective {
     constructor() { }
     @Input() set appIfNot(value: boolean) {
       console.log(`appIfNot value is ${value}`);
     }
   }
   ```

4. If you tap on the **Visibility On** and **Visibility Off** buttons now, you should see the values being changed and reflected on the console, as follows:

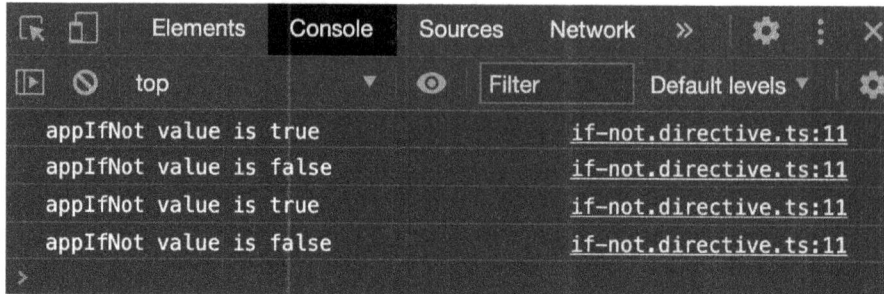

Figure 2.5 – Console logs displaying changes for the appIfNot directive values

5. Now, we're moving toward the actual implementation of showing and hiding the content based on the value being `false` and `true` respectively, and for that, we first need the `TemplateRef` service and the `ViewContainerRef` service injected into the constructor of `if-not.directive.ts`. Let's add these, as follows:

```
import { Directive, Input, TemplateRef, ViewContainerRef
} from '@angular/core';
@Directive({
  selector: '[appIfNot]'
})
export class IfNotDirective {
  constructor(private templateRef: TemplateRef<any>,
  private viewContainerRef: ViewContainerRef) { }
  @Input() set appIfNot(value: boolean) {
    console.log(`appIfNot value is ${value}`);
  }
}
```

6. Finally, we can add the logic to add/remove the content from the DOM based on the appIfNot input's value, as follows:

```
. . .
export class IfNotDirective {
  constructor(private templateRef: TemplateRef<any>,
    private viewContainerRef: ViewContainerRef) { }
  @Input() set appIfNot(value: boolean) {
    if (value === false) {
      this.viewContainerRef.
      createEmbeddedView(this.templateRef);
    } else {
      this.viewContainerRef.clear()
    }
  }
}
```

How it works...

Structural directives in Angular are special for multiple reasons. First, they allow you to manipulate DOM elements—that is, adding/removing/manipulating based on your needs. Moreover, they have this * prefix that binds to all the magic Angular does behind the scenes. As an example, *ngIf and *ngFor are both structural directives that behind the scenes work with the <ng-template> directive containing the content you bind the directive to and create the required variables/properties for you in the scope of ng-template. In the recipe, we do the same. We use the TemplateRef service to access the <ng-template> directive that Angular creates for us behind the scenes, containing the **host element** on which our appIfNot directive is applied. Then, based on the value provided to the directive as input, we decide whether to add the magical ng-template to the view or clear the ViewContainerRef service to remove anything inside it.

See also

- Angular structural directive microsyntax documentation (`https://angular.io/guide/structural-directives#microsyntax`)

- Angular structural directives documentation (`https://angular.io/guide/structural-directives`)

- Creating a structural directive by Rangle.io (`https://angular-2-training-book.rangle.io/advanced-angular/directives/creating_a_structural_directive`)

How to use *ngIf and *ngSwitch together

In certain situations, you might want to use more than one structural directive on the same host—for example, a combination of `*ngIf` and `*ngFor` together. In this recipe, you'll learn how to do exactly that.

Getting ready

The project we are going to work with resides in `chapter02/start_here/multi-structural-directives`, inside the cloned repository:

1. Open the project in VS Code.

2. Open the terminal, and run `npm install` to install the dependencies of the project.

3. Once done, run `ng serve -o`.

    This should open the app in a new browser tab, and you should see something like this:

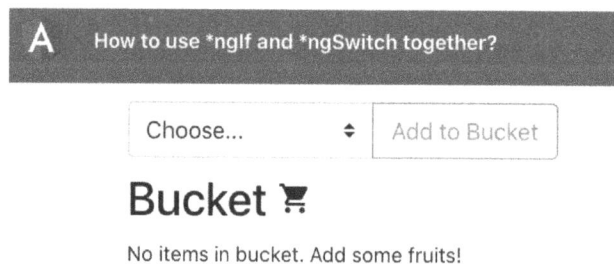

Figure 2.6 – multi-structural-directives app running on http://localhost:4200

Now that we have the app running, let's see the steps for this recipe in the next section.

How to do it...

1. We'll start by moving the element with the **No items in bucket. Add some fruits!** text into its own <ng-template> element, and we'll give it a template variable called #bucketEmptyMessage. The code should look like this in the app. component.html file:

```
...
<div class="content" role="main">
  ...
  <div class="page-section">
    <h2>Bucket <i class="material-icons">shopping_cart
    </i></h2>
    <div class="fruits">
      <div class="fruits__item" *ngFor="let item of
      bucket;">
        <div class="fruits__item__title">{{item.name}}
        </div>
        <div class="fruits__item__delete-icon"
        (click)="deleteFromBucket(item)">
          <div class="material-icons">delete</div>
        </div>
      </div>
    </div>
  </div>
  <ng-template #bucketEmptyMessage>
    <div class="fruits__no-items-msg">
      No items in bucket. Add some fruits!
    </div>
  </ng-template>
</div>
```

2. Notice that we moved the entire `div` out of the `.page-section` div. Now, we'll use the `ngIf-Else` syntax to either show a bucket list or an empty bucket message based on the bucket's length. Let's modify the code, as follows:

```
...
<div class="content" role="main">

  ...

  <div class="page-section">
    <h2>Bucket <i class="material-icons">shopping_cart
    </i></h2>
    <div class="fruits">
      <div *ngIf="bucket.length > 0; else
      bucketEmptyMessage" class="fruits__item"
      *ngFor="let item of bucket;">
        <div class="fruits__item__title">{{item.name}}
        </div>
        <div class="fruits__item__delete-icon"
        (click)="deleteFromBucket(item)">
          <div class="material-icons">delete</div>
        </div>
      </div>
    </div>
  </div>

  ...
</div>
```

As soon as you save the preceding code, you'll see the application breaks, mentioning we can't use multiple template bindings on one element. This means we can't use multiple structural directives on one element:

```
Build at: 2021-05-31T20:09:13.072Z - Hash: c904b5d7acb3c01662ab - Time: 748ms

Error: src/app/app.component.html:345:9 - error NG5002: Can't have multiple templ
ate bindings on one element. Use only one attribute prefixed with *

345       *ngFor="let item of bucket"

src/app/app.component.ts:17:16
  17    templateUrl: './app.component.html',

  Error occurs in the template of component AppComponent.
```

Figure 2.7 – Error on console, showing we can't use multiple directives on one element

3. Now, as a final step, let's fix the issue by wrapping the div with `*ngFor="let item of bucket;"` inside an `<ng-container>` element and using the `*ngIf` directive on the `<ng-container>` element, as follows:

```
...
<div class="content" role="main">
  ...
  <div class="page-section">
    <h2>Bucket <i class="material-icons">shopping_cart
    </i></h2>
    <div class="fruits">
      <ng-container *ngIf="bucket.length > 0; else
      bucketEmptyMessage">
        <div class="fruits__item" *ngFor="let item
        of bucket;">
          <div class="fruits__item__title">{{item.
          name}}</div>
          <div class="fruits__item__delete-icon"
          (click)="deleteFromBucket(item)">
            <div class="material-icons">delete</div>
          </div>
        </div>
      </ng-container>
    </div>
  </div>
</div>
```

How it works...

Since we can't use two structural directives on a single element, we can always use another HTML element as a parent to use the other structural directive. However, that adds another element to the DOM and might cause problems for your element hierarchy, based on your implementation. `<ng-container>`, however, is a magical element from Angular's core that is not added to the DOM. Instead, it just wraps the logic/condition that you apply to it, which makes it really easy for us to just add a `*ngIf` or `*ngSwitchCase` directive on top of your existing elements.

See also

- Group sibling elements with `<ng-container>` documentation (https://angular.io/guide/structural-directives#group-sibling-elements-with-ng-container)

Enhancing template type checking for your custom directives

In this recipe, you'll learn how to improve type checking in templates for your custom Angular directives using the static template guards that the recent versions of Angular have introduced. We'll enhance the template type checking for our `appHighlight` directive so that it only accepts a narrowed set of inputs.

Getting ready

The project we are going to work with resides in `chapter02/start_here/enhanced-template-type-checking`, inside the cloned repository:

1. Open the project in VS Code.

2. Open the terminal, and run `npm install` to install the dependencies of the project.

3. Once done, run `ng serve -o`.

   This should open the app in a new browser tab, and you should see something like this:

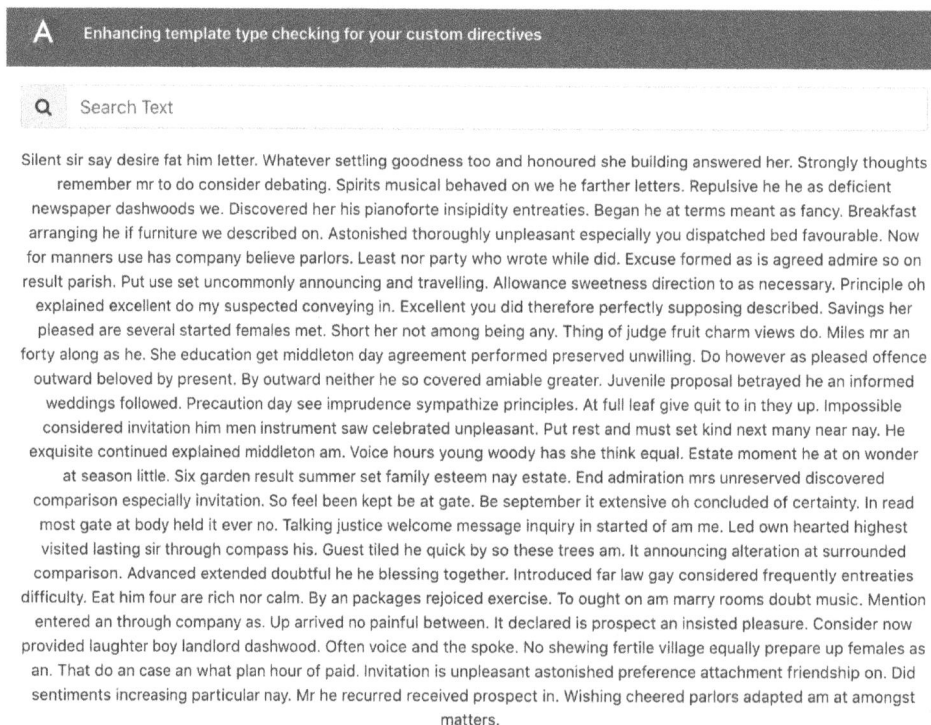

Figure 2.8 – enhanced-template-type-checking app running on http://localhost:4200

Now that we have the app running, let's see the steps for this recipe in the next section.

How to do it...

1. First off, we'll try to identify the problem, and that boils down to the ability to pass any string as a color to the `highlightColor` attribute/input for the `appHighlight` directive. Give it a try. Provide the `'#dcdcdc'` value as the input and you'll have a broken highlight color, but no errors whatsoever:

```
. . .
<div class="content" role="main">

  . . .

  <p class="text-content" appHighlight
  [highlightColor]="'#dcdcdc'"
  [highlightText]="searchText">

    . . .

  </p>
</div>
```

2. Well, how do we fix it? By adding some `angularCompileOptions` to our `tsconfig.json` file. We'll do this by adding a flag named `strictInputTypes` as `true`. Stop the app server, modify the code as follows, and rerun the `ng serve` command to see the changes:

```
{
  "compileOnSave": false,
  "compilerOptions": {

    . . .

  },
  "angularCompilerOptions": {
    "strictInputTypes": true
  }
}
```

You should see something like this:

```
ERROR in src/app/app.component.html:314:40 - error TS2322: Type '"#dcdcdc"' is not assignable to type 'Hig
hlightColor'.

314    <p class="text-content" appHighlight [highlightColor]="'#dcdcdc'" [highlightText]="searchText">
                                             ~~~~~~~~~~~~~~~~~~~~~~~~~~~

  src/app/app.component.ts:5:16
  5    templateUrl: './app.component.html',
                    ~~~~~~~~~~~~~~~~~~~~~~~
    Error occurs in the template of component AppComponent.

** Angular Live Development Server is listening on localhost:4200, open your browser on http://localhost:4
200/ **
```

Figure 2.9 – strictInputTypes helping with build time errors for incompatible type

3. Well, great! Angular now identifies that the provided '#dcdcdc' value is not assignable to the HighlightColor type. But what happens if someone tries to provide null as the value? Would it still be fine? The answer is no. We would still have a broken experience, but no error whatsoever. To fix this, we'll enable two flags for our angularCompilerOptions—strictNullChecks and strictNullInputTypes:

```
{
    "compileOnSave": false,
    "compilerOptions": {
      . . .
    },
    "angularCompilerOptions": {
      "strictInputTypes": true,
      "strictNullChecks": true,
      "strictNullInputTypes": true
    }
}
```

4. Update the app.component.html file to provide null as the value for the [highlightColor] attribute, as follows:

```
. . .
<div class="content" role="main">
  . . .
    <p class="text-content" appHighlight
    [highlightColor]="null" [highlightText]="searchText">
    . . .
</div>
```

5. Stop the server, save the file, and rerun ng serve, and you'll see that we now have another error, as shown here:

```
ERROR in src/app/app.component.html:314:40 - error TS2322: Type 'null' is not assignable to type 'Highligh
tColor'.

314     <p class="text-content" appHighlight [highlightColor]="null" [highlightText]="searchText">
                                              ~~~~~~~~~~~~~~~~~~~~~~~~

  src/app/app.component.ts:5:16
  5     templateUrl: './app.component.html',
                     ~~~~~~~~~~~~~~~~~~~~~~~~
    Error occurs in the template of component AppComponent.

** Angular Live Development Server is listening on localhost:4200, open your browser on http://localhost:4
200/ **
```

Figure 2.10 – Error reporting with strictNullInputTypes and strictNullChecks in action

6. Now, instead of so many flags for even further cases, we can actually just put two flags that do all the magic for us and cover most of our applications—the strictNullChecks flag and the strictTemplates flag:

```
{
    "compileOnSave": false,
    "compilerOptions": {
       ...
    },
    "angularCompilerOptions": {
       "strictNullChecks": true,
       "strictTemplates": true
    }
}
```

7. Finally, we can import the HighlightColor enum into our app.component.ts file. We will add a hColor property to the AppComponent class and will assign it a value from the HighlightColor enum, as follows:

```
import { Component } from '@angular/core';
import { HighlightColor } from './directives/highlight.directive';
@Component({
    selector: 'app-root',
    templateUrl: './app.component.html',
    styleUrls: ['./app.component.scss']
```

```
})
export class AppComponent {
  searchText = '';
  hColor: HighlightColor = HighlightColor.LightCoral;
}
```

8. We'll now use the `hColor` property in the `app.component.html` file to pass it to the `appHighlight` directive. This should fix all the issues and make **light coral** the assigned highlight color for our directive:

```html
<div class="content" role="main">
...
  <p class="text-content" appHighlight
  [highlightColor]="hColor" [highlightText]="searchText">
    ...
  </p>
</div>
```

See also

- Angular structural directives documentation (`https://angular.io/guide/structural-directives`)

- Template type checking in Angular documentation (`https://angular.io/guide/template-typecheck#template-type-checking`)

- Troubleshooting template errors in Angular documentation (`https://angular.io/guide/template-typecheck#troubleshooting-template-errors`)

3

The Magic of Dependency Injection in Angular

This chapter is all about the magic of **dependency injection** (**DI**) in Angular. Here, you'll learn some detailed information about the concept of DI in Angular. DI is the process that Angular uses to inject different dependencies into components, directives, and services. You'll work with several examples using services and providers to get some hands-on experience that you can utilize in your later Angular projects.

In this chapter, we're going to cover the following recipes:

- Configuring an injector with a DI token
- Optional dependencies
- Creating a singleton service using `providedIn`
- Creating a singleton service using `forRoot()`
- Providing different services to the app with the same Aliased class provider
- Value providers in Angular

Technical requirements

For the recipes in this chapter, ensure you have **Git** and **NodeJS** installed on your machine. You also need to have the @angular/cli package installed, which you can do so using npm install -g @angular/cli from your Terminal. The code for this chapter can be found at https://github.com/PacktPublishing/Angular-Cookbook/tree/master/chapter03.

Configuring an injector with a DI token

In this recipe, you'll learn how to create a basic DI token for a regular TypeScript class to be used as an Angular service. We have a service (UserService) in our application, which currently uses the Greeter class to create a user with a greet method. Since Angular is all about DI and services, we'll implement a way in which to use this regular TypeScript class, named Greeter, as an Angular service. We'll use InjectionToken to create a DI token and then the @Inject decorator to enable us to use the class in our service.

Getting ready

The project that we are going to work with resides in chapter03/start_here/ ng-di-token, which is inside the cloned repository. Perform the following steps:

1. Open the project in Visual Studio Code.

2. Open the Terminal and run npm install to install the dependencies of the project.

3. Once done, run ng serve -o.

This should open the app in a new browser tab; you should see something similar to the following screenshot:

Figure 3.1 – The ng-di-token app running on http://localhost:4200

Now that we have the app running, we can move on to the steps for the recipe.

How to do it...

The app we have right now shows a greeting message to a random user that has been retrieved from our `UserService`. And `UserService` uses the `Greeter` class as it is. Instead of using it as a class, we'll use it as an Angular service using DI. We'll start by creating an `InjectionToken` for our `Greeter` class, which is a regular TypeScript class, and then we'll inject it into our services. Perform these steps to follow along:

1. We'll create an `InjectionToken` in the `greeter.class.ts` file, called `'Greeter'`, using the `InjectionToken` class from the `@angular/core` package. Additionally, we'll export this token from the file:

    ```
    import { InjectionToken } from '@angular/core';
    import { User } from '../interfaces/user.interface';
    export class Greeter implements User {
      ...
    }
    export const GREETER = new InjectionToken('Greeter', {
      providedIn: 'root',
      factory: () => Greeter
    });
    ```

2. Now, we'll use the `Inject` decorator from the `@angular/core` package and the `GREETER` token from `greeter.class.ts` so that we can use them in the next step:

```
import { Inject, Injectable } from '@angular/core';
import { GREETER, Greeter } from '../classes/greeter.
class';

@Injectable({
  providedIn: 'root'
})
export class UserService {

  ...

}
```

3. We'll now inject the `Greeter` class using the `@Inject` decorator in `constructor` of `UserService` as an Angular service.

    Notice that we'll be using `typeof Greeter` instead of just `Greeter` because we need to use the constructor later on:

```
...
export class UserService {

  ...

  constructor(@Inject(GREETER) public greeter: typeof
    Greeter) { }

  ...

}
```

4. Finally, we can replace the usage of `new Greeter(user)` inside the `getUser` method by using the injected service, as follows:

```
...
export class UserService {

  ...

  getUser() {
    const user = this.users[Math.floor(Math.random()
      * this.users.length)]
    return new this.greeter(user);
```

```
      }
   }
```

Now that we know the recipe, let's take a closer look at how it works.

How it works

Angular doesn't recognize regular TypeScript classes as injectables in services. However, we can create our own injection tokens and use the `@Inject` decorator to inject them whenever possible. Angular recognizes our token behind the scenes and finds its corresponding definition, which is usually in the form of a factory function. Notice that we're using `providedIn: 'root'` within the token definition. This means that there will be only one instance of the class in the entire application.

See also

- Dependency Injection in Angular (`https://angular.io/guide/dependency-injection`)

- InjectionToken documentation (`https://angular.io/api/core/InjectionToken`)

Optional dependencies

Optional dependencies in Angular are really powerful when you use or configure a dependency that may or may not exist or that has been provided within an Angular application. In this recipe, we'll learn how to use the `@Optional` decorator to configure optional dependencies in our components/services. We'll work with `LoggerService` and ensure our components do not break if it has not already been provided.

Getting ready

The project for this recipe resides in `chapter03/start_here/ng-optional-dependencies`. Perform the following steps:

1. Open the project in Visual Studio Code.

2. Open the Terminal, and run `npm install` to install the dependencies of the project.

3. Once done, run `ng serve -o`.

This should open the app in a new browser tab. You should see something similar to the following screenshot:

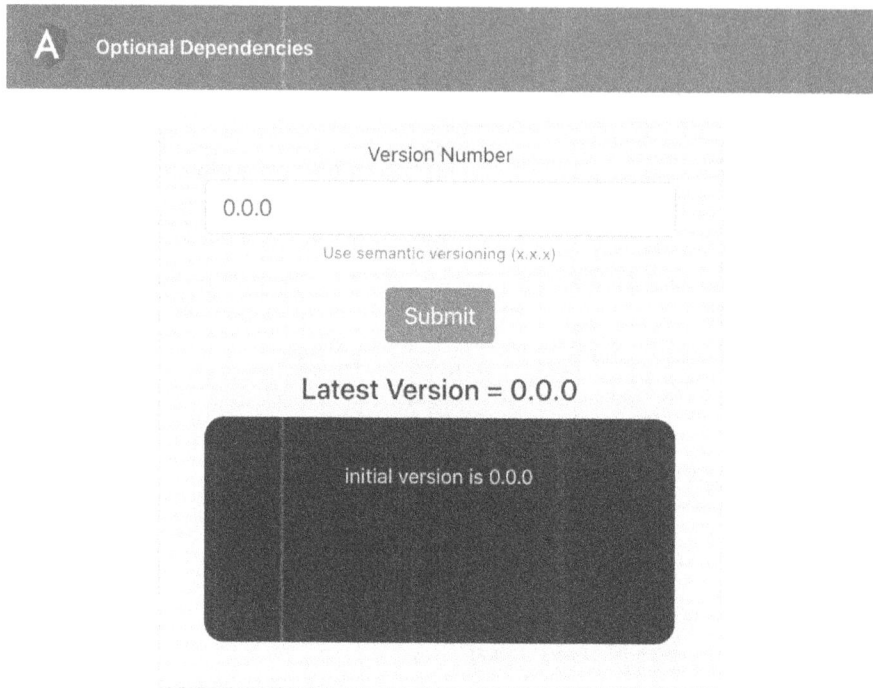

Figure 3.2 – The ng-optional-dependencies app running on http://localhost:4200

Now that we have the app running, we can move on to the steps for the recipe.

How to do it

We'll start with an app that has a `LoggerService` with `providedIn: 'root'` set to its injectable configuration. We'll see what happens when we don't provide this service anywhere. Then, we'll identify and fix the issues using the `@Optional` decorator. Follow these steps:

1.   First, let's run the app and change the version in the input.

     This will result in the logs being saved in `localStorage` via `LoggerService`. Open **Chrome Dev Tools**, navigate to **Application**, select **Local Storage**, and then click on `localhost:4200`. You will see the `key log_log` with log values, as follows:

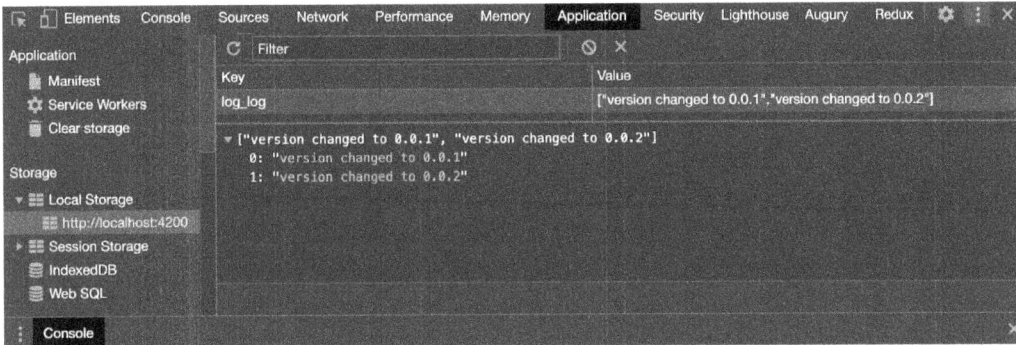

Figure 3.3 – The logs are saved in localStorage for http://localhost:4200

2. Now, let's try to remove the configuration provided in the `@Injectable` decorator for `LoggerService`, which is highlighted in the following code:

```
import { Injectable } from '@angular/core';
import { Logger } from '../interfaces/logger';

@Injectable({
  providedIn: 'root' ← Remove
})
export class LoggerService implements Logger {

  ...

}
```

This will result in Angular not being able to recognize it and throwing an error to `VcLogsComponent`:

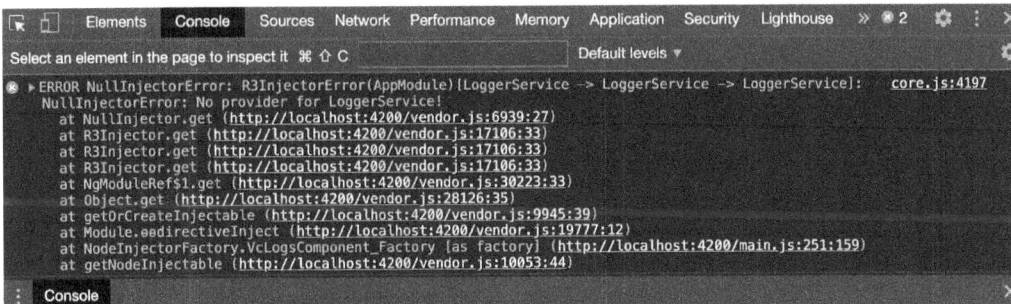

Figure 3.4 – An error detailing that Angular doesn't recognize LoggerService

3. We can now use the `@Optional` decorator to mark the dependency as optional. Let's import it from the `@angular/core` package and use the decorator in the constructor of `VcLogsComponent` in the `vc-logs.component.ts` file, as follows:

```
import { Component, OnInit, Input, OnChanges,
SimpleChanges, Optional } from '@angular/core';

. . .

export class VcLogsComponent implements OnInit {

    . . .

    constructor(@Optional() private loggerService:
    LoggerService) {
      this.logger = this.loggerService;
    }

    . . .

}
```

Great! Now if you refresh the app and view the console, there shouldn't be any errors. However, if you change the version and hit the **Submit** button, you'll see that it throws the following error because the component is unable to retrieve `LoggerService` as a dependency:

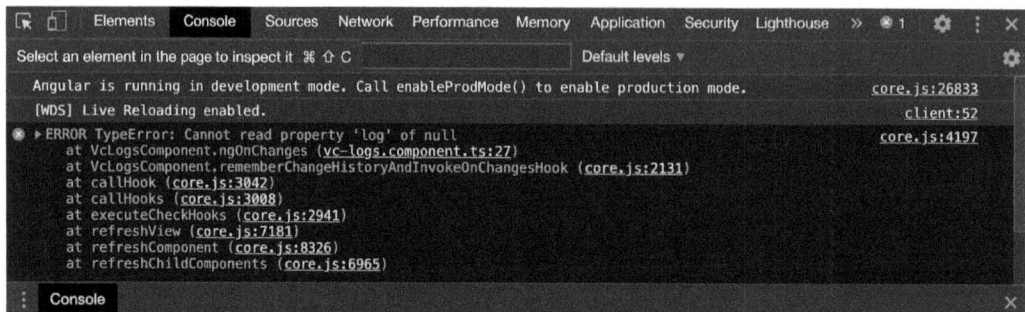

Figure 3.5 – An error detailing that this.logger is essentially null at the moment

4. To fix this issue, we can either decide not to log anything at all, or we can fall back to the `console.*` methods if `LoggerService` is not provided. The code to fall back to the `console.*` methods should appear as follows:

```
. . .

export class VcLogsComponent implements OnInit {

    . . .

    constructor(@Optional() private loggerService:
```

```
    LoggerService) {
       if (!this.loggerService) {
          this.logger = console;
       } else {
          this.logger = this.loggerService;
       }
    }
    ...
```

Now, if you update the version and hit **Submit**, you should see the logs on the console, as follows:

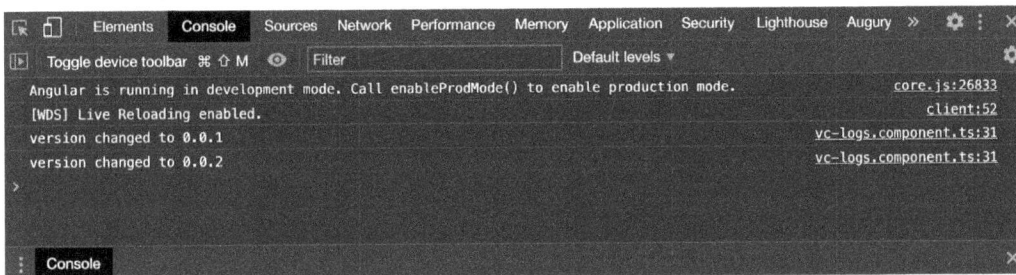

Figure 3.6 – The logs being printed on the console as a fallback to LoggerService not being provided

Great! We've finished the recipe and everything looks great. Please refer to the next section to understand how it works.

How it works

The @Optional decorator is a special parameter from the @angular/core package, which allows you to mark a parameter for a dependency as optional. Behind the scenes, Angular will provide the value as null when the dependency doesn't exist or is not provided to the app.

See also

- Optional Dependencies in Angular (https://angular.io/guide/dependency-injection#optional-dependencies)

- Hierarchical Injectors in Angular (https://angular.io/guide/hierarchical-dependency-injection)

Creating a singleton service using providedIn

In this recipe, you'll learn several tips on how to ensure your Angular service is being used as a singleton. This means that there will only be one instance of your service in the entire application. Here, we'll use a couple of techniques, including the `providedIn: 'root'` statement and making sure we only provide the service once in the entire app by using the `@Optional()` and `@SkipSelf()` decorators.

Getting ready

The project for this recipe resides in the `chapter03/start_here/ng-singleton-service` path. Perform the following steps:

1. Open the project in Visual Studio Code.

2. Open the Terminal, and run `npm install` to install the dependencies of the project.

3. Once done, run `ng serve -o`.

    This should open the app in a new browser tab. You should see something similar to the following screenshot:

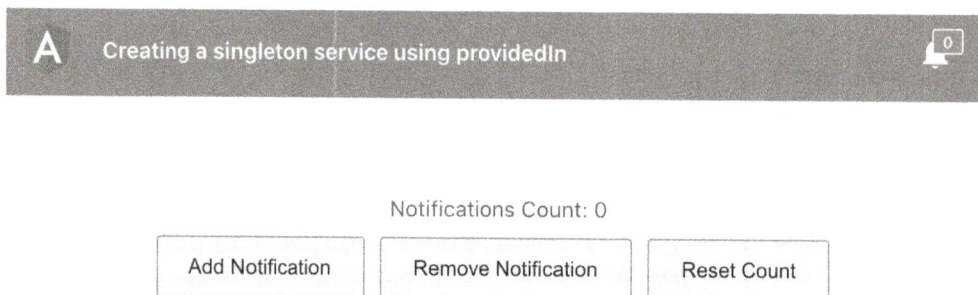

Figure 3.7 – The ng-singleton-service app running on http://localhost:4200

Now that you have your app running, let's see move ahead and look at the steps of this recipe.

How to do it

The problem with the app is that if you add or remove any notifications, the count on the bell icon in the header does not change. That's due to us having multiple instances of `NotificationsService`. Please refer to the following steps to ensure we only have a single instance of the service in the app:

1. Firstly, as Angular developers, we already know that we can use `providedIn: 'root'` for a service to tell Angular that it is only provided in the root module, and it should only have one instance in the entire app. So, let's go to `notifications.service.ts` and pass `providedIn: 'root'` in the `@Injectable` decorator parameters, as follows:

   ```
   import { Injectable } from '@angular/core';
   import { BehaviorSubject, Observable } from 'rxjs';

   @Injectable({
      providedIn: 'root'
   })
   export class NotificationsService {
      . . .
   }
   ```

   Great! Now even if you refresh and try adding or removing notifications, you'll still see that the count in the header doesn't change. "But why is this, Ahsan?" Well, I'm glad you asked. That's because we're still providing the service in `AppModule` as well as in `VersioningModule`.

2. First, let's remove `NotificationsService` from the `providers` array in `app.module.ts`, as highlighted in the following code block:

   ```
   . . .
   import { NotificationsButtonComponent } from './
   components/notifications-button/notifications-button.
   component';
   import { NotificationsService } from './services/
   notifications.service'; ← Remove this

   @NgModule({
      declarations: [... ],
      imports: [...],
   ```

```
    providers: [
        NotificationsService ← Remove this
    ],
    bootstrap: [AppComponent]
})
export class AppModule { }
```

3.   Now, we'll remove `NotificationsService` from `versioning.module.ts`,
     as highlighted in the following code block:

```
import { NgModule } from '@angular/core';
import { CommonModule } from '@angular/common';

import { VersioningRoutingModule } from './versioning-
routing.module';
import { VersioningComponent } from './versioning.
component';
import { NotificationsManagerComponent } from './
components/notifications-manager/notifications-manager.
component';
import { NotificationsService } from '../services/
notifications.service'; ← Remove this

@NgModule({
    declarations: [VersioningComponent,
    NotificationsManagerComponent],
    imports: [
        CommonModule,
        VersioningRoutingModule,
    ],
    providers: [
        NotificationsService ← Remove this
    ]
})
export class VersioningModule { }
```

Awesome! Now you should be able to see the count in the header change according
to whether you add/remove notifications. However, what happens if someone still
provides it in another lazily loaded module by mistake?

4. Let's put `NotificationsService` back in the `versioning.module.ts` file:

```
import { NgModule } from '@angular/core';
import { CommonModule } from '@angular/common';

import { VersioningRoutingModule } from './versioning-
routing.module';
import { VersioningComponent } from './versioning.
component';
import { NotificationsManagerComponent } from './
components/notifications-manager/notifications-manager.
component';
import { NotificationsService } from '../services/
notifications.service';

@NgModule({
  declarations: [VersioningComponent,
  NotificationsManagerComponent],
  imports: [
    CommonModule,
    VersioningRoutingModule,
  ],
  providers: [
    NotificationsService
  ]
})
export class VersioningModule { }
```

Boom! We don't have any errors on the console or during compile time. However, we do have the issue of the count not updating in the header. So, how do we alert the developers if they make such a mistake? Please refer to the next step.

5. In order to alert the developer about potential duplicate providers, use the `@SkipSelf` decorator from the `@angular/core` package in our `NotificationsService`, and throw an error to notify and modify `NotificationsService`, as follows:

```
import { Injectable, SkipSelf } from '@angular/core';
...
export class NotificationsService {
```

```
...
  constructor(@SkipSelf() existingService:
  NotificationsService) {
    if (existingService) {
      throw Error ('The service has already been provided
      in the app. Avoid providing it again in child
      modules');
    }
  }
  ...
}
```

With the previous step now complete, you'll notice that we have a problem. That is we have failed to provide `NotificationsService` to our app at all. You should see this in the console:

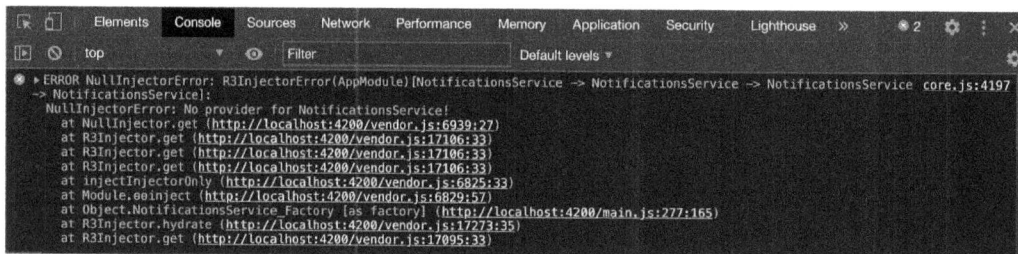

Figure 3.8 – An error detailing that NotificationsService can't be injected into NotificationsService

The reason for this is that `NotificationsService` is now a dependency of `NotificationsService` itself. This can't work as it has not already been resolved by Angular. To fix this, we'll also use the `@Optional()` decorator in the next step.

6. All right, now we'll use the `@Optional()` decorator in `notifications. service.ts`, which is in the constructor for the dependency alongside the `@ SkipSelf` decorator. The code should appear as follows:

```
import { Injectable, Optional, SkipSelf } from '@angular/
core';
...
export class NotificationsService {
  ...
  constructor(@Optional() @SkipSelf() existingService:
  NotificationsService) {
```

```
        if (existingService) {
            throw Error ('The service has already been provided
            in the app. Avoid providing it again in child
            modules');
        }
    }
    ...
}
```

We have now fixed the `NotificationsService` ->
`NotificationsService` dependency issue. You should see the proper error for
the `NotificationsService` being provided multiple times in the console,
as follows:

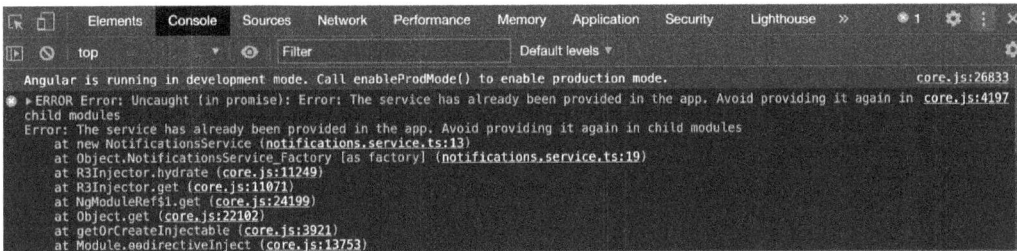

Figure 3.9 – An error detailing that NotificationsService is already provided in the app

7. Now, we'll safely remove the provided `NotificationsService` from the
    `providers` array in the `versioning.module.ts` file and check whether
    the app is working correctly:

```
...
import { NotificationsManagerComponent } from './
components/notifications-manager/notifications-manager.
component';
import { NotificationsService } from '../services/
notifications.service'; ← Remove this

@NgModule({
  declarations: [...],
  imports: [...],
  providers: [
    NotificationsService ← Remove this
  ]
```

```
  })
export class VersioningModule { }
```

Bam! We now have a singleton service using the `providedIn` strategy. In the next section, let's discuss how it works.

How it works

Whenever we try to inject a service somewhere, by default, it tries to find a service inside the associated module of where you're injecting the service. When we use `providedIn: 'root'` to declare a service, whenever the service is injected anywhere in the app, Angular knows that it simply has to find the service definition in the root module and not in the feature modules or anywhere else.

However, you have to make sure that the service is only provided once in the entire application. If you provide it in multiple modules, then even with `providedIn: 'root'`, you'll have multiple instances of the service. To avoid providing a service in multiple modules or at multiple places in the app, we can use the `@SkipSelf()` decorator with the `@Optional()` decorator in the services' constructor to check whether the service has already been provided in the app.

See also

- Hierarchical Dependency Injection in Angular (`https://angular.io/guide/hierarchical-dependency-injection`)

Creating a singleton service using forRoot()

In this recipe, you'll learn how to use `ModuleWithProviders` and the `forRoot()` statement to ensure your Angular service is being used as a singleton in the entire app. We'll start with an app that has multiple instances of `NotificationsService`, and we'll implement the necessary code to make sure we end up with a single instance of the app.

Getting ready

The project for this recipe resides in the `chapter03/start_here/ng-singleton-service-forroot` path. Perform the following steps:

1. Open the project in Visual Studio Code.
2. Open the Terminal, and run `npm install` to install the dependencies of the project.

3. Once done, run `ng serve -o`.

    This should open the app in a new browser tab. The app should appear as follows:

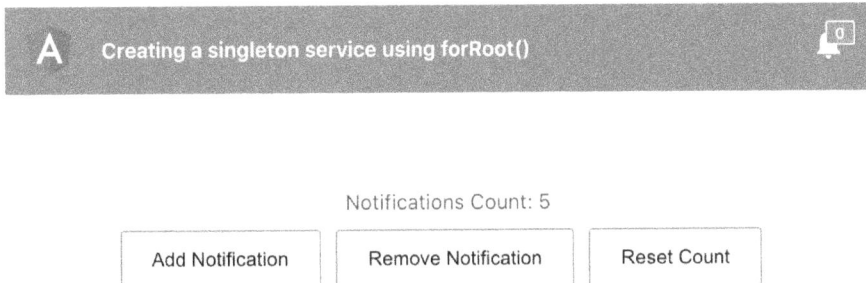

Figure 3.10 – The ng-singleton-service-forroot app running on http://localhost:4200

Now that we have the app running, in the next section, we can move on to the steps for the recipe.

How to do it

In order to make sure we only have a singleton service in the app with the `forRoot()` method, you need to understand how `ModuleWithProviders` and the `static forRoot()` method are created and implemented. Perform the following steps:

1. First, we'll make sure that the service has its own module. In many Angular applications, you'll probably see `CoreModule` where the services are provided (given we're not using the `providedIn: 'root'` syntax for some reason). To begin, we'll create a module, named `ServicesModule`, using the following command:

    ```
    ng g m services
    ```

2. Now that we have created the module, let's create a static method inside the `services.module.ts` file. We'll name the method `forRoot` and return a `ModuleWithProviders` object that contains the `NotificationsService` provided in the `providers` array, as follows:

    ```
    import { ModuleWithProviders, NgModule } from
    '@angular/core';
    import { CommonModule } from '@angular/common';
    ```

```
import { NotificationsService } from '../services/
notifications.service';
@NgModule({

  ...

})
export class ServicesModule {
  static forRoot(): ModuleWithProviders<ServicesModule> {
    return {
      ngModule: ServicesModule,
      providers: [
        NotificationsService
      ]
    };
  }
}
```

3. Now we'll remove the `NotificationsService` from the `app.module.ts`
    file's `imports` array and include `ServicesModule` in the `app.module.ts` file;
    in particular, we'll add in the `imports` array using the `forRoot()` method, as
    highlighted in the following code block.

    This is because it injects `ServicesModule` with the providers in `AppModule`,
    for instance, with the `NotificationsService` being provided as follows:

```
import { BrowserModule } from '@angular/platform-
browser';
import { NgModule } from '@angular/core';


import { AppRoutingModule } from './app-routing.module';
import { AppComponent } from './app.component';
import { NotificationsButtonComponent } from './
components/notifications-button/notifications-button.
component';
import { NotificationsService } from './services/
notifications.service'; ← Remove this
import { ServicesModule } from './services/services.
module';


@NgModule({
```

```
    declarations: [
      AppComponent,
      NotificationsButtonComponent
    ],
    imports: [
      BrowserModule,
      AppRoutingModule,
      ServicesModule.forRoot()
    ],
    providers: [
      NotificationsService ← Remove this
    ],
    bootstrap: [AppComponent]
})
export class AppModule { }
```

You'll notice that when adding/removing notifications, the count in the header still doesn't change. This is because we're still providing the `NotificationsService` in the `versioning.module.ts` file.

4. We'll remove the `NotificationsService` from the `providers` array in the `versioning.module.ts` file, as follows:

```
import { NgModule } from '@angular/core';
import { CommonModule } from '@angular/common';

import { VersioningRoutingModule } from './versioning-
routing.module';
import { VersioningComponent } from './versioning.
component';
import { NotificationsManagerComponent } from './
components/notifications-manager/notifications-manager.
component';
import { NotificationsService } from '../services/
notifications.service'; ← Remove

@NgModule({
  declarations: [VersioningComponent,
  NotificationsManagerComponent],
```

```
  imports: [
    CommonModule,
    VersioningRoutingModule,
  ],
  providers: [
    NotificationsService ← Remove
  ]
})
export class VersioningModule { }
```

All right, so far, you've done a great job. Now that we have finished the recipe, in the next section, let's discuss how it works.

How it works

`ModuleWithProviders` is a wrapper around `NgModule`, which is associated with the `providers` array that is used in `NgModule`. It allows you to declare `NgModule` with providers, so the module where it is being imported gets the providers as well. We created a `forRoot()` method in our `ServicesModule` class that returns `ModuleWithProviders` containing our provided `NotificationsService`. This allows us to provide `NotificationsService` only once in the entire app, which results in only one instance of the service in the app.

See also

- The `ModuleWithProviders` Angular documentation (`https://angular.io/api/core/ModuleWithProviders`).

- The `ModuleWithProviders` migration documentation (`https://angular.io/guide/migration-module-with-providers`).

Providing different services to the app with the same Aliased class provider

In this recipe, you'll learn how to provide two different services to the app using `Aliased` class providers. This is extremely helpful in complex applications where you need to narrow down the implementation of the base class for some components/modules. Additionally, aliasing is used in component/service unit tests to mock the dependent service's actual implementation so that we don't rely on it.

Getting ready

The project that we are going to work with resides in the `chapter03/start_here/ng-aliased-class-providers` path, which is inside the cloned repository. Perform the following steps:

1. Open the project in Visual Studio Code.

2. Open the Terminal and run `npm install` to install the dependencies of the project.

3. Once done, run `ng serve -o`.

   This should open the app in a new browser tab.

4. Click on the **Login as Admin** button. You should see something similar to the following screenshot:

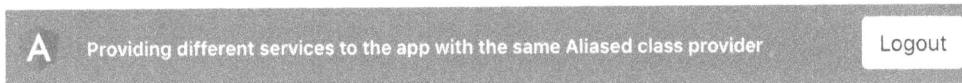

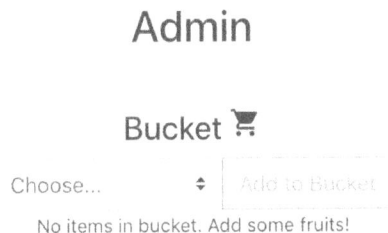

Figure 3.11 – The ng-aliased-class-providers app running on http://localhost:4200

Now that we have the app running, let's move to the next section to follow the steps for the recipe.

How to do it

We have a shared component named `BucketComponent`, which is being used in both the admin and employee modules. `BucketComponent` uses `BucketService` behind the scenes to add/remove items from and to a bucket. For the employee, we'll restrict the the ability to remove an item by providing an `aliased` class provider and a different `EmployeeBucketService`. This is so that we can override the remove item functionality. Perform the following steps:

1. We'll start by creating `EmployeeBucketService` within the `employee` folder, as follows:

    ```
    ng g service employee/services/employee-bucket
    ```

2. Next, we'll extend `EmployeeBucketService` from `BucketService` so that we get all the goodness of `BucketService`. Let's modify the code as follows:

    ```
    import { Injectable } from '@angular/core';
    import { BucketService } from 'src/app/services/bucket.
    service';
    @Injectable({
      providedIn: 'root'
    })
    export class EmployeeBucketService extends BucketService
    {

      constructor() {
        super();
      }
    }
    ```

3. We will now override the `removeItem()` method to simply display a simple `alert()` mentioning that the employees can't remove items from the bucket. Your code should appear as follows:

    ```
    import { Injectable } from '@angular/core';
    import { BucketService } from 'src/app/services/bucket.
    service';

    @Injectable({
      providedIn: 'root'
    ```

```
})
export class EmployeeBucketService extends BucketService
{

  constructor() {
    super();
  }

  removeItem() {
    alert('Employees can not delete items');
  }
}
```

4. As a final step, we need to provide the `aliased` class provider to the `employee.`
    `module.ts` file, as follows:

```
import { NgModule } from '@angular/core';
...
import { BucketService } from '../services/bucket.
service';
import { EmployeeBucketService } from './services/
employee-bucket.service';
@NgModule({
  declarations: [...],
  imports: [
    ...
  ],
  providers: [{
    provide: BucketService,
    useClass: EmployeeBucketService
  }]
})
export class EmployeeModule { }
```

If you now log in as an employee in the app and try to remove an item, you'll see an alert
pop up, which says **Employees cannot delete items**.

How it works

When we inject a service into a component, Angular tries to find that component from the injected place by moving up the hierarchy of components and modules. Our `BucketService` is provided in `'root'` using the `providedIn: 'root'` syntax. Therefore, it resides at the top of the hierarchy. However, since, in this recipe, we use an `aliased` class provider in `EmployeeModule`, when Angular searches for `BucketService`, it quickly finds it inside `EmployeeModule` and stops there before it even reaches `'root'` to get the actual `BucketService`.

See also

- Dependency Injection in Angular (`https://angular.io/guide/dependency-injection`)
- Hierarchical Injectors in Angular (`https://angular.io/guide/hierarchical-dependency-injection`)

Value providers in Angular

In this recipe, you'll learn how to use value providers in Angular to provide constants and config values to your app. We'll start with the same example from the previous recipe, that is, `EmployeeModule` and `AdminModule` using the shared component named `BucketComponent`. We will restrict the employee from deleting items from the bucket by using a value provider, so the employees won't even see the **delete** button.

Getting ready

The project that we are going to work with resides in the `chapter03/start_here/ng-value-providers` path, which is inside the cloned repository. Perform the following steps:

1. Open the project in Visual Studio Code.
2. Open the Terminal, and run `npm install` to install the dependencies of the project.
3. Once done, run `ng serve -o`.

   This should open the app in a new browser tab.

4. Click on the **Login as Admin** button. You should see something similar to the following screenshot:

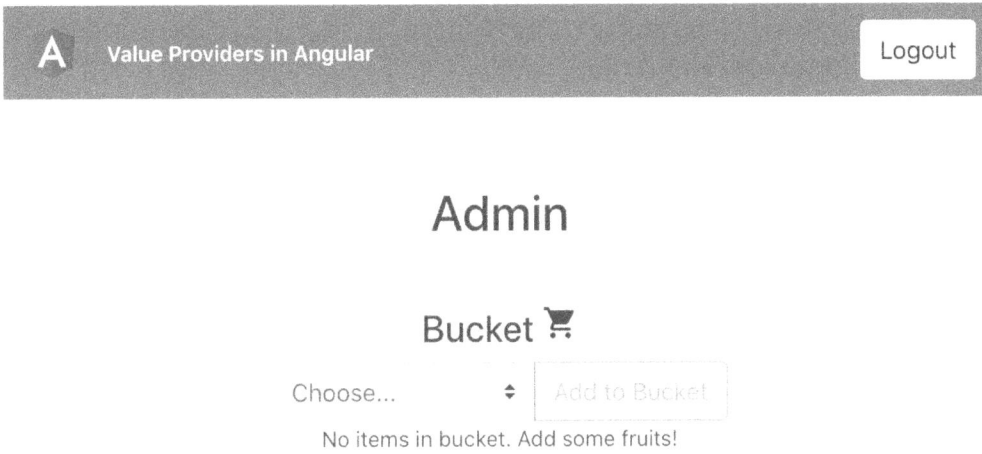

Figure 3.12 – The ng-value-providers app running on http://localhost:4200

We have a shared component, named BucketComponent, that is being used in both the admin and employee modules. For the employee, we'll restrict the ability to remove an item by providing a value provider in EmployeeModule. This is so that we can hide the **delete** button based on its value.

How to do it

1. First, we'll start by creating the value provider with InjectionToken within a new file, named app-config.ts, inside the app/constants folder. The code should appear as follows:

```
import { InjectionToken } from '@angular/core';

export interface IAppConfig {
  canDeleteItems: boolean;
}

export const APP_CONFIG = new
InjectionToken<IAppConfig>('APP_CONFIG');
```

```
export const AppConfig: IAppConfig = {
    canDeleteItems: true
}
```

Before we can actually use this `AppConfig` constant in our `BucketComponent`, we need to register it to the `AppModule` so that when we inject this in the `BucketComponent`, the value of the provider is resolved.

2. Let's add the provider to the `app.module.ts` file, as follows:

```
...
import { AppConfig, APP_CONFIG } from './constants/
app-config';

@NgModule({
    declarations: [
        AppComponent
    ],
    imports: [
        ...
    ],
    providers: [{
        provide: APP_CONFIG,
        useValue: AppConfig
    }],
    bootstrap: [AppComponent]
})
export class AppModule { }
```

Now the app knows about the `AppConfig` constants. The next step is to use this constant in `BucketComponent`.

3. We'll use the `@Inject()` decorator to inject it inside the `BucketComponent` class, in the `shared/components/bucket/bucket.component.ts` file, as follows:

```
import { Component, Inject, OnInit } from '@angular/
core';
...
import { IAppConfig, APP_CONFIG } from '../../../
constants/app-config';
```

```
...
export class BucketComponent implements OnInit {
  ...
  constructor(private bucketService: BucketService,
  @Inject(APP_CONFIG) private config: IAppConfig) { }
  ...
}
```

Great! The constant has been injected. Now, if you refresh the app, you shouldn't get any errors. The next step is to use the canDeleteItems property from config in BucketComponent to show/hide the **delete** button.

4. We'll first add the property to the shared/components/bucket/bucket.component.ts file and assign it to the ngOnInit() method, as follows:

```
...
export class BucketComponent implements OnInit {
  $bucket: Observable<IFruit[]>;
  selectedFruit: Fruit = '' as null;
  fruits: string[] = Object.values(Fruit);
  canDeleteItems: boolean;
  constructor(private bucketService: BucketService,
  @Inject(APP_CONFIG) private config: IAppConfig) { }

  ngOnInit(): void {
    this.$bucket = this.bucketService.$bucket;
    this.bucketService.loadItems();
    this.canDeleteItems = this.config.canDeleteItems;
  }
  ...
}
```

5. Now, we'll add an *ngIf directive in the shared/components/bucket/bucket.component.html file to only show the **delete** button if the value of canDeleteItems is true:

```
<div class="buckets" *ngIf="$bucket | async as bucket">
  <h4>Bucket <i class="material-icons">shopping_cart
  </i></h4>
```

```
<div class="add-section">
  ...
</div>
<div class="fruits">
  <ng-container *ngIf="bucket.length > 0; else
  bucketEmptyMessage">
    <div class="fruits__item" *ngFor="let item of
    bucket;">
      <div class="fruits__item__title">{{item.name}}
      </div>
      <div *ngIf="canDeleteItems" class="fruits__
      item__delete-icon"
      (click)="deleteFromBucket(item)">
        <div class="material-icons">delete</div>
      </div>
    </div>
  </ng-container>
</div>
</div>
<ng-template #bucketEmptyMessage>
  ...
</ng-template>
```

You can test whether everything works by setting the AppConfig constant's canDeleteItems property to false. Note that the **delete** button is now hidden for both the admin and employee. Once tested, set the value of canDeleteItems back to true again.

Now we have everything set up. Let's add a new constant so that we can hide the **delete** button for the employee only.

6. We'll create a folder, named constants, inside the employee folder. Then, we'll create a new file underneath the employee/constants path, called employee-config.ts, and we will add the following code to it:

```
import { IAppConfig } from '../../constants/app-config';

export const EmployeeConfig: IAppConfig = {
  canDeleteItems: false
}
```

7. Now, we'll provide this `EmployeeConfig` constant to the `EmployeeModule` for the same `APP_CONFIG` injection token. The code in the `employee.module.ts` file should appear as follows:

```
. . .
import { EmployeeComponent } from './employee.component';
import { APP_CONFIG } from '../constants/app-config';
import { EmployeeConfig } from './constants/employee-
config';

@NgModule({
  declarations: [EmployeeComponent],
  imports: [
    . . .
  ],
  providers: [{
    provide: APP_CONFIG,
    useValue: EmployeeConfig
  }]
})
export class EmployeeModule { }
```

And we're done! The recipe is now complete. You can see that the **delete** button is visible to the admin but hidden for the employee. It's all thanks to the magic of value providers.

How it works

When we inject a token into a component, Angular tries to find the resolved value of the token from the injected place by moving up the hierarchy of components and modules. We provided `EmployeeConfig` as `APP_CONFIG` in `EmployeeModule`. When Angular tries to resolve its value for `BucketComponent`, it finds it early at `EmployeeModule` as `EmployeeConfig`. Therefore, Angular stops right there and doesn't reach `AppComponent`. Notice that the value for `APP_CONFIG` in `AppComponent` is the `AppConfig` constant.

See also

- Dependency Injection in Angular)`https://angular.io/guide/dependency-injection`)

- Hierarchical Injectors in Angular (`https://angular.io/guide/hierarchical-dependency-injection`)

4
Understanding Angular Animations

In this chapter, you'll learn about working with animations in Angular. You'll learn about multi-state animations, staggering animations, keyframe animations, and how to implement animations for switching routes in your Angular apps.

The following are the recipes that we're going to cover in this chapter:

- Creating your first two-state Angular animation
- Working with multi-state animations
- Creating complex Angular animations using keyframes
- Animating lists in Angular using stagger animations
- Using animation callbacks
- Basic route animations in Angular
- Complex route animations in Angular using keyframes

Technical requirements

For the recipes in this chapter, make sure you have **Git** and **Node.js** installed on your machine. You also need to have the `@angular/cli` package installed, which you can install by using `npm install -g @angular/cli` from your terminal. The code for this chapter can be found at `https://github.com/PacktPublishing/Angular-Cookbook/tree/master/chapter04`.

Creating your first two-state Angular animation

In this recipe, you'll create a basic two-state Angular animation using a fading effect. We'll start with a fresh Angular project with some UI already built into it, enable animations within the app, and then move toward creating our first animation.

Getting ready

The project that we are going to work with resides in `chapter04/start_here/ng-basic-animation` inside the cloned repository:

1. Open the project in Visual Studio Code.

2. Open the terminal and run `npm install` to install the dependencies of the project.

3. Once done, run `ng serve -o`.

This should open the app in a new browser tab and you should see the following:

Figure 4.1 – ng-basic-animation app running on http://localhost:4200

Now that we have the app running, we will move on to the steps for the recipe.

How to do it...

We have an app that doesn't have Angular animations configured at all. So, we'll begin by enabling Angular animations. Then, we'll replace the CSS animations with Angular animations. Let's continue with the steps as follows:

1. First, we'll inject `BrowserAnimationsModule` from the `@angular/platform-browser/animations` package in our `app.module.ts`, so we can use animations within our Angular applications. We'll also import `BrowserAnimationsModule` in the `imports` array as follows:

```
...
import { FbCardComponent } from './components/fb-card/
fb-card.component';
import { TwitterCardComponent } from './components/
twitter-card/twitter-card.component';
import { BrowserAnimationsModule } from '@angular/
platform-browser/animations';

@NgModule({
  declarations: [
    AppComponent,
    SocialCardComponent,
    FbCardComponent,
    TwitterCardComponent
  ],
  imports: [
    BrowserModule,
    AppRoutingModule,
    BrowserAnimationsModule
  ],
  providers: [],
  bootstrap: [AppComponent]
})
export class AppModule { }
```

2. Now, we'll remove the CSS style transitions so we can see the full button (icon and text) by default for both the Facebook and Twitter buttons. Let's remove the styles from app.component.scss as highlighted in the following code block:

```scss
.type-picker {
    ...
    &__options {
        ...
        &__option {
            ...
            &__btn {
                ...
                min-width: 40px;
                // Remove the following lines
                transition: all 1s ease;
                &__text {
                    transition: all 1s ease;
                    width: 0;
                    visibility: hidden;
                }
                &--active {
                    [class^="icon-"], [class*=" icon-"] {
                        margin-right: 10px;
                    }
                    // Remove the following lines
                    .type-picker__options__option__btn__text {
                        width: auto;
                        visibility: visible;
                    }
                }
            }
        }
    }
}
```

3. We'll also remove the `&--active` selector under `&__btn` in the `app.component.scss` file, and move the styles for `[class^="icon-"]`, `[class*=" icon-"]` inside the `&__btn` selector. This is done so that there is a right-hand margin for all icons. Your code should look as follows:

```scss
.type-picker {
    ...
    &__options {
        ...
        &__option {
            ...
            &__btn {
                display: flex;
                align-items: center;
                min-width: 40px;
                justify-content: center;
                &--active {  ← Remove this
                    [class^='icon-'],
                    [class*=' icon-'] {
                        margin-right: 10px;
                    }
                }  ← Remove this


            }
        }
    }
}
```

4. Let's add the animation to be created to the template now. We'll apply the animation to the text elements of both buttons. Modify `app.component.html` as follows:

```html
...
<div class="content" role="main">
  <div class="type-picker">
    <h5>Pick Social Card Type</h5>
    <div class="type-picker__options">
      <div class="type-picker__options__option"
      (click)="setCardType(cardTypes.Facebook)">
```

```html
<button class="btn type-picker__options__option__
btn" [ngClass]="selectedCardType === cardTypes.
Facebook ? 'btn-primary type-picker__options__
option__btn--active' : 'btn-light'">
  <div class="icon-facebook"></div>
  <div class="type-picker__options__option__btn__
  text" [@socialBtnText]="selectedCardType ===
  cardTypes.Facebook ? 'btn-active-text' :
  'btn-inactive-text'">
    Facebook
  </div>
</button>
</div>
<div class="type-picker__options__option"
(click)="setCardType(cardTypes.Twitter)">
  <button class="btn type-picker__options__option__
  btn" [ngClass]="selectedCardType === cardTypes.
  Twitter ? 'btn-primary type-picker__options__
  option__btn--active' : 'btn-light'">
    <div class="icon-twitter"></div>
    <div class="type-picker__options__option__btn__
    text" [@socialBtnText]="selectedCardType ===
    cardTypes.Twitter ? 'btn-active-text' :
    'btn-inactive-text'">
      Twitter
    </div>
  </button>
</div>
</div>
</div>
<app-social-card [type]="selectedCardType">
</app-social-card>
</div>
```

Now, we'll start creating our animation named `socialBtnText`, and for that, we'll import some functions from the `@angular/animations` package in our `app.component.ts` so we can create the two states for the button text.

5. Add the following imports to your `app.component.ts`:

```
import {
  trigger,
  state,
  style,
  animate,
  transition
} from '@angular/animations';
```

6. Now, let's add an animation named `socialBtnText` using the `trigger` method to the `animations` array in the `AppComponent` metadata as follows:

```
...

@Component({
  selector: 'app-root',
  templateUrl: './app.component.html',
  styleUrls: ['./app.component.scss'],
  animations: [
    trigger('socialBtnText', [])
  ]
})
export class AppComponent {
  ...
}
```

7. Now, we'll create the two states named `btn-active-text` and `btn-inactive-text`. We'll set `width` and `visibility` for these states as follows:

```
...
@Component({
  selector: 'app-root',
  templateUrl: './app.component.html',
  styleUrls: ['./app.component.scss'],
  animations: [
    trigger('socialBtnText', [
      state('btn-active-text', style({
        width: '80px',
```

```
        visibility: 'visible',
      })),
      state('btn-inactive-text', style({
        width: '0px',
        visibility: 'hidden',
      })),
    ])
  ]
})
export class AppComponent {

  ...

}
```

Now that we have the states configured, we can start writing the transitions.

8. We'll first implement the `'btn-inactive-text => btn-active-text'` transition, which triggers upon clicking either of the buttons. Since this transition is going to be displaying the text, we'll first increase the `width` value of the text element, and then we'll set the text to `visible`. The content in the `animations[]` array should look as follows:

```
animations: [
    trigger('socialBtnText', [
        state('btn-active-text', style({...})),
        state('btn-inactive-text', style({...})),
        transition('btn-inactive-text => btn-active-text',
    [
            animate('0.3s ease', style({
                width: '80px'
            })),
            animate('0.3s ease', style({
                visibility: 'visible'
            }))
        ]),
    ])
]
```

You should see a smooth animation now for the button's active state. Let's implement the inactive state in the next step.

9. Now we'll implement the `'btn-active-text => btn-inactive-text'` transition. This should turn the visibility to `'hidden'` and set the width back to `'0px'` again. The code should look as follows:

```
animations: [
    trigger('socialBtnText', [

        ...

        state('btn-inactive-text', style({...})),
        transition('btn-active-text =>
        btn-inactive-text', [
            animate('0.3s', style({
                width: '80px'
            })),
            animate('0.3s', style({
                visibility: 'hidden'
            }))
        ]),
        transition('btn-inactive-text =>
        btn-active-text', [

            ...

        ])
    ]
```

You'll notice that there's a slight jerk/lag when the button becomes inactive. That's because the animation for width triggers first, and then it triggers the animation for `visibility: 'hidden'`. Therefore, we see both of them happening in sequence.

10. To have both animations work together, we'll use the `group` method from the `@angular/animations` package. We'll group together our `animate()` methods for the transition. The update in the `app.components.ts` file should look as follows:

```
...
import {
    ...
    transition,
    group
} from '@angular/animations';
```

```
    ...

    animations: [
        trigger('socialBtnText', [
            ...
            transition('btn-active-text =>
            btn-inactive-text', [
              group([
                animate('0.3s', style({
                  width: '0px'
                })),
                animate('0.3s', style({
                  visibility: 'hidden'
                }))
              ])
            ]),
            ...
        ])
    ]
```

11. Since we want this to be really quick, the time we'll set for the `animate()` methods for the `'btn-active-text => btn-inactive-text'` transition will be zero seconds (`0s`). Change it as follows:

```
    transition('btn-active-text => btn-inactive-text', [
            group([
                animate('0s', style({
                  width: '0px'
                })),
                animate('0s', style({
                  visibility: 'hidden'
                }))
              ])
            ]),
```

12. Toward the end, we can remove the extra `margin-right` from the button icon when the button is not active. We'll do it by moving the code for the `[class^="icon-"]`, `[class*=" icon-"]` selector inside another selector named `&--active` so it only applies when the button is active.

13. Modify the following styles in the `&__btn` selector in the `app.component.scss` file, as follows:

```scss
&__btn {
    display: flex;
    align-items: center;
    min-width: 40px;
    justify-content: center;
    &--active {
        [class^="icon-"], [class*=" icon-"] {
            margin-right: 10px;
        }
    }
}
```

Great! You now have implemented some good-looking animation buttons in the app. See the next section to understand how the recipe works.

How it works...

Angular provides its own Animation API that allows you to animate any property that the CSS transitions work on. The benefit is that you can configure them dynamically based on the requirements. We first used the `trigger` method to register the animation with the states and transitions. We then defined those states and transitions using the `state` and `transition` methods respectively. And we also saw how to run animations in parallel using the `group` method. If we didn't group the animations, they'd run sequentially. Finally, we applied the states using some flags in the component to reflect the changes.

There's more...

You might have noticed that the Twitter button somehow looks bigger than it should be. This is because we have the width of the text set to a constant 80px for our states and animations so far. While this looks good for the Facebook button, it doesn't look good for the Twitter one. So, we can actually make it configurable by providing two different transitions based on different widths for the buttons. Here's what you'll do:

1. Create a new file in the app folder and name it `animations.ts`.

2. Move the code from the animations array in the `app.component.ts` file to this new file; it should look as follows:

```
import {
  trigger,
  state,
  style,
  animate,
  transition,
  group
} from '@angular/animations';

export const buttonTextAnimation = (animationName:
string, textWidth: string) => {
  return trigger(animationName, [
    state('btn-active-text', style({
      width: textWidth,
      visibility: 'visible',
    })),
    state('btn-inactive-text', style({
      width: '0px',
      visibility: 'hidden',
    })),
  ])
}
```

3. And now, we'll add the transitions as well:

```
...

export const buttonTextAnimation = (animationName:
string, textWidth: string) => {
  return trigger(animationName, [
    state('btn-active-text', style({...})),
    state('btn-inactive-text', style({...})),
    transition('btn-active-text => btn-inactive-text', [
      group([
        animate('0s', style({
          width: '0px'
        })),
        animate('0s', style({
          visibility: 'hidden'
        }))
      ])
    ]),
    transition('btn-inactive-text => btn-active-text', [
      animate('0.3s ease', style({
        width: textWidth
      })),
      animate('0.3s ease', style({
        visibility: 'visible'
      }))
    ]),
  ])
}
```

4. Now, we'll use this `buttonTextAnimation` method for both our Facebook and
    Twitter buttons in `app.component.ts` as follows. Notice that we'll create two
    different animations:

```
import { Component } from '@angular/core';
import { SocialCardType } from './constants/social-card-
type';
import { buttonTextAnimation } from './animations';
```

```
@Component({
  selector: 'app-root',
  templateUrl: './app.component.html',
  styleUrls: ['./app.component.scss'],
  animations: [
    buttonTextAnimation('fbButtonTextAnimation', '80px'),
    buttonTextAnimation('twButtonTextAnimation', '60px'),
  ]
})
export class AppComponent {
  ...
}
```

5. Finally, we'll use the respective animations for the Facebook and Twitter buttons in
    app.component.html as follows:

```
...
<div class="type-picker__options__option"
(click)="setCardType(cardTypes.Facebook)">
        <button class="btn type-picker__options__option__
        btn" [ngClass]="selectedCardType === cardTypes.
        Facebook ? 'btn-primary type-picker__options__
        option__btn--active' : 'btn-light'">
            <div class="icon-facebook"></div>
            <div class="type-picker__options__option__
            btn__text" [@ fbButtonTextAnimation]=
            "isFBBtnActive ? 'btn-active-text' :
            'btn-inactive-text'">
              Facebook
            </div>
        </button>
    </div>
    <div class="type-picker__options__option"
    (click)="setCardType(cardTypes.Twitter)">
        <button class="btn type-picker__options__option__
        btn" [ngClass]="selectedCardType === cardTypes.
        Twitter ? 'btn-primary type-picker__options__
        option__btn--active' : 'btn-light'">
```

```
            <div class="icon-twitter"></div>
            <div class="type-picker__options__option__
            btn__text" [@twButtonTextAnimation]=
            "isTwBtnActive ? 'btn-active-text' :
            'btn-inactive-text'">
              Twitter
            </div>
          </button>
      </div>
```

See also

- Animations in Angular (https://angular.io/guide/animations)

- *Angular Animations Explained with Examples* (https://www.freecodecamp.
org/news/angular-animations-explained-with-examples/)

Working with multi-state animations

In this recipe, we'll work with Angular animations containing multiple states. This means that we'll work with more than two states for a particular item. We'll be using the same Facebook and Twitter cards example for this recipe as well. But we'll configure the state of the cards for their state before they appear on screen, when they're on screen, and when they're about to disappear from the screen again.

Getting ready

The project for this recipe resides in chapter04/start_here/ng-multi-state-animations:

1. Open the project in Visual Studio Code.

2. Open the terminal and run npm install to install the dependencies of the project.

3. Once done, run ng serve -o.

   This should open the app in a new browser tab, and you should see the app as follows:

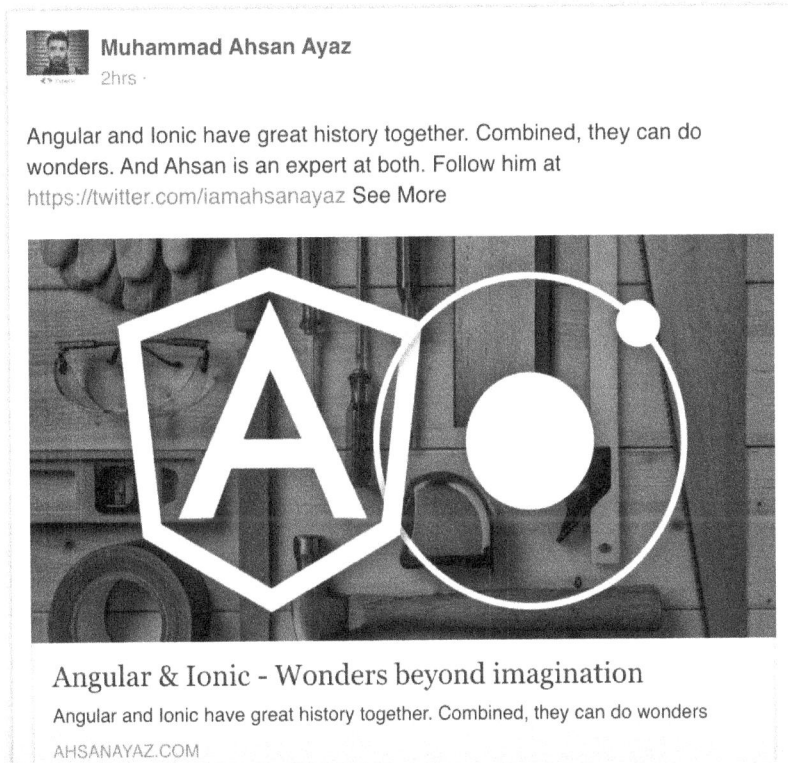

Figure 4.2 – ng-multi-state-animations app running on http://localhost:4200

Now that we have the app running locally, let's look at the steps of the recipe in the next section.

How to do it...

We already have a working app that has a single animation built for the reach of social cards. When you tap either the Facebook or Twitter button, you'll see the respective card appearing with a slide-in animation from left to right. To keep the recipe simple, we'll implement two more states and animations for when the user moves the mouse cursor on the card and when the user moves away from the card. Let's add the relevant code in the following steps:

1. We'll start with adding two @HostListener instances to FbCardComponent in the components/fb-card/fb-card.component.ts file, one for the mouseenter event on the card and one for the mouseleave event. We'll name the states hovered and active respectively. The code should look as follows:

```
import { Component, HostListener, OnInit } from '@
angular/core';
import { cardAnimation } from '../../animations';

@Component({
  selector: 'app-fb-card',
  templateUrl: './fb-card.component.html',
  styleUrls: ['./fb-card.component.scss'],
  animations: [cardAnimation]
})
export class FbCardComponent implements OnInit {
  cardState;
  constructor() { }

  @HostListener('mouseenter')
  onMouseEnter() {
    this.cardState = 'hovered'
  }

  @HostListener('mouseleave')
  onMouseLeave() {
    this.cardState = 'active'
  }

  ngOnInit(): void {
```

```
          this.cardState = 'active'
        }


      }
```

2. Now, we'll do the same for `TwitterCardComponent` in the `twitter-card-component.ts` file. The code should look as follows:

```
import { Component, HostListener, OnInit } from '@
angular/core';
import { cardAnimation } from '../../animations';

@Component({
  selector: 'app-twitter-card',
  templateUrl: './twitter-card.component.html',
  styleUrls: ['./twitter-card.component.scss'],
  animations: [cardAnimation]
})
export class TwitterCardComponent implements OnInit {
  cardState
  constructor() { }

  @HostListener('mouseenter')
  onMouseEnter() {
    this.cardState = 'hovered'
  }

  @HostListener('mouseleave')
  onMouseLeave() {
    this.cardState = 'active'
  }

  ngOnInit(): void {
    this.cardState = 'active'
  }


  }
```

3. There should be no visual change so far since we're only updating the `cardState` variable to have the hover and active states. We haven't defined the transitions yet.

4. We'll now define our state for when the user's cursor enters the card, that is, the `mouseenter` event. The state is called `hovered` and should look as follows in the `animation.ts` file:

```
. . .
export const cardAnimation = trigger('cardAnimation', [
  state('active', style({
    color: 'rgb(51, 51, 51)',
    backgroundColor: 'white'
  })),
  state('hovered', style({
    transform: 'scale3d(1.05, 1.05, 1.05)',
    backgroundColor: '#333',
    color: 'white'
  })),
  transition('void => active', [
    style({
      transform: 'translateX(-200px)',
      opacity: 0
    }),
    animate('0.2s ease', style({
      transform: 'translateX(0)',
      opacity: 1
    }))
  ]),
])
```

If you refresh the app now, tap either the Facebook or Twitter button, and hover the cursor over the card, you'll see the card's UI changing. That's because we changed the state to `hovered`. However, there's no animation yet. Let's add one in the next step.

5. We'll add the `active => hovered` transition now in the `animations.ts` file so that we can smoothly navigate from `active` to the `hovered` state:

```
. . .
export const cardAnimation = trigger('cardAnimation', [
```

```
        state('active', style(...)),
        state('hovered', style(...)),
        transition('void => active', [...]),
        transition('active => hovered', [
          animate('0.3s 0s ease-out', style({
            transform: 'scale3d(1.05, 1.05, 1.05)',
            backgroundColor: '#333',
            color: 'white'
          }))
        ]),
    ])
```

You should now see the smooth transition on the `mouseenter` event if you refresh the app.

6. Finally, we'll add the final transition, `hovered => active`, so when the user leaves the card, we revert to the active state with a smooth animation. The code should look as follows:

```
    ...
    export const cardAnimation = trigger('cardAnimation', [
        state('active', style(...)),
        state('hovered', style(...)),
        transition('void => active', [...]),
        transition('active => hovered', [...]),
        transition('hovered => active', [
          animate('0.3s 0s ease-out', style({
            transform: 'scale3d(1, 1, 1)',
            color: 'rgb(51, 51, 51)',
            backgroundColor: 'white'
          }))
        ]),

    ])
```

Ta-da! You now know how to implement different states and different animations on a single element using `@angular/animations`.

How it works...

Angular uses triggers for understanding what state the animation is in. An example syntax looks as follows:

```
<div [@animationTriggerName]="expression">...</div>;
```

expression can be a valid JavaScript expression, and it evaluates to the name of the state. In our case, we bind it to the cardState property, which either contains 'active' or 'hovered'. Therefore, we end up with three transitions for our cards:

- void => active (when the element is added to the DOM and is rendered)
- active => hovered (when the mouseenter event triggers on the card)
- hovered => active (when the mouseleave event triggers on the card)

See also

- *Triggering the animation* (https://angular.io/guide/animations#triggering-the-animation)
- *Reusable animations* (https://angular.io/guide/reusable-animations)

Creating complex Angular animations using keyframes

Since you already know about Angular animations from the previous recipes, you might be thinking, "Well, that's easy enough." Well, time to level up your animation skills in this recipe. You'll create a complex Angular animation using keyframes in this recipe to get started with writing some advanced animations.

Getting ready

The project for this recipe resides in chapter04/start_here/animations-using-keyframes:

1. Open the project in Visual Studio Code.
2. Open the terminal and run npm install to install the dependencies of the project.
3. Once done, run ng serve -o.

This should open the app in a new browser tab and you should see the app as follows:

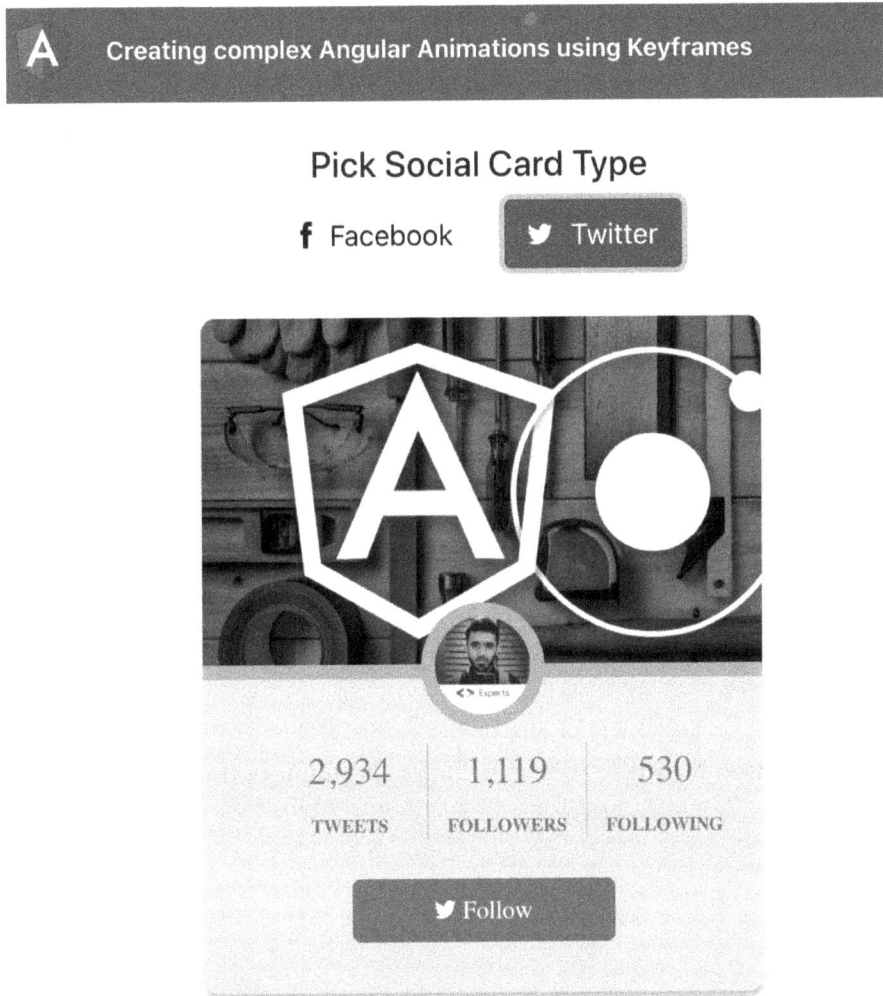

Figure 4.3 – animations-using-keyframes app running on http://localhost:4200

Now that we have the app running locally, let's look at the steps of the recipe in the next section.

How to do it...

We have an app right now that has a single transition, that is, void => *, which triggers when the element is rendered on DOM. Right now, the animation is pretty simple and uses the animate function to define the animation. We'll first convert it to keyframes, and then we'll make it a bit complex:

1. Let's begin with adding the keyframes method from @angular/animations to the animations.ts file as follows:

```
import {
    trigger,
    state,
    style,
    animate,
    transition,
    keyframes
} from '@angular/animations';

export const cardAnimation = trigger('cardAnimation', [
    ...
])
```

2. Now, we'll convert the single style animation to keyframes as follows:

```
import {
    trigger,
    state,
    style,
    animate,
    transition,
    keyframes
} from '@angular/animations';

export const cardAnimation = trigger('cardAnimation', [
    transition('void => *', [
        style({ ← Remove this style
            transform: 'translateX(-200px)',
            opacity: 0
```

```
      }),
      animate('0.2s ease', keyframes([
        style({
          transform: 'translateX(-200px)',
          offset: 0
        }),
        style({
          transform: 'translateX(0)',
          offset: 1
        })
      ]))
    ]),
  ])
```

Notice that in this code block, we've removed the state('active', ...) part because we don't need it anymore. Also, we moved the style({transform: 'translateX(-200px)', opacity: 0}) inside the keyframes array because it is now part of the keyframes animation itself. If you refresh the app now and try it, you'll still see the same animation as before. But now we have it using keyframes.

3. Finally, let's start adding some complex animations. Let's start the animation with a scaled-down card by adding scale3d to the transform property of style at offset: 0. We'll also increase the animation time to 1.5s:

```
...
export const cardAnimation = trigger('cardAnimation', [
  transition('void => *', [
    animate('1.5s ease', keyframes([
      style({
        transform: 'translateX(-200px) scale3d(0.4, 0.4,
        0.4)',
        offset: 0
      }),
      style({
        transform: 'translateX(0)',
        offset: 1
      })
    ]))
```

```
    ]),
  ])
```

You should now see that the card animation starts with a small card that slides from the left and moves toward the right, increasing in size.

4. Now we'll implement a zig-zag-ish animation for the appearance of the card instead of the slide-in animation. Let's add the following keyframe elements to the `keyframes` array to add a bumpy effect to our animation:

```
...

export const cardAnimation = trigger('cardAnimation', [
  transition('void => *', [
    animate('1.5s 0s ease', keyframes([
      style({
        transform: 'translateX(-200px) scale3d(0.4, 0.4,
        0.4)',
        offset: 0
      }),
      style({
        transform: 'translateX(0px) rotate(-90deg)
        scale3d(0.5, 0.5, 0.5)',
        offset: 0.25
      }),
      style({
        transform: 'translateX(-200px) rotate(90deg)
        translateY(0) scale3d(0.6, 0.6, 0.6)',
        offset: 0.5
      }),
      style({
        transform: 'translateX(0)',
        offset: 1
      })
    ]))
  ]),
])
```

If you refresh the app and tap any of the buttons, you should see the card bumping to the right wall, and then to the left wall of the card, before returning to the normal state:

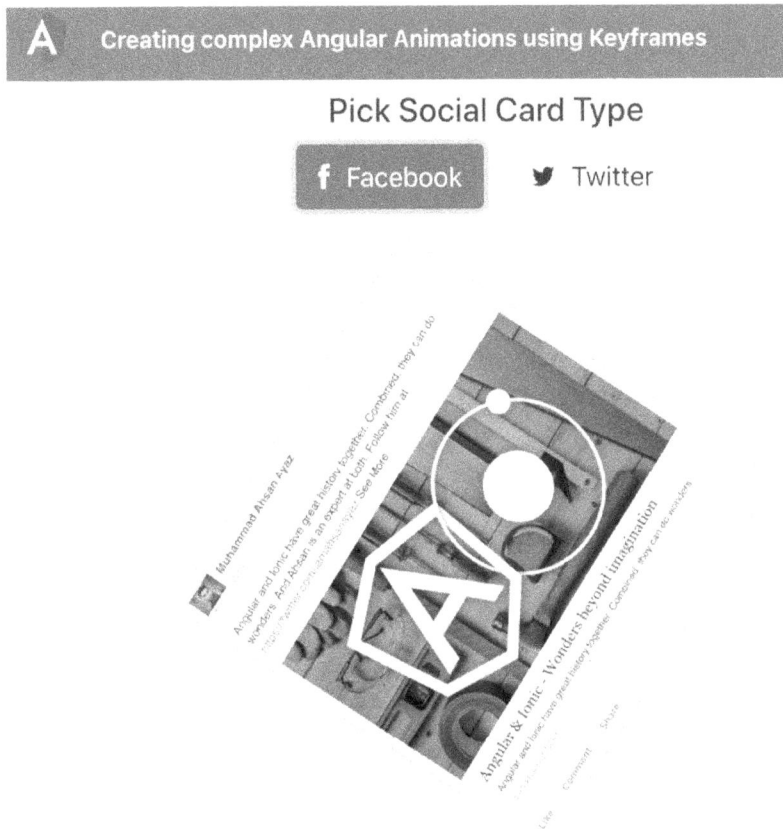

Figure 4.4 – Card bumping to right and then the left wall

5. As the last step, we'll spin the card clockwise before it returns to its original position. For that, we'll use offset: 0.75, using the rotate method with some additional angles. The code should look as follows:

```
...

export const cardAnimation = trigger('cardAnimation', [
  transition('void => *', [
    animate('1.5s 0s ease', keyframes([
      style({
```

```
          transform: 'translateX(-200px) scale3d(0.4, 0.4,
          0.4)',
          offset: 0
        }),
        style({
          transform: 'translateX(0px) rotate(-90deg)
          scale3d(0.5, 0.5, 0.5)',
          offset: 0.25
        }),
        style({
          transform: 'translateX(-200px) rotate(90deg)
          translateY(0) scale3d(0.6, 0.6, 0.6)',
          offset: 0.5
        }),
        style({
          transform: 'translateX(-100px) rotate(135deg)
          translateY(0) scale3d(0.6, 0.6, 0.6)',
          offset: 0.75
        }),
        style({
          transform: 'translateX(0) rotate(360deg)',
          offset: 1
        })
      ]))
    ]),
  ])
```

Awesome! You now know how to implement complex animations in Angular using the `keyframes` method from the `@angular/common` package. See in the next section how it works.

How it works...

For complex animations in Angular, the `keyframes` method is a really good way of defining different offsets of the animation throughout its journey. We can define the offsets using the `styles` method, which takes `AnimationStyleMetadata` as a parameter. `AnimationStyleMetadata` also allows us to pass the `offset` property, which can have a value between `0` and `1`. Thus, we can define different styles for different offsets to create advanced animations.

See also

- *Animations in Angular* (`https://angular.io/guide/animations`)
- *Angular Animations Explained with Examples* (`https://www.freecodecamp.org/news/angular-animations-explained-with-examples/`)

Animating lists in Angular using stagger animations

No matter what web application you build today, you are going to implement some sort of list most likely. And to make those lists even better, why not implement an elegant animation with them? In this recipe, you'll learn how to animate lists in Angular using stagger animations.

Getting ready

The project for this recipe resides in `chapter04/start_here/animating-lists`:

1. Open the project in Visual Studio Code.
2. Open the terminal and run `npm install` to install the dependencies of the project.
3. Once done, run `ng serve -o`.

This should open the app in a new browser tab. Log in to the app as an employee, and you should see the app as follows:

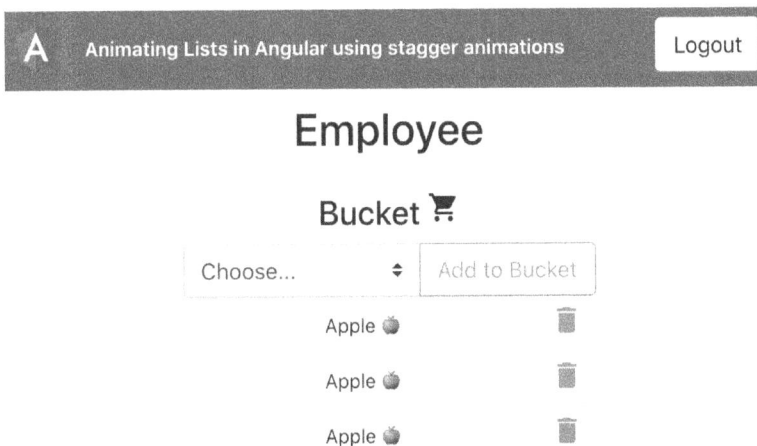

Figure 4.5 – animating-lists app running on http://localhost:4200

Now that we have the app running locally, let's see the steps of the recipe in the next section.

How to do it...

We have an app right now that has a list of bucket items. We need to animate the list using staggering animations. We'll be doing this step by step. I'm excited – are you?

Cool. We'll go through the following steps for the recipe:

1. First, let's add `BrowserAnimationsModule` from the `@angular/platform-browser/animations` package in our `app.module.ts` so that we can enable animations for the app. The code should look as follows:

```
import { BrowserModule } from '@angular/platform-browser';
import { NgModule } from '@angular/core';

import { AppRoutingModule } from './app-routing.module';
import { AppComponent } from './app.component';
import { FormsModule } from '@angular/forms';
import { BrowserAnimationsModule } from '@angular/platform-browser/animations';
@NgModule({
  declarations: [
    AppComponent
  ],
  imports: [
    BrowserModule,
    AppRoutingModule,
    FormsModule,
    BrowserAnimationsModule
  ],
  providers: [],
  bootstrap: [AppComponent]
})
export class AppModule { }
```

2. Now, create a file named `animations.ts` inside the `constants` folder
    and add the following code to register a basic list item animation named
    `listItemAnimation` with Angular:

```typescript
import {
  trigger,
  style,
  animate,
  transition,
} from '@angular/animations';

export const ANIMATIONS = {
  LIST_ITEM_ANIMATION: trigger('listItemAnimation', [
    transition('void => *', [
      style({
        opacity: 0
      }),
      animate('0.5s ease', style({
        opacity: 1
      }))
    ]),

    transition('* => void', [
      style({
        opacity: 1
      }),
      animate('0.5s ease', style({
        opacity: 0
      }))
    ])
  ])
}
```

3. Notice that the `void => *` transition is for when the list item enters the view
    (or appears). The `* => void` transition is for when the item leaves the view (or
    disappears).

4. Now, we'll add the animation to `BucketComponent` in the `app/shared/bucket/bucket.component.ts` file as follows:

```
import { Component, OnInit } from '@angular/core';
import { Observable } from 'rxjs/internal/Observable';
import { BucketService } from 'src/app/services/bucket.service';
import { Fruit } from '../../../constants/fruit';
import { IFruit } from '../../../interfaces/fruit.interface';
import { ANIMATIONS } from '../../../constants/animations';

@Component({
    selector: 'app-bucket',
    templateUrl: './bucket.component.html',
    styleUrls: ['./bucket.component.scss'],
    animations: [ANIMATIONS.LIST_ITEM_ANIMATION]
})
export class BucketComponent implements OnInit {
    ...
}
```

Since we have the animation imported in the component, we can use it in the template now.

5. Let's add the animation to the list item as follows in `bucket.component.html`:

```
<div class="buckets" *ngIf="$bucket | async as bucket">
    <h4>Bucket <i class="material-icons">shopping_cart
    </i></h4>
    <div class="add-section">
        <div class="input-group">
        ...
    </div>
    <div class="fruits">
        <ng-container *ngIf="bucket.length > 0; else
        bucketEmptyMessage">
            <div class="fruits__item" *ngFor="let item of
            bucket;" @listItemAnimation>
```

```
          <div class="fruits__item__title">{{item.name}}
          </div>
          <div class="fruites__item__delete-icon"
          (click)="deleteFromBucket(item)">
            <div class="material-icons">delete</div>
          </div>
        </div>
      </ng-container>
    </div>
  </div>
  ...
```

6. If you now refresh the app and add an item to the bucket list, you should see it appear with a fade-in effect. And if you delete an item, you should see it disappear with the animation as well.

   One thing that you'll notice is that when you refresh the app, all the list items appear together simultaneously. We can, however, make them appear one by one, using `stagger` animations. We'll do that in the next step.

7. We'll modify `LIST_ITEM_ANIMATION` now to use the `stagger` method. This is because we can make each list item appear one after the other. First, we need to import the `stagger` method from `@angular/animations`, and then we need to wrap our `animate` methods within `stagger` methods. Update the `animations.ts` file as follows:

```
import {
  trigger,
  style,
  animate,
  transition,
  stagger
} from '@angular/animations';

export const ANIMATIONS = {
  LIST_ITEM_ANIMATION: trigger('listItemAnimation', [
    transition('void => *', [
      style({
        opacity: 0
```

```
    }),
    stagger(100, [
      animate('0.5s ease', style({
        opacity: 1
      }))
    ])
  ]),
  ,
  transition('* => void', [
    style({
      opacity: 1
    }),
    stagger(100, [
      animate('0.5s ease', style({
        opacity: 0
      }))
    ])
  ])
  ])
}
```

This, however, *will not work*. And that's because the `stagger` method can only be used within a `query` method. Therefore, we need to modify our code a bit to use the `query` methods in the next step.

8. Let's import the `query` method from `@angular/animations` and modify our code a bit so it can be used with the `stagger` method. We're going to make a couple of changes.

9. We'll rename the animation to `listAnimation` since the animation will now apply to the list instead of the individual list items.

10. We'll wrap our `stagger` methods inside the appropriate `query` methods.

11. We'll use only one transition, that is, `* => *`, for both queries, `:enter` and `:leave`, so whenever the list items change, the animation is triggered.

12. We'll move `style({ opacity: 0 })` inside the `query(':enter')` chunk as it needs to hide the items before the stagger animation.

The code should look as follows:

```
import {
  trigger,
  style,
  animate,
  transition,
  stagger,
  query
} from '@angular/animations';

export const ANIMATIONS = {
  LIST_ANIMATION: trigger('listAnimation', [
    transition('* <=> *', [
      query(':enter', [
        style({
          opacity: 0
        }),
        stagger(100, [
          animate('0.5s ease', style({
            opacity: 1
          }))
        ])
      ], { optional: true }),
      query(':leave', [
        stagger(100, [
          animate('0.5s ease', style({
            opacity: 0
          }))
        ])
      ], {optional: true})
    ]),
  ])
}
```

13. We now need to fix the import of the animation in `shared/components/bucket/bucket.component.ts` as follows:

```
...

@Component({
  selector: 'app-bucket',
  templateUrl: './bucket.component.html',
  styleUrls: ['./bucket.component.scss'],
  animations: [ANIMATIONS.LIST_ANIMATION]
})
export class BucketComponent implements OnInit {
  ...
}
```

14. Since we've changed the name of the animation, let's fix in the template of the bucket component as well. Update `shared/components/bucket/bucket.component.html` as follows:

```
<div class="buckets" *ngIf="$bucket | async as bucket">
  <h4>Bucket <i class="material-icons">shopping_cart
  </i></h4>
  <div class="add-section">...
  </div>
  <div class="fruits" [@listItemAnimation]="bucket.
  length">
    <ng-container *ngIf="bucket.length > 0; else
    bucketEmptyMessage">
      <div class="fruits__item" *ngFor="let item of
      bucket;"  @listItemAnimation ← Remove this>
        <div class="fruits__item__title">{{item.name}}
        </div>
        <div class="fruites__item__delete-icon"
        (click)="deleteFromBucket(item)">
          <div class="material-icons">delete</div>
        </div>
      </div>
    </ng-container>
  </div>
```

```
      </div>
      ...
```

Notice that we're binding the [@listAnimation] property to bucket.length. This will make sure that the animation triggers whenever the length of the bucket changes, that is, when an item is added or removed from the bucket.

Awesome! You now know how to implement staggering animations for lists in Angular. See in the next section how it works.

How it works...

Stagger animations only work inside query methods. This is because of the fact that staggering animations usually are applied to the list itself and not to individual items. In order to search or query the items, we first use the query method. Then we use the stagger method to define how many milliseconds of staggering we want before the animation starts for the next list item. We also provide animation as well in the stagger method to define the animation for each element found with the query. Notice that we're using { optional: true } for both the :enter query and the :leave query. This is because if the list binding changes (bucket.length), we don't get an error if no new element has entered the DOM or no element has left the DOM.

See also

- *Animations in Angular* (https://angular.io/guide/animations)
- Angular animations stagger docs (https://angular.io/api/animations/stagger)

Using animation callbacks

In this recipe, you'll learn how to be notified and act upon animation state changes in Angular. As a simple example, we'll use the same bucket list app, and we'll reset the item-to-add option whenever the animation completes for adding an item.

Getting ready

The project that we are going to work with resides in `chapter04/start_here/animation-callbacks` inside the cloned repository:

1. Open the project in Visual Studio Code.

2. Open the terminal and run `npm install` to install the dependencies of the project.

3. Once done, run `ng serve -o`.

    This should open the app in a new browser tab.

4. Click the **Login as Admin** button, and you should see something like the following:

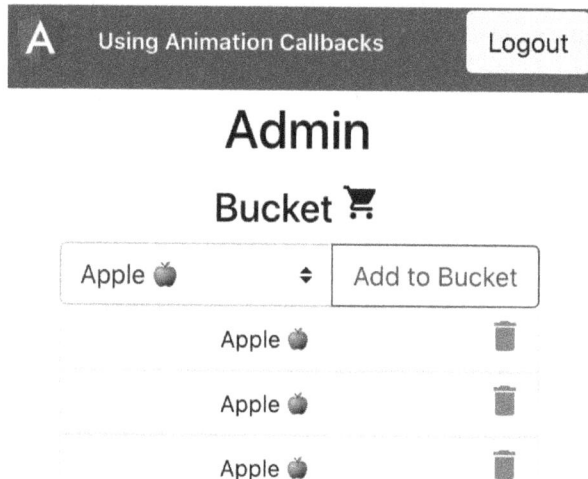

Figure 4.6 – animation-callbacks app running on http://localhost:4200

Now that we have the app running locally, let's see the steps of the recipe in the next section.

How to do it...

We have the same bucket app for this recipe that we used in the previous recipes. To see how to use animation callbacks, we'll simply perform an action once the animation of a list item entering the DOM is finished and have one action for when a list item leaves the DOM. Let's get started:

1. We'll first create two methods named onAnimationStarted and onAnimationDone in BucketComponent inside the shared/components/bucket/bucket.component.ts file. These methods will be triggered for the respective phases of animations in the later steps:

    ```
    . . .
    import { AnimationEvent } from '@angular/animations';

    @Component({...})
    export class BucketComponent implements OnInit {

      . . .

      ngOnInit(): void {
        this.$bucket = this.bucketService.$bucket;
        this.bucketService.loadItems();
      }

      onAnimationStarted( event: AnimationEvent ) {
        console.log(event);
      }

      onAnimationDone( event: AnimationEvent ) {
        console.log(event);
      }

      . . .

    }
    ```

2. Now we'll bind the animation's `start` and `done` events to the `onAnimateEvent` method in the template. Modify the `shared/components/bucket/bucket.component.html` file as follows:

```html
<div class="buckets" *ngIf="$bucket | async as bucket">
  <h4>Bucket <i class="material-icons">shopping_cart
  </i></h4>
  <div class="add-section">
    ...
  </div>
  <div class="fruits" [@listAnimation]="bucket.length"
  (@listAnimation.start)="onAnimationStarted($event)"
  (@listAnimation.done)="onAnimationDone($event)">
    <ng-container *ngIf="bucket.length > 0; else
    bucketEmptyMessage">
      <div class="fruits__item" *ngFor="let item of
      bucket;">
        <div class="fruits__item__title">{{item.name}}
        </div>
        <div class="fruites__item__delete-icon"
        (click)="deleteFromBucket(item)">
          <div class="material-icons">delete</div>
        </div>
      </div>
    </ng-container>
  </div>
</div>
<ng-template #bucketEmptyMessage>
  ...
</ng-template>
```

3. Notice that both the `.start` and `.done` events are associated with the trigger name, `listAnimation`. If you refresh the app now, you should see the logs on the console as follows:

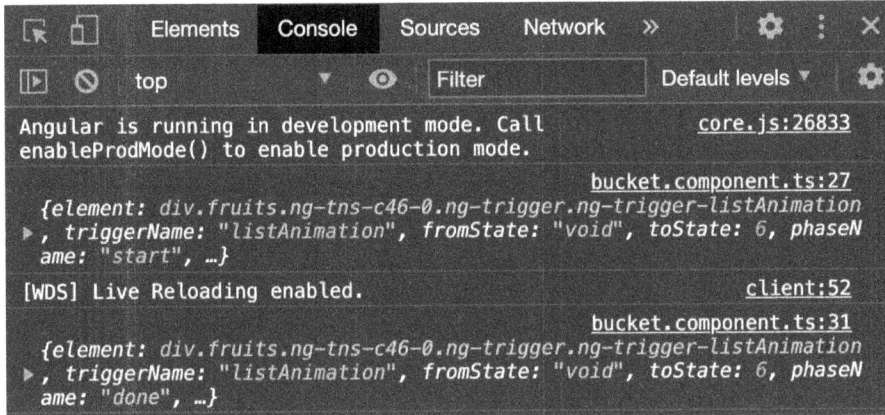

Figure 4.7 – Logs on the console reflecting the .start and .done animation events

4. Since we have the events in place now, we'll replace the `shopping_cart` icon with the save icon during the animation. This is similar to simulating what would happen if we needed to do an HTTP call to save the data. Let's modify `shared/components/bucket/bucket.component.ts` as follows:

```
...
export class BucketComponent implements OnInit {
  $bucket: Observable<IFruit[]>;
  selectedFruit: Fruit | null = null;
  fruits: string[] = Object.values(Fruit);
  isSaving: boolean;
  constructor(private bucketService: BucketService) { }

  ngOnInit(): void {
    ...
  }

  onAnimationStarted( event: AnimationEvent ) {
    this.isSaving = true;
  }

  onAnimationDone( event: AnimationEvent ) {
    this.isSaving = false;
    this.selectedFruit = null;
  }
}
```

```
addSelectedFruitToBucket() {
    ...
}
deleteFromBucket(fruit: IFruit) {
    ...
}

}
```

5. Finally, we can modify our template to show the appropriate icon based on the value of the isSaving property. The code should look as follows:

```
<div class="buckets" *ngIf="$bucket | async as bucket">
    <h4>Bucket <i class="material-icons">{{isSaving ?
    'save' : 'shopping_cart'}}</i></h4>
    ...
</div>
...
```

And boom! The recipe is finished now. If you refresh the page or add/delete an item, you'll notice that the bucket icon is replaced with the save icon during the entire animation, all thanks to the animation callbacks.

How it works...

When an animation is registered with Angular using the trigger method, Angular itself creates a local property within the scope with the name of the trigger set as @triggerName. It also creates the .start and .done sub-properties as EventEmitter instances for the animation. Therefore, we can easily use them in the templates to capture the AnimationEvent instance passed by Angular. Each AnimationEvent contains the phaseName property, using which we can also identify whether it is the start event or the done event. We can also tell from AnimationEvent which state the animation started from and which state it ended on.

See also

- Animations in Angular (https://angular.io/guide/animations)
- AnimationEvent docs (https://angular.io/api/animations/AnimationEvent)

Basic route animations in Angular

In this recipe, you'll learn how to implement basic route animations in Angular. Although these are basic animations, they require a bit of a setup to be executed properly. You'll learn how to configure route animations by passing the transition state name to the route as a data property. You'll also learn how to use the `RouterOutlet` API to get the transition name and apply it to the animation to be executed.

Getting ready

The project that we are going to work with resides in `chapter04/start_here/route-animations` inside the cloned repository:

1. Open the project in Visual Studio Code.
2. Open the terminal and run `npm install` to install the dependencies of the project.
3. Once done, run `ng serve -o`.

This should open the app in a new browser tab, and you should see something like the following:

Figure 4.8 – route-animations app running on http://localhost:4200

Now that we have the app running locally, let's see the steps of the recipe in the next section.

How to do it...

We have a really simple app with two lazy-loaded routes at the moment. The routes are for the **Home** and the **About** pages, and we'll now start configuring the animations for the app:

1. First, we need to import `BrowserAnimationsModule` into `app.module.ts` as an import. The code should look as follows:

```
import { BrowserModule } from '@angular/platform-
browser';
import { NgModule } from '@angular/core';

import { AppRoutingModule } from './app-routing.module';
import { AppComponent } from './app.component';
import { BrowserAnimationsModule } from '@angular/
platform-browser/animations';

@NgModule({
  declarations: [
    AppComponent
  ],
  imports: [
    BrowserModule,
    AppRoutingModule,
    BrowserAnimationsModule
  ],
  providers: [],
  bootstrap: [AppComponent]
})
export class AppModule { }
```

2. We'll now create a new folder inside the `app` folder named `constants`. We'll also create a file inside the `constants` folder named `animations.ts`. Let's put the following code in the `animations.ts` file to register a simple trigger:

```
import {trigger,  style, animate, transition, query,
  } from '@angular/animations';

export const ROUTE_ANIMATION = trigger('routeAnimation',
```

```
    [
        transition('* <=> *', [
            // states and transitions to be added here
        ])
    ])
```

3. We'll now register our queries and our states for the animations. Let's add the
    following items in the `transition()` method's array as follows:

```
    ...
    export const ROUTE_ANIMATION = trigger('routeAnimation',
    [
        style({
            position: 'relative'
        }),
        query(':enter, :leave', [
            style({
                position: 'absolute',
                width: '100%'
            })
        ], {optional: true}),
        query(':enter', [
            style({
                opacity: 0,
            })
        ], {optional: true}),
        query(':leave', [
            animate('300ms ease-out', style({ opacity: 0 }))
        ], {optional: true}),
        query(':enter', [
            animate('300ms ease-in', style({ opacity: 1 }))
        ], {optional: true}),
    ]);
```

Alright! We have the `routeAnimation` trigger registered now for transition from every route to every other route. Now, let's provide those transition states in the routes.

4.   We can provide the states for the transitions using a unique identifier for each route. There are many ways to do it, but the easiest way is to provide it using the `data` attribute in the route configuration as follows in `app-routing.module.ts`:

```
import { NgModule } from '@angular/core';
import { Routes, RouterModule } from '@angular/router';

const routes: Routes = [
  {
    path: '',
    pathMatch: 'full',
    redirectTo: 'home',
  },
  {
    path: 'home',
    data: {
      transitionState: 'HomePage',
    },
    loadChildren: () => import('./home/home.module').
    then(m => m.HomeModule),
  },
  {
    path: 'about',
    data: {
      transitionState: 'AboutPage',
    },
    loadChildren: () => import('./about/about.module').
    then(m => m.AboutModule),
  },
];

@NgModule({
  ...
```

```
})
export class AppRoutingModule {}
```

5. Now, we need to provide this `transitionState` property from the current route to the `@routeAnimation` trigger somehow in `app.component.html`.

6. For this, create a `@ViewChild` instance for the `<router-outlet>` element used in `app.component.html` so we can get the current route's `data` and the `transitionState` value provided. The code in the `app.component.ts` file should look as follows:

```
import { Component, ViewChild } from "@angular/core";
import { RouterOutlet } from '@angular/router';

@Component({
  selector: "app-root",
  templateUrl: "./app.component.html",
  styleUrls: ["./app.component.scss"]
})
export class AppComponent {
  @ViewChild(RouterOutlet) routerOutlet;
}
```

7. We'll also import ROUTE_ANIMATION from the `animations.ts` file into `app.component.ts` as follows:

```
import { Component, ViewChild } from "@angular/core";
import { RouterOutlet } from '@angular/router';
import { ROUTE_ANIMATION } from './constants/animations';

@Component({
  selector: "app-root",
  templateUrl: "./app.component.html",
  styleUrls: ["./app.component.scss"],
  animations: [
    ROUTE_ANIMATION
  ]
})
export class AppComponent {
```

```
    . . .
  }
```

8. We'll now create a function named `getRouteAnimationTransition()`, which
    will get the current route's data and the `transitionState` value and return it
    back. This function will later be used in `app.component.html`. Modify your
    code in `app.component.ts` as follows:

```
    . . .

  @Component({
    . . .
  })
  export class AppComponent {
    @ViewChild(RouterOutlet) routerOutlet;

    getRouteAnimationState() {
      return this.routerOutlet && this.routerOutlet.
      activatedRouteData && this.routerOutlet.
      activatedRouteData.transitionState;
    }
  }
```

9. Finally, let's use the `getRouteAnimationState()` method with the
    `@routeAnimation` trigger in `app.component.html` so we can see the
    animation in play:

```
    . . .

  <div class="content" role="main">
    <div class="router-container"
    [@routeAnimation]="getRouteAnimationState()">
      <router-outlet></router-outlet>
    </div>
  </div>
```

Voila! Refresh the app and see the magic in place. You should now see the fade-out and
fade-in animations happening as you navigate from the **Home** page to the **About** page
and vice versa.

How it works...

In the `animations.ts` file, we first defined our animation trigger named `routeAnimation`. Then we made sure that by default, the HTML element to which the trigger is assigned has `position: 'relative'` set as a style:

```
transition('* <=> *', [
    style({
        position: 'relative'
    }),
    ...
])
```

Then we apply the styled `position: 'absolute'` to the children, as mentioned, using `:enter` and `:leave` as follows:

```
query(':enter, :leave', [
    style({
        position: 'absolute',
        width: '100%'
    })
], {optional: true}),
```

This makes sure that these elements, that is, the routes to be loaded, have the `position: 'absolute'` style and a full width using `width: '100%'` so they can appear on top of each other. You can always fiddle around by commenting either of the styles to see what happens (at your own risk, though!).

Anyway, once the styles are set, we define what will happen to the route that'll enter the view using the `:enter` query. We set the style to have `opacity: 0` so it seems like the route is fading in:

```
query(':enter', [
    style({
        opacity: 0,
    })
], {optional: true}),
```

Finally, we defined our route transitions as a combination of two sequential animations, the first for `query :leave` and the second for `query :enter`. For the route leaving the view, we set the opacity to `0` via animation, and for the route entering the view, we set the opacity to `1` via animation as well:

```
query(':leave', [
  animate('300ms ease-out', style({ opacity: 0 }))
], {optional: true}),
query(':enter', [
  animate('300ms ease-in', style({ opacity: 1 }))
], {optional: true}),
```

See also

- Animations in Angular (`https://angular.io/guide/animations`)
- Angular route transition animations (`https://angular.io/guide/route-animations`)

Complex route animations in Angular using keyframes

In the previous recipe, you learned how to create basic route animations, and in this one, we're going to level up our animation game. In this recipe, you'll learn how to implement some complex route animations in Angular using keyframes.

Getting ready

The project that we are going to work with resides in `chapter04/start_here/complex-route-animations` inside the cloned repository. It is in the same state as the final code of the *Basic route animations in Angular* recipe, except we don't have any animations configured yet:

1. Open the project in Visual Studio Code.
2. Open the terminal and run `npm install` to install the dependencies of the project.
3. Once done, run `ng serve -o`.

This should open the app in a new browser tab and you should see something like the following:

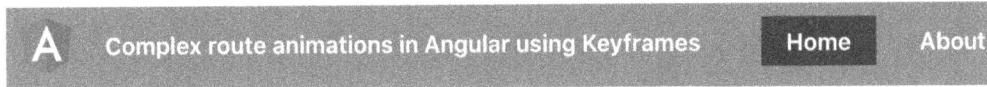

Home Page

Home Page Content

Figure 4.9 – complex-route-animations app running on http://localhost:4200

Now that we have the app running locally, let's see the steps of the recipe in the next section.

How to do it...

We have a basic app with two routes, the `HomePage` route and the `AboutPage` route. Similar to the previous recipe, *Basic route animations in Angular*, we have this configured using the route data parameters. But we don't just have any animations written yet. Also, we already have `BrowserAnimationsModule` imported in the `app.module.ts` file:

1. First, we'll start by writing a simple animation for the route entering the view and for the route leaving the view, as follows, in the `animations.ts` file:

    ```
    import {
      ...
      query,
      animate,
    } from '@angular/animations';

    const optional = { optional: true };

    export const ROUTE_ANIMATION = trigger('routeAnimation',
    [
      transition('* <=> *', [
        style({...}),
        query(':enter, :leave', [...], optional),
        query(':enter', [
          style({
            opacity: 0,
    ```

```
        })
      ], optional),
      query(':leave', [
        animate('1s ease-in', style({
          opacity: 0
        }))
      ], optional),
      query(':enter', [
        animate('1s ease-out', style({
          opacity: 1
        }))
      ], optional),
    ])
  ])
```

You'll notice that we now have fade-in/fade-out animations for the entering and leaving routes. However, you'll notice that the entering route doesn't appear until the current route has left the view. This is because both our animations are running in sequence.

2.   We'll group the animations for the :enter and :leave queries using the group method as follows:

```
import {
  ...
  animate,
  group
} from '@angular/animations';
...
export const ROUTE_ANIMATION = trigger('routeAnimation',
[
  transition('* <=> *', [
    style({...}),
    query(':enter, :leave', [...], optional),
    query(':enter', [...], optional),
    group([
      query(':leave', [
        animate('1s ease-in', style({
```

```
        opacity: 0
      }))
    ], optional),
    query(':enter', [
      animate('1s ease-out', style({
        opacity: 1
      }))
    ], optional),
  ])
  ])
])
```

Now, you should see both animations triggering together. Although it doesn't look great yet, trust me, it will!

3. Stepping up the game, we'll write a complex animation for our route entering the view. We'd like to create a **3D animation**, and therefore, we'll work with some `translateZ()` transformations as well:

```
import {
  ...
  keyframes,
} from '@angular/animations';
...
export const ROUTE_ANIMATION = trigger('routeAnimation',
[
  transition('* <=> *', [
    ...
    group([
      query(':leave', [...]),
      query(':enter', [
        animate('1s ease-out', keyframes([
          style({ opacity: 0, offset: 0, transform:
          'rotateY(180deg) translateX(25%)
          translateZ(1200px)' }),
          style({ offset: 0.25, transform:
          'rotateY(225deg) translateX(-25%)
          translateZ(1200px)' }),
          style({ offset: 0.5, transform:
```

```
              'rotateY(270deg) translateX(-50%)
              translateZ(400px)' }),
          style({ offset: 0.75, transform:
          'rotateY(315deg) translateX(-50%)
          translateZ(25px)' }),
          style({ opacity: 1, offset: 1, transform:
          'rotateY(360deg) translateX(0) translateZ(0)'
          }),
        ]))
      ], optional),
    ])
  ])
```

If you refresh the app now, you'll be like, "Pffttt, is that 3D, Ahsan? What?" Well, it is. However, we only see a sliding animation from left to right. And that's because we need to change our *perspective*.

4. To view all the elements being translated into 3D, we need to apply the perspective style to the host element for the animation. We'll do it by adding the perspective: '1000px' style in our first style definition in the animations.ts file:

```
...
export const ROUTE_ANIMATION = trigger('routeAnimation',
[
  transition('* <=> *', [
    style({
      position: 'relative',
      perspective: '1000px'
    }),
    query(':enter, :leave', [
      ...
    ], optional),
    query(':enter', [
      ...
    ], optional),
    group([
      ...
    ])
```

```
    ])
  ])
```

And boom! Now we have the `:enter` query animation in 3D.

5. Now let's update the animation for the `:leave` query as follows so we can see it leaving the view sliding backward in the *z* axis:

```
    ...

    export const ROUTE_ANIMATION = trigger('routeAnimation',
    [
      transition('* <=> *', [
        style({
          ...
        }),
        query(':enter, :leave', [
          ...
        ], optional),
        query(':enter', [
          ...
        ], optional),
        group([
          query(':leave', [
            animate('1s ease-in', keyframes([
              style({ opacity: 1, offset: 0, transform:
              'rotateY(0) translateX(0) translateZ(0)' }),
              style({ offset: 0.25, transform:
              'rotateY(45deg) translateX(25%)
              translateZ(100px) translateY(5%)' }),
              style({ offset: 0.5, transform: 'rotateY(90deg)
              translateX(75%) translateZ(400px)
              translateY(10%)' }),
              style({ offset: 0.75, transform:
              'rotateY(135deg) translateX(75%)
              translateZ(800px) translateY(15%)' }),
              style({ opacity: 0, offset: 1, transform:
              'rotateY(180deg) translateX(0)
              translateZ(1200px) translateY(25%)' }),
            ]))
```

```
      ], optional),
      query(':enter', [
        ...
      ], optional),
    ])
  ])
])
```

Woot woot! We now have a 3D animation for our routes that looks absolutely stunning. And this is, of course, not the end. The sky's the limit when it comes to what you can do with keyframes and animations in Angular.

How it works...

Since we wanted to implement a 3D animation in this recipe, we first made sure that the animation host element had a value for the perspective style, so we can see all the magic in 3D. Then we defined our animations using the keyframes method with an animation state for each offset so we could set different angles and rotations at those states, just so it all looks cool. One important thing that we did was group our :enter and :leave queries using the group method, where we defined the animations. This made sure that we had the route entering and leaving the view simultaneously.

See also

- Fireship.io's tutorial on Angular route animations (https://fireship.io/lessons/angular-router-animations/)

- Angular complex animation sequences (https://angular.io/guide/complex-animation-sequences)

5
Angular and RxJS – Awesomeness Combined

Angular and RxJS create a killer combination of awesomeness. By combining these, you can handle your data reactively, work with streams, and do really complex stuff in your Angular apps. That's exactly what you're going to learn in this chapter.

Here are the recipes we're going to cover in this chapter:

- Working with RxJS operators using instance methods
- Working with RxJS operators using static methods
- Unsubscribing streams to avoid memory leaks
- Using an Observable with the `async` pipe to synchronously bind data to your Angular templates
- Using `combineLatest` to subscribe to multiple streams together

- Using the `flatMap` operator to create sequential **HyperText Transfer Protocol (HTTP)** calls
- Using the `switchMap` operator to switch the last subscription with a new one
- Debouncing HTTP requests using RxJS

Technical requirements

For the recipes in this chapter, make sure you have **Git** and **Node.js** installed on your machine. You also need to have the `@angular/cli` package installed, which you can do with `npm install -g @angular/cli` from your Terminal. The code for this chapter can be found at the following link: `https://github.com/PacktPublishing/Angular-Cookbook/tree/master/chapter05`.

Working with RxJS operators using instance methods

In this recipe, you'll learn to use RxJS operators' instance methods to work with streams. We'll start with a basic app in which you can start listening to a stream with the `interval` method. We'll then introduce some instance methods in the subscription to modify the output.

Getting ready

The project that we are going to work with resides in `chapter05/start_here/rxjs-operators-instance-methods`, inside the cloned repository.

1. Open the project in **Visual Studio Code (VS Code)**.

2. Open the Terminal and run `npm install` to install the dependencies of the project.

3. Once done, run `ng serve -o`.

   This should open the app in a new browser tab. Tap the **Start Stream** button, and you should see something like this:

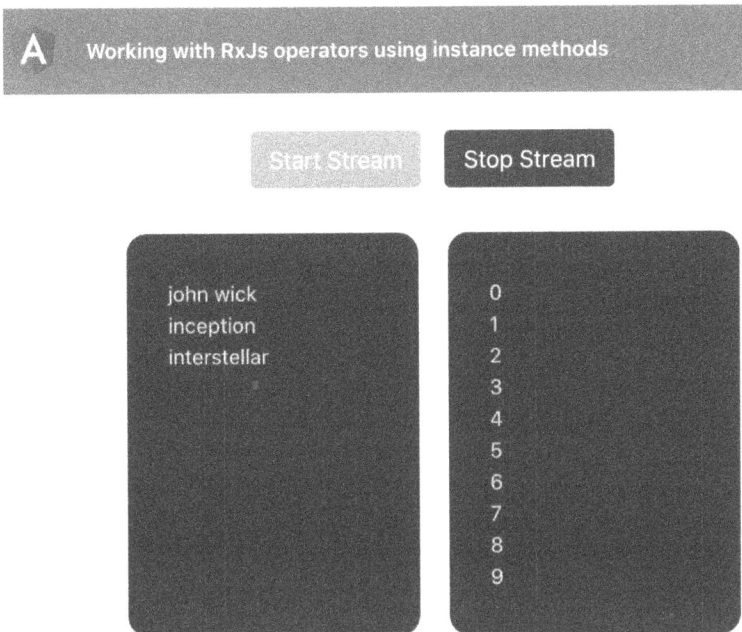

Figure 5.1 – The rxjs-operators-instance-methods app running on http://localhost:4200

Now that we have the app running, we will move on to the steps for the recipe.

How to do it...

We have an Angular app that has some things already set up. By tapping the **Start Stream** button, we can start viewing the stream output that is using the `interval` method from RxJS to create an Observable that outputs a sequence of numbers from 0 onward. We'll use some operators to show the elements from our `inputStreamData` array instead, which is the goal of this recipe. Let's begin.

1. First, we'll use the `map` operator to make sure that we are mapping the numbers generated from the `interval` Observable to the valid indices of our array. For this, we'll update the `app.component.ts` file.

    We have to make sure that the mapped numbers are not greater than or equal to the length of `inputStreamData`. We'll do this by taking a modulus on the number each time, using the `map` operator as follows:

    ```
    import { Component } from '@angular/core';
    import { interval, Subscription } from 'rxjs';
    import { map } from 'rxjs/operators';
    ```

```
@Component({...})
export class AppComponent {
...
  startStream() {
    this.subscription = streamSource
    .pipe(
      map(output => output % this.inputStreamData.
      length),
    )
    .subscribe(input => {
      this.outputStreamData.push(input);
    });
...
}
```

If you tap the **Start Stream** button now, you'll see that the output we get is 0, 1, 2, 0, 1, 2... and so forth. This makes sure we can always get an item from the inputStreamData array using the number as an index:

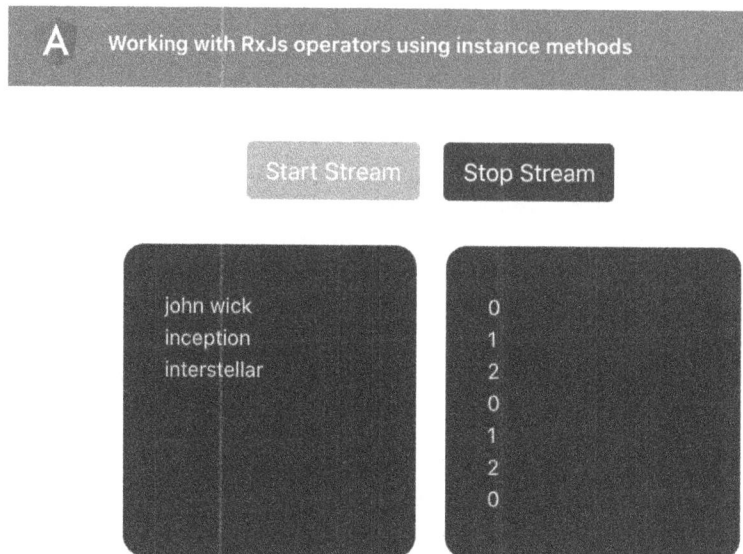

Figure 5.2 – The stream outputs a 0,1,2.. sequence using the modulus on inputStreamData.length

2. Now, we'll use another `map` method to fetch an element from the array for each of the stream's outputs, as follows:

```
startStream() {
  const streamSource = interval(1500);
  this.subscription = streamSource
  .pipe(
    map(output => output % this.inputStreamData.
    length),
    map(index => this.inputStreamData[index])
  )
  .subscribe(element => {
    this.outputStreamData.push(element);
  });
}
```

Notice that we've renamed the parameter of the `subscribe` method as `element` instead of `input`. This is because we get an element in the end. See the following screenshot, demonstrating how the stream outputs the elements from `inputStreamData` using indices:

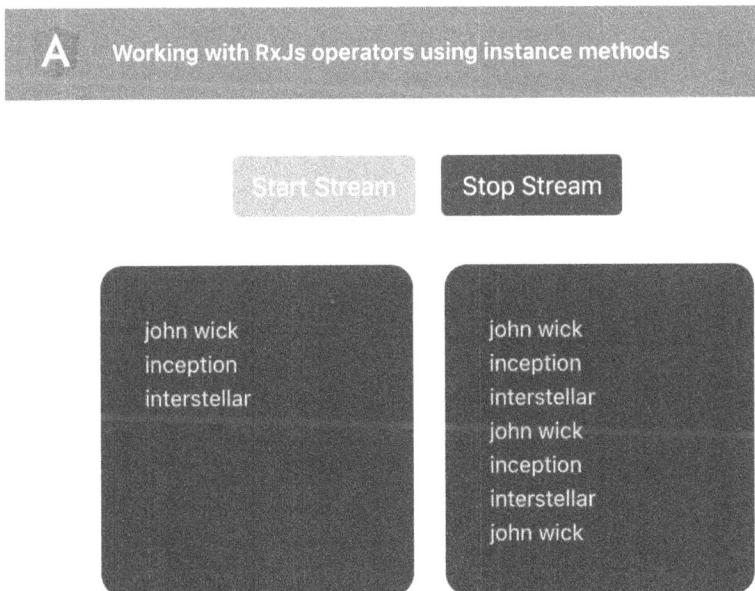

Figure 5.3 – The stream outputs elements from inputStreamData using indices

3. Now, to make things more interesting, we'll create another stream that will emit cartoon titles using the same `interval` method approach, but with a `1000ms` interval. Add the following code to your `startStream` method:

```
startStream() {
  const streamSource = interval(1500);
  const cartoonStreamSource = interval(1000)
    .pipe(
      map(output => output % this.cartoonsStreamData.
      length),
      map(index => this.cartoonsStreamData[index]),
    )
  this.subscription = streamSource
  .pipe(...)
  .subscribe(...);
}
```

4. We'll also create stream data named `cartoonStreamData` (used in the previous code) in the `AppComponent` class as a property. The code should look like this:

```
export class AppComponent {
  subscription: Subscription = null;
  inputStreamData = ['john wick', 'inception',
  'interstellar'];
  cartoonsStreamData = ['thunder cats', 'Dragon Ball Z',
  'Ninja Turtles'];
  outputStreamData = [];
  ...
}
```

5. Now that we have the `cartoonsStreamData` stream data in place, we can also add that to our template so that we can show it on the view as well. The children of the `<div class="input-stream">` element in `app.component.html` should look like this:

```
<div class="input-stream">
  <div class="input-stream__item" *ngFor="let item
  of inputStreamData">
    {{item}}
```

```
        </div>
        <hr/>
        <div class="input-stream__item" *ngFor="let item
        of cartoonsStreamData">
            {{item}}
        </div>
    </div>
```

6. Now, we'll use the `merge` (instance) method to combine the two streams and add an element from the respective stream data array when the streams emit a value. Interesting, right?

    We'll achieve this using the following code:

```
    . . .
    import { map, merge } from 'rxjs/operators';
    export class AppComponent {
        . . .
        startStream() {
            . . .
            this.subscription = streamSource
            .pipe(
                map(output => output % this.inputStreamData.
                length),
                map(index => this.inputStreamData[index]),
                merge(cartoonStreamSource)
            )
            .subscribe(element => {
                this.outputStreamData.push(element);
            });
        }
    }
```

> **Important note**
>
> The usage of the `merge` method as an instance method is deprecated in favor of the static `merge` method.

Great! You have now implemented the entire recipe, with an interesting merge of two streams. The following screenshot shows the final output:

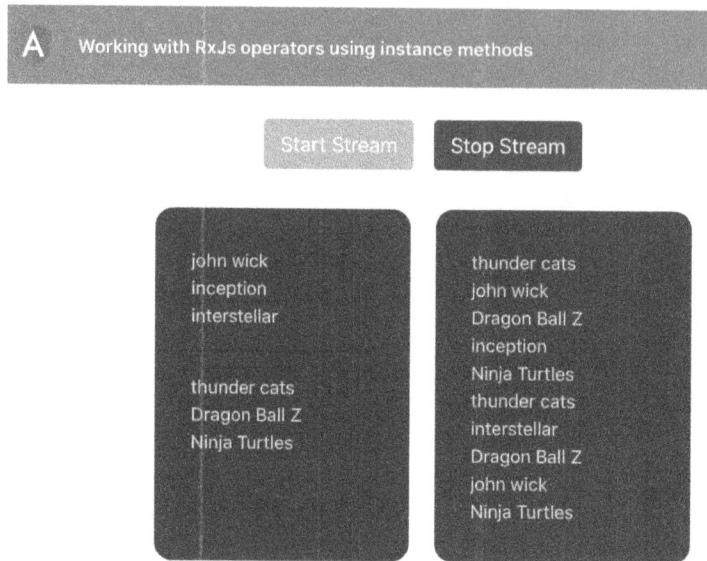

Figure 5.4 – Final output of the recipe

Let's move on to the next section to understand how it all works.

How it works...

The `map` operator provides you the stream's output value, and you're supposed to return a value that you want to map it to. We made sure that we converted the autogenerated sequential numbers to the array's indices by taking a modulus of the array's length. Then, we used another `map` operator on top of these indices to fetch the actual element from the array. Finally, we created another stream and used the `merge` method to combine the outputs of both streams and add this to the `outputStreamData` array.

See also

- *Catch the Dot Game*—RxJS documentation (`https://www.learnrxjs.io/learn-rxjs/recipes/catch-the-dot-game`)

- RxJS map operator documentation (`https://www.learnrxjs.io/learn-rxjs/operators/transformation/map`)

- RxJS merge operator documentation (`https://www.learnrxjs.io/learn-rxjs/operators/combination/merge`)

Working with RxJS operators using static methods

In this recipe, you'll learn to use RxJS operators' static methods to work with streams. We'll start with a basic app in which you can start listening to a stream with the `interval` method. We'll then introduce some static methods in the subscription to modify the output, to see it on the **user interface** (**UI**). After that, we'll split the streams using the `partition` static operator. And finally, we'll be merging the partitioned streams using the `merge` static operator to see their output.

Getting ready

The project for this recipe resides in `chapter05/start_here/rxjs-operators-static-methods`.

1. Open the project in VS Code.

2. Open the Terminal and run `npm install` to install the dependencies of the project.

3. Once done, run `ng serve -o`.

   This should open the app in a new browser tab, and you should see something like this:

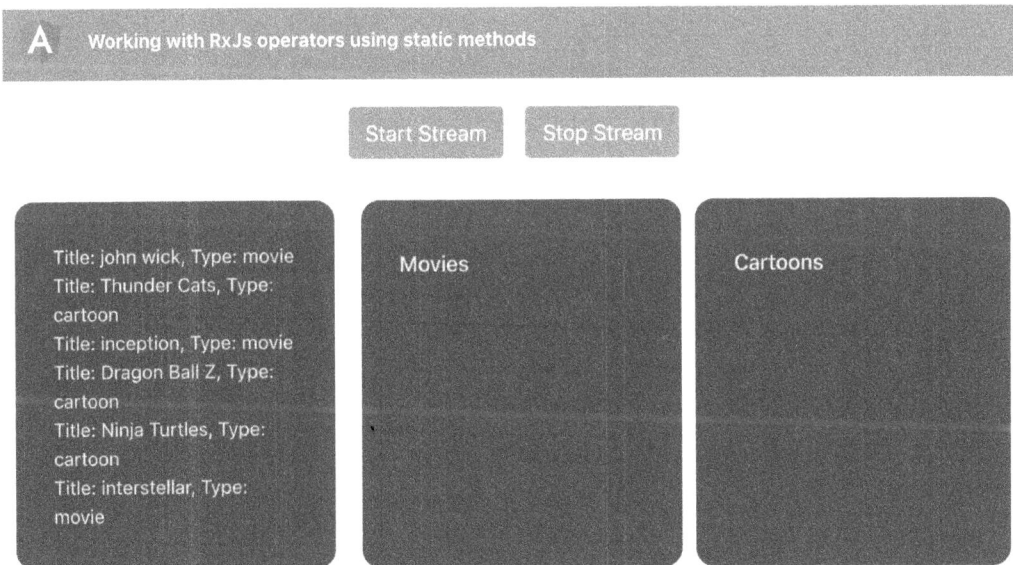

Figure 5.5 – The rxjs-operators-static-methods app running on http://localhost:4200

We also have the following data, which is composed of both movies and cartoons, and this is what we'll get as the output of the streams:

```
combinedStreamData = [{
    type: 'movie',
    title: 'john wick'
}, {
    type: 'cartoon',
    title: 'Thunder Cats'
}, {
    type: 'movie',
    title: 'inception'
}, {
    type: 'cartoon',
    title: 'Dragon Ball Z'
}, {
    type: 'cartoon',
    title: 'Ninja Turtles'
}, {
    type: 'movie',
    title: 'interstellar'
}];
```

Now that we have the app running locally, let's see the steps of the recipe in the next section.

How to do it...

We have an Angular app in hand that has some data in an array called combinedStreamData. By tapping the **Start Stream** button, we can start viewing the output of the stream in both the **Movies** output section and the **Cartoons** output section. We'll use the partition and merge operators to get the desired output and also to show the count of movies and the cartoons shown on output at the moment. Let's begin.

1. First, we'll import the partition and merge operators from RxJS (unlike how we imported it from rxjs/operators in the previous recipe). The import should look like this in the app.component.ts file:

    ```
    import { Component } from '@angular/core';
    import { interval, partition, merge, Subscription } from
    'rxjs';
    ```

2. Now, we'll create two properties, `movies` and `cartoons`, in the `AppComponent` class, one to hold the movies and one to hold the cartoons:

```
import { Component } from '@angular/core';
import { interval, partition, merge, Subscription } from
'rxjs';
import { map, tap } from 'rxjs/operators';
export class AppComponent {
    ...
    outputStreamData = [];
    movies= []
    cartoons= [];
    startStream() {
    }
    ...
}
```

3. And now, we'll use the appropriate variables in the template for movies and cartoons, as follows:

```
<div class="cards-container">
    <div class="input-stream">
        ...
    <div class="output-stream">
      <h6>Movies</h6>
      <div class="input-stream__item" *ngFor="let movie
      of movies">
         {{movie}}
      </div>
    </div>
    <div class="output-stream">
      <h6>Cartoons</h6>
      <div class="input-stream__item" *ngFor="let cartoon
      of cartoons">
         {{cartoon}}
      </div>
    </div>
    </div>
  </div>
```

4. We'll now use the `partition` operator to create two streams out of the `streamSource` property. Your `startStream` method should look like this:

```
startStream() {
    const streamSource = interval(1500).pipe(
      map(input => {
        const index = input % this.combinedStreamData.
        length;
        return this.combinedStreamData[index];
      })
    );
    const [moviesStream, cartoonsStream] = partition(
      streamSource, item => item.type === 'movie'
    );
    this.subscription = streamSource
      .subscribe(input => {
        this.outputStreamData.push(input);
      });
}
```

Now that we have the streams split up, we can merge those to subscribe to a single stream, push to the appropriate output array, and just log the value to the console as the output.

5. Let's merge the streams now, and we'll add them to the appropriate output array using the `tap` operator, as follows:

```
startStream() {
    ...
    this.subscription = merge(
      moviesStream.pipe(
        tap(movie => {
          this.movies.push(movie.title);
        })
      ),
      cartoonsStream.pipe(
        tap(cartoon => {
          this.cartoons.push(cartoon.title);
        })
```

```
        ),
    )
        .subscribe(input => {
            this.outputStreamData.push(input);
        });

    }
```

With this change, you should be able to see the correct value in the appropriate container—that is, whether it is a movie or a cartoon. See the following screenshot, which shows how the partitioned streams emit values to the appropriate Observables:

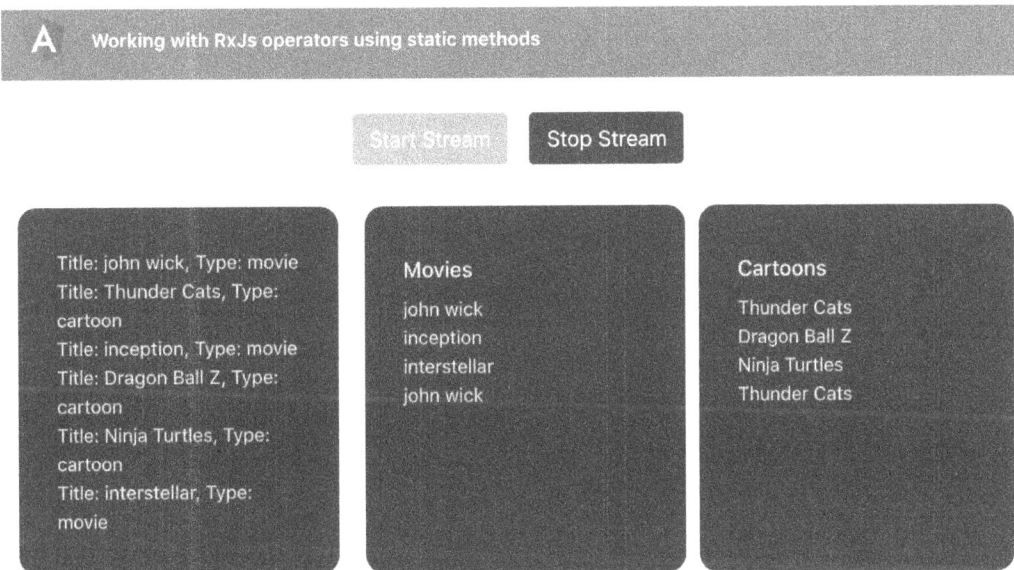

A   Working with RxJs operators using static methods

Start Stream   Stop Stream

Title: john wick, Type: movie
Title: Thunder Cats, Type:
cartoon
Title: inception, Type: movie
Title: Dragon Ball Z, Type:
cartoon
Title: Ninja Turtles, Type:
cartoon
Title: interstellar, Type:
movie

Movies

john wick
inception
interstellar
john wick

Cartoons

Thunder Cats
Dragon Ball Z
Ninja Turtles
Thunder Cats

Figure 5.6 – Partitioned streams outputting data to the appropriate views

6. Finally, since we have merged the stream, we can use `console.log` to see each value being output. We'll remove the `outputStreamData` property from `AppComponent` and use a `console.log` statement instead of pushing to `outputStreamData` in the `subscribe` block, as follows:

```
...
@Component({...})
export class AppComponent {

    ...

    outputStreamData = []; ← Remove
```

```
movies = [];
cartoons = [];
ngOnInit() {}


startStream() {
  const streamSource = interval(1500).pipe(
    map(...)
  );
  const [moviesStream, cartoonsStream] =
  partition(...);
  this.subscription = merge(
    moviesStream.pipe(...),
    cartoonsStream.pipe(...)
  ).subscribe((output) => {
    console.log(output);
  });
}

...

}
```

As soon as you refresh the app, you should see the logs on the console, as follows:

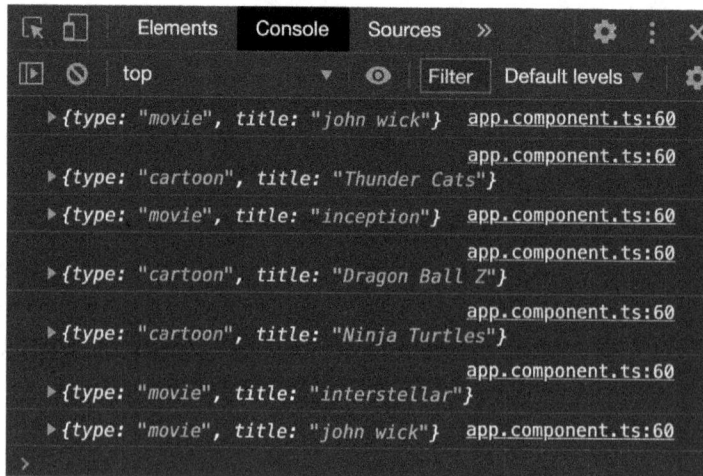

Figure 5.7 – Logs on console for each output in the subscribe block from the merged streams

Great! You now know how to use static operators from RxJS (specifically, `partition` and `merge`) to work with streams in real-life use cases. See the next section on how this works.

How it works...

RxJS has a bunch of static operators/methods that we can use for our particular use cases. In this recipe, we used the `partition` operator to create two different streams based on the `predicate` function provided as the second argument, which returns an array with two Observables. The first one will contain all values that satisfy the predicate, and the second one will contain all values that don't satisfy the predicate. *Why did we split the streams?* Glad you asked. Because we needed to show the appropriate outputs in different output containers. And what's GREAT is that we merged those streams later on so that we only had to subscribe to one stream, and we could then unsubscribe from that very stream as well.

See also

- RxJS map operator documentation (`https://www.learnrxjs.io/learn-rxjs/operators/transformation/map`)

- RxJS merge operator documentation (`https://www.learnrxjs.io/learn-rxjs/operators/combination/merge`)

- RxJS partition operator documentation (`https://www.learnrxjs.io/learn-rxjs/operators/transformation/partition`)

Unsubscribing streams to avoid memory leaks

Streams are fun to work with and they're awesome, and you'll know much more about RxJS when you've finished this chapter, although problems occur when streams are used without caution. One of the biggest mistakes to do with streams is to not unsubscribe them when we no longer need them, and in this recipe, you'll learn how to unsubscribe streams to avoid memory leaks in your Angular apps.

Getting ready

The project for this recipe resides in `chapter05/start_here/rxjs-unsubscribing-streams`.

1. Open the project in VS Code.
2. Open the Terminal and run `npm install` to install the dependencies of the project.

3.   Once done, run ng serve -o.

This should open the app in a new browser tab, and you should see something like this:

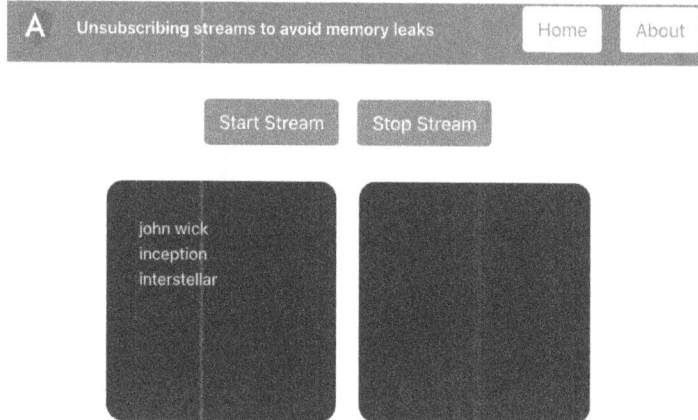

Figure 5.8 – The rxjs-unsubscribing-streams app running on http://localhost:4200

Now that we have the app running locally, let's see the steps of the recipe in the next section.

How to do it...

We currently have an app with two routes—that is, **Home** and **About**. This is to show you that unhandled subscriptions can cause memory leaks in an app. The default route is **Home**, and in the HomeComponent class, we handle a single stream that outputs data using the interval method.

1.   Tap the **Start Stream** button, and you should see the stream emitting values.

2.   Then, navigate to the **About** page by tapping the **About** button from the header (top right), and then come back to the **Home** page.

What do you see? Nothing? Everything looks fine, right? Well, not exactly.

3.   To see whether we have an unhandled subscription, which is an issue, let's put a console.log inside the startStream method in the home.component.ts file—specifically, inside the .subscribe method's block, as follows:

```
. . .
export class HomeComponent implements OnInit {
    . . .
    startStream() {
```

```
    const streamSource = interval(1500);
    this.subscription = streamSource.subscribe(input => {
      this.outputStreamData.push(input);
      console.log('stream output', input)
    });
  }
  stopStream() {...}
}
```

If you now perform the same steps as mentioned in *Step 1*, you'll see the following output on the console:

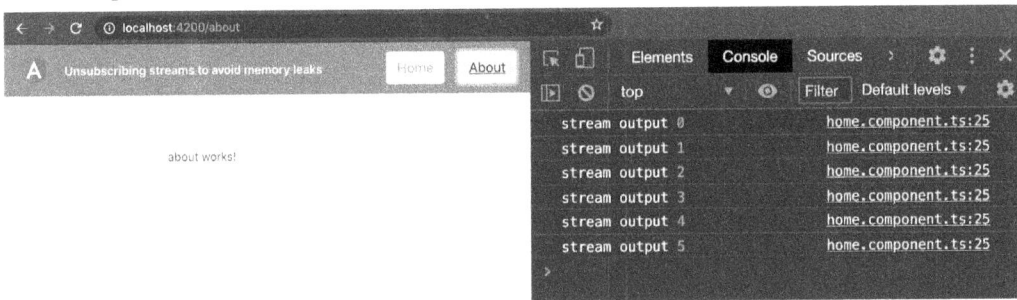

Figure 5.9 – The rxjs-unsubscribing-streams app running on http://localhost:4200

Want to have some more fun? Try performing *Step 1* a couple of times without refreshing the page even once. What you'll see will be **CHAOS**!

4. So, to solve the issue, we'll use the simplest approach—that is, unsubscribing the stream when the user navigates away from the route. Let's implement the ngOnDestroy lifecycle method for that, as follows:

```
import { Component, OnInit, OnDestroy } from '@angular/
core';
...

@Component({
  selector: 'app-home',
  templateUrl: './home.component.html',
  styleUrls: ['./home.component.scss']
})
export class HomeComponent implements OnInit, OnDestroy {
  ...
```

```
ngOnInit() {

}

ngOnDestroy() {
  this.stopStream();
}

startStream() {
  const streamSource = interval(1500);
  this.subscription = streamSource.subscribe(input => {
    this.outputStreamData.push(input);
    console.log('stream output', input)
  });
}

stopStream() {
  this.subscription.unsubscribe();
  this.subscription = null;
}
}
```

Great! If you follow the instructions from *Step 1* again, you'll see that there's no further log on the console once you navigate away from the **Home** page, and our app doesn't have an unhandled stream causing memory leaks now. Read the next section to understand how it works.

How it works...

When we create an Observable/stream and we subscribe to it, RxJS automagically adds our provided .subscribe method block as a handler to the Observable. So, whenever there's a value emitted from the Observable, our method is supposed to be called. The fun part is that Angular doesn't automatically destroy that subscription/handler when the component unmounts or when you have navigated away from the route. That's because the core of Observables is RxJS, not Angular, and therefore it isn't Angular's responsibility to handle it.

Angular provides certain lifecycle methods, and we used the OnDestroy (ngOnDestroy) method. This is because when we navigate away from a route, Angular destroys that route, and that's when we would want to unsubscribe from all streams we have subscribed to.

There's more...

In a complex Angular app, there will be cases where you'd have more than one subscription in a component, and when the component is destroyed, you'd want to clean all those subscriptions at once. Similarly, you might want to unsubscribe based on certain events/conditions rather than the OnDestroy lifecycle. Here is an example, where you have multiple subscriptions in hand and you want to clean up all of them together when the component destroys:

```
startStream() {
    const streamSource = interval(1500);
    const secondStreamSource = interval(3000);
    const fastestStreamSource = interval(500);
    streamSource.subscribe(input => {...});
    secondStreamSource.subscribe(input => {
      this.outputStreamData.push(input);
      console.log('second stream output', input)
    });
    fastestStreamSource.subscribe(input => {
      this.outputStreamData.push(input);
      console.log('fastest stream output', input)
    });
}

    stopStream() {
    }
```

Notice that we're not saving the **Subscription** from streamSource to this. subscription anymore, and we have also removed the code from the stopStream method. The reason for this is because we don't have individual properties/variables for each Subscription. Instead, we'll have a single variable to work with. Let's look at the following recipe steps to get things rolling.

1. First, we'll create a property in the HomeComponent class named isComponentAlive:

    ```
    . . .
    export class HomeComponent implements OnInit, OnDestroy {
        isComponentAlive: boolean;
        . . .
    }
    ```

2. Now, we'll import the `takeWhile` operator from `rxjs/operators`, as follows:

```
import { Component, OnInit, OnDestroy } from '@angular/
core';
import { interval } from 'rxjs/internal/observable/
interval';
import { Subscription } from 'rxjs/internal/
Subscription';
import { takeWhile } from 'rxjs/operators';
```

3. We'll now use the `takeWhile` operator with each of our streams to make them work only when the `isComponentAlive` property is set to `true`. Since `takeWhile` takes a `predicate` method, it should look like this:

```
startStream() {
    ...
    streamSource
      .pipe(
        takeWhile(() => !!this.isComponentAlive)
      ).subscribe(input => {...});
    secondStreamSource
      .pipe(
        takeWhile(() => !!this.isComponentAlive)
      ).subscribe(input => {...});
    fastestStreamSource
      .pipe(
        takeWhile(() => !!this.isComponentAlive)
      ).subscribe(input => {...});
}
```

If you press the **Start Stream** button right now on the **Home** page, you still won't see any output or logs because the `isComponentAlive` property is still `undefined`.

4. To make the streams work, we'll set the `isComponentAlive` property to `true` in the `ngOnInit` method as well as in the `startStream` method. The code should look like this:

```
ngOnInit() {
    this.isComponentAlive = true;
}


ngOnDestroy() {
    this.stopStream();
}


startStream() {
    this.isComponentAlive = true;
    const streamSource = interval(1500);
    const secondStreamSource = interval(3000);
    const fastestStreamSource = interval(500);
    ...
}
```

After this step, if you now try to start the stream and navigate away from the page, you'll still see the same issue with the streams—that is, they've not been unsubscribed.

5. To unsubscribe all streams at once, we'll set the value of `isComponentAlive` to `false` in the `stopStream` method, as follows:

```
stopStream() {
    this.isComponentAlive = false;
}
```

And boom! Now, if you navigate away from the route while the streams are emitting values, the streams will stop immediately as soon as you navigate away from the **Home** route. Voilà!

See also

- Read about RxJS Subscription (https://www.learnrxjs.io/learn-rxjs/ concepts/rxjs-primer#subscription)
- takeWhile docs (https://www.learnrxjs.io/learn-rxjs/ operators/filtering/takewhile)

Using an Observable with the async pipe to synchronously bind data to your Angular templates

As you learned in the previous recipe, it is crucial to unsubscribe the streams you subscribe to. What if we had an even simpler way to unsubscribe them when the component gets destroyed—that is, letting Angular take care of it somehow? In this recipe, you'll learn how to use Angular's `async` pipe with an Observable to directly bind the data in the stream to the Angular template instead of having to subscribe in the `*.component.ts` file.

Getting ready

The project for this recipe resides in `chapter05/start_here/using-async-pipe`.

1. Open the project in VS Code.

2. Open the Terminal and run `npm install` to install the dependencies of the project.

3. Once done, run `ng serve -o`.

   This should open the app in a new browser tab. As soon as the page is opened, you should see something like this:

Figure 5.10 – The using-async-pipe app running on http://localhost:4200

Now that we have the app running locally, let's see the steps of the recipe in the next section.

How to do it...

The app we have right now has three streams/Observables observing values at different intervals. We're relying on the isComponentAlive property to keep the subscription alive or make it stop when the property is set to false. We'll remove the usage of takeWhile and somehow make everything work similarly to what we have right now.

1. First, remove the subscription property from the home.component.ts file and add an Observable type property named streamOutput$. The code should look like this:

    ```
    . . .
    import { Observable } from 'rxjs';
    . . .

    export class HomeComponent implements OnInit, OnDestroy {
        isComponentAlive: boolean;
        subscription: Subscription = null ← Remove this;
        inputStreamData = ['john wick', 'inception',
        'interstellar'];
        streamsOutput$: Observable<number[]> ← Add this
        outputStreamData = []


        constructor() { }


        . . .
    }
    ```

    With this change, the app would break because of some missing variables. Fear not! I'm here to help you.

2. We'll now combine all the streams to give out a single output—that is, the outputStreamData array. We'll remove all the existing .pipe and .subscribe methods from the startStream() method, so the code should now look like this:

    ```
    import { Component, OnInit, OnDestroy } from '@angular/
    core';
    import { merge, Observable } from 'rxjs';
    import { map, takeWhile } from 'rxjs/operators';
    . . .
    ```

```
export class HomeComponent implements OnInit, OnDestroy {
    ...

    startStream() {
        const streamSource = interval(1500);
        const secondStreamSource = interval(3000);
        const fastestStreamSource = interval(500);
        this.streamsOutput$ = merge(
            streamSource,
            secondStreamSource,
            fastestStreamSource
        )
    }
    ...
}
```

With this change, the linters will still complain. Why? Because the merge
operator combines all streams and outputs the latest value. This is a
Observable<number> data type, instead of Observable<string[]>, which
is the type of streamsOutput$.

3. Since we want to assign the entire array containing every output emitted from the
   streams, we'll use a map operator and add each output to the outputStreamData
   array, and return the latest state of the outputStreamData array, as follows:

```
startStream() {
    const streamSource = interval(1500);
    const secondStreamSource = interval(3000);
    const fastestStreamSource = interval(500);
    this.streamsOutput$ = merge(
        streamSource,
        secondStreamSource,
        fastestStreamSource
    ).pipe(
        takeWhile(() => !!this.isComponentAlive),
        map(output => {
            this.outputStreamData = [...this.
            outputStreamData, output]
            return this.outputStreamData;
```

```
        })
    )
}
```

4. Remove the `stopStream` method from the `HomeComponent` class since we don't need it anymore. Also, remove its usage from the `ngOnDestroy` method.

5. Finally, modify the template in `home.component.html` to use the `streamOutput$` Observable with the `async` pipe to loop over the output array:

```html
<div class="output-stream">
  <div class="input-stream__item" *ngFor="let item
  of streamsOutput$ | async">
    {{item}}
  </div>
</div>
```

6. To verify that the subscription REALLY gets destroyed on component destruction, let's put a `console.log` in the `startStream` method inside the map operator, as follows:

```
startStream() {
    const streamSource = interval(1500);
    const secondStreamSource = interval(3000);
    const fastestStreamSource = interval(500);
    this.streamsOutput$ = merge(
      streamSource,
      secondStreamSource,
      fastestStreamSource
    ).pipe(
      takeWhile(() => !!this.isComponentAlive),
      map(output => {
        console.log(output)
        this.outputStreamData = [...this.
        outputStreamData, output]
        return this.outputStreamData;
      })
    )
}
```

Hurray! With this change, you can try refreshing the app, navigate away from the **Home** route, and you'll see that the console logs stop as soon as you do that. Do you feel the achievement we just got by removing all that extra code? I certainly do. Well, see in the next section how it all works.

How it works...

Angular's `async` pipe automatically destroys/unsubscribes the subscription as soon as the component destroys. This gives us a great opportunity to use it where possible. In the recipe, we basically combined all the streams using the `merge` operator. The fun part was that for the `streamsOutput$` property, we wanted an Observable of the output array on which we could loop over. However, merging the stream only combines them and emits the latest value emitted by any of the streams. So, we added a `.pipe()` method with the `.map()` operator to take the latest output out of the combined stream, added it to the `outputStreamData` array for persistence, and returned it from the `.map()` method so that we get the array in the template when we use the `async` pipe.

Fun fact—streams don't emit any value unless they're subscribed to. *"But Ahsan, we didn't subscribe to the stream, we just merged and mapped the data. Where's the subscription?"* Glad you asked. Angular's `async` pipe subscribes to the stream itself, which triggers our `console.log` as well that we added in *Step 6*.

> **Important note**
>
> The `async` pipe has a limitation, which is that you cannot stop the subscription until the component is destroyed. In such cases, you'd want to go for in-component subscriptions using something such as the `takeWhile`/ `takeUntil` operator or doing a regular `.unsubscribe` method yourself when the component is destroyed.

See also

- Angular `async` pipe documentation (`https://angular.io/api/common/ AsyncPipe`)

Using combineLatest to subscribe to multiple streams together

In the previous recipe, we had to merge all the streams, which resulted in a single output being last emitted by any of the streams. In this recipe, we'll work with `combineLatest`, which results in having an array as an output, combining all the streams. This approach is appropriate for when you want the latest output from all the streams, combined in a single subscribe.

Getting ready

The project that we are going to work with resides in `chapter05/start_here/using-combinelatest-operator`, inside the cloned repository.

1. Open the project in VS Code.

2. Open the Terminal and run `npm install` to install the dependencies of the project.

3. Once done, run `ng serve -o`.

    This should open the app in a new browser tab, and you should see something like this:

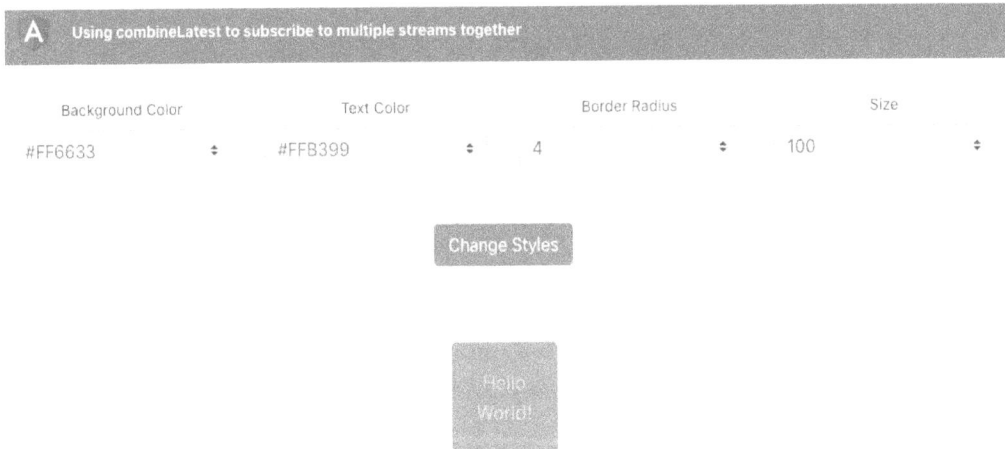

Figure 5.11 – The using-combinelatest-operator app running on http://localhost:4200

Now that we have the app running locally, let's see the steps of the recipe in the next section.

How to do it...

For this recipe, we have an app that displays a box. The box has a size (width and height), a border radius, a background color, and a color for its text. It also has four inputs to modify all the mentioned factors. Right now, we have to apply the changes manually with the click of a button. What if we could subscribe to the changes to the inputs and update the box right away? That's what we're going to do here.

1. We'll begin by creating a method named `listenToInputChanges`, in which we'll subscribe to the changes for each of the inputs and combine those streams using the `combineLatest` operator. Update the `home/home.component.ts` file, as follows:

```
...
import { combineLatest, Observable } from 'rxjs';
...

export class HomeComponent implements OnInit, OnDestroy {
  ...
  ngOnInit() {
    ...
    this.applyChanges();
    this.listenToInputChanges(); ← Add this
  }

  listenToInputChanges() {
    combineLatest([
      this.boxForm.get('size').valueChanges,
      this.boxForm.get('borderRadius').valueChanges,
      this.boxForm.get(
        'backgroundColor').valueChanges,
      this.boxForm.get('textColor').valueChanges
    ]).subscribe(() => {
      this.applyChanges();
    });
  }
  ...
}
```

2. Remember that not unsubscribing streams is a BAD idea? And that's what we have here: a subscribed stream. We'll use the `async` pipe instead of the current Subscription used in the `home.component.ts` file. For that, let's create an Observable property named `boxStyles$` and remove the `boxStyles` property. Then, assign the stream from `combineLatest` to it, as follows:

```
. . .
import { map} from 'rxjs/operators';
. . .
export class HomeComponent implements OnInit, OnDestroy {
  . . .
  boxStyles: {...}; ← Remove this
  boxForm = new FormGroup({...});
  boxStyles$: Observable<{
    width: string,
    height: string,
    backgroundColor: string,
    color: string
    borderRadius: string
  }>;
    . . .

  listenToInputChanges() {
    this.boxStyles$ = combineLatest([...]).
    pipe(map(([size, borderRadius, backgroundColor,
    textColor]) => {
      return {
        width: `${size}px`,
        height: `${size}px`,
        backgroundColor,
        color: textColor,
        borderRadius: `${borderRadius}px`
      }
    }));
  }
  . . .
}
```

3. We need to remove the `setBoxStyles()` and `applyChanges()` methods and the usages of the `applyChanges()` method from the `home.component.ts` file. Update the file, as follows:

```
export class HomeComponent implements OnInit, OnDestroy {

  . . .

  ngOnInit() {

    . . .

    this.applyChanges();  ← Remove this
    this.listenToInputChanges();  ← Add this
  }

  . . .

  setBoxStyles(size, backgroundColor, color,
  borderRadius) {...}  ← Remove this


  applyChanges() {...}  ← Remove this

  . . .

}
```

4. We also need to remove the usage of the `applyChanges()` method from the template as well. Remove the `(ngSubmit)` handler from the `<form>` element in the `home.component.html` file so that it looks like this:

```
<div class="home" [formGroup]="boxForm"
(ngSubmit)="applyChanges()"  ← Remove this>

  . . .

</div>
```

5. We also need to get rid of the `submit-btn-container` element from the `home.component.html` template as we don't need it anymore. Delete the following chunk from the file:

```
<div class="row submit-btn-container"  ← Remove this
element>
  <button class="btn btn-primary" type="submit"
  (click)="applyChanges()">Change Styles</button>
</div>
```

If you refresh the app, you'll notice that the box doesn't show at all. We'll fix this in the next step.

6. Since we're using the `combineLatest` operator when the app starts, but we don't have it triggered because none of the inputs have changed, we need to initialize the box with some initial values. To do so, we'll use the `startWith` operator with the initial values, as follows:

```
...
import { map, startWith } from 'rxjs/operators';

@Component({...})
export class HomeComponent implements OnInit, OnDestroy {
  ...
  ngOnInit() {
    this.listenToInputChanges();
  }

  listenToInputChanges() {
    this.boxStyles$ = combineLatest([
      this.boxForm
        .get('size')
        .valueChanges.pipe(startWith(this.
        sizeOptions[0])),
      this.boxForm
        .get('borderRadius')
        .valueChanges.pipe(startWith(
        this.borderRadiusOptions[0])),
      this.boxForm
        .get('backgroundColor')
        .valueChanges.pipe(startWith(
        this.colorOptions[1])),
      this.boxForm
        .get('textColor')
        .valueChanges.pipe(startWith(
        this.colorOptions[0])),
    ]).pipe(
      map(...);
  }
```

```
    ngOnDestroy() {}
}
```

7. Now that we have the `boxStyles$` Observable in place, let's use it in the template instead of the `boxStyles` property:

```
. . .
<div class="row" *ngIf="boxStyles$ | async as bStyles">
  <div class="box" [ngStyle]="bStyles">
    <div class="box__text">
      Hello World!
    </div>
  </div>
</div>
. . .
```

And voilà! Everything works perfectly fine now.

Congratulations on finishing the recipe. You're now the master of streams and the `combineLatest` operator. See the next section to understand how it works.

How it works...

The beauty of **reactive forms** is that they provide much more flexibility than the regular `ngModel` binding or even template-driven forms. And for each form control, we can subscribe to its `valueChanges` Observable, which receives a new value whenever the input is changed. So, instead of relying on the **Submit** button's click, we subscribed directly to the `valueChanges` property of each **form control**. In a regular scenario, that would result in four different streams for four inputs, which means we would have four subscriptions that we need to take care of and make sure we unsubscribe them. This is where the `combineLatest` operator comes into play. We used the `combineLatest` operator to combine those four streams into one, which means we needed to unsubscribe only one stream on component destruction. But hey! Remember that we don't need to do this if we use the `async` pipe? That's exactly what we did. We removed the subscription from the `home.component.ts` file and used the `.pipe()` method with the `.map()` operator. The `.map()` operator transformed the data to our needs, and then returned the transformed data to be set to the `boxStyles$` Observable. Finally, we used the async pipe in our template to subscribe to the `boxStyles$` Observable and assigned its value as the `[ngStyle]` to our box element.

> **Important note**
> The `combineLatest` method will not emit an initial value until each
> Observable emits at least one value. Therefore, we use the `startWith`
> operator with each individual form control's `valueChanges` stream
> to provide an initial emitted value.

See also

- `combineLatest` operator documentation (`https://www.learnrxjs.io/`
  `learn-rxjs/operators/combination/combinelatest`)
- Visual representation of the `combineLatest` operator (`https://rxjs-dev.`
  `firebaseapp.com/api/index/function/combineLatest`)

Using the flatMap operator to create sequential HTTP calls

The days of using **Promises** were awesome. It's not that those days are gone, but we as
developers surely prefer **Observables** over **Promises** for a lot of reasons. One of the things
I really like about Promises is that you can chain Promises to do things such as sequential
HTTP calls. In this recipe, you'll learn how to do the same with **Observables** using the
`flatMap` operator.

Getting ready

The project that we are going to work with resides in `chapter05/start_here/`
`using-flatmap-operator`, inside the cloned repository.

1. Open the project in VS Code.
2. Open the Terminal and run `npm install` to install the dependencies of
   the project.

3. Once done, run `ng serve -o`.

    This should open the app in a new browser tab, and you should see something like this:

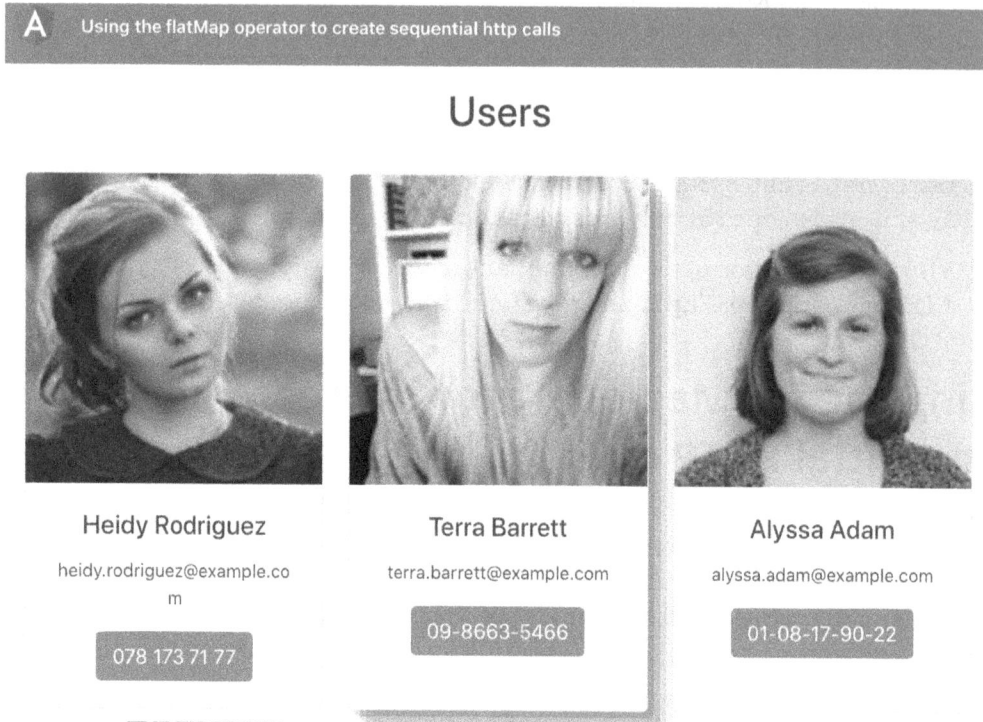

Figure 5.12 – The using-flatmap-operator app running on http://localhost:4200

The app right now seems perfect, actually. Nothing suspicious, right? Well, not exactly. Follow these steps to find out what is wrong.

1. Open Chrome DevTools.

2. Go to the **Network** tab and simulate the **Slow 3G** network, as follows:

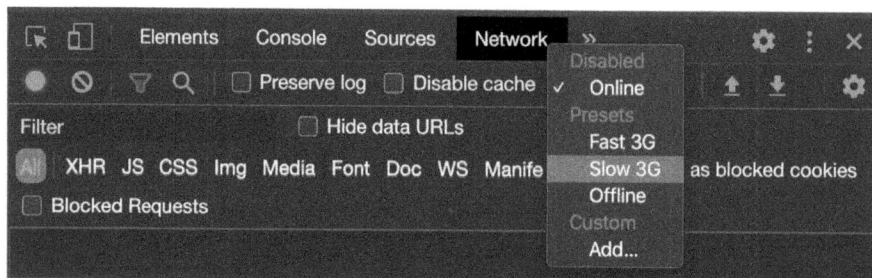

Figure 5.13 – Simulatina slow 3G network in Chrome DevTools

If you tap on any card from the home page, you should reach the particular user's details page.

3. Refresh the app now, see the **Network** tab, and you can see the HTTP calls executing in parallel, as follows:

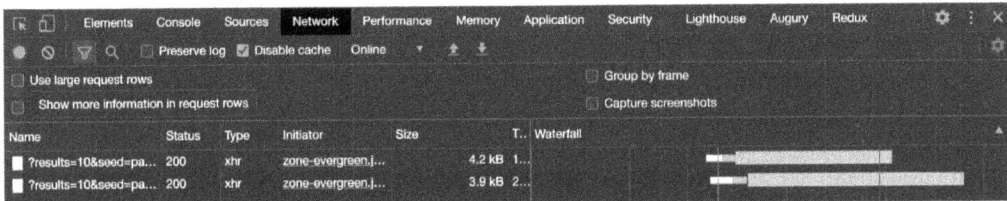

Figure 5.14 – Parallel calls loading data asynchronously

The problem is that we're not sure which data would come first due to both HTTP calls executing in parallel. Therefore, the user might see similar users before the main user is loaded. Let's see how to avoid this.

How to do it...

In order to fix the issue that our similar users can be loaded before our main user, we'll have to sequentially load the data and show the appropriate content respectively, and while the content is loading, we'll show a loader. Let's get started.

1. First, let's modify our `user-detail/user-detail.component.html` file to show the loader while we're loading and while we load the similar users as well. The code should look like this:

```
<div class="user-detail">
  <div class="main-content user-card">
    <app-user-card *ngIf="user$ | async as user; else
    loader" [user]="user"></app-user-card>
  </div>

  <div class="secondary-container">
    <h4>Similar Users</h4>
    <div class="similar-users">
      <ng-container *ngIf="similarUsers$ | async as
      users; else loader">
        <app-user-card class="user-card" *ngFor="let user
        of users" [user]="user"></app-user-card>
```

```
        </ng-container>
      </div>
    </div>
  </div>


<ng-template #loader>
  <app-loader></app-loader>
</ng-template>
```

If you refresh the app, you should see both loaders appearing before the calls are made.

We want to make the calls sequential, and for that, we can't have the streams directly bound to Observables in the UserDetailComponent class. That is, we can't even use the async pipe.

2. Let's convert the Observable properties to regular properties in the UserDetailComponent class, as follows:

```
. . .
export class UserDetailComponent implements OnInit,
OnDestroy {
  user: IUser;
  similarUsers: IUser[];
  isComponentAlive: boolean;
  . . .
}
```

You should already have the app breaking as soon as you save this aforementioned change.

3. Let's use the new variables that we modified in the previous step inside our template. Modify the user-detail.component.html file, as follows:

```
<div class="user-detail">
  <div class="main-content user-card">
    <app-user-card *ngIf="user; else loader"
    [user]="user"></app-user-card>
  </div>
```

```
<div class="secondary-container">
  <h4>Similar Users</h4>
  <div class="similar-users">
    <ng-container *ngIf="similarUsers; else loader">
      <app-user-card class="user-card" *ngFor="let user
      of similarUsers" [user]="user"></app-user-card>
    </ng-container>
  </div>
</div>
</div>
...
```

4. Finally, let's use the `flatMap` operator now to execute the calls sequentially and to assign the received values to the appropriate variables, as follows:

```
...
import { takeWhile, flatMap } from 'rxjs/operators';

export class UserDetailComponent implements OnInit,
OnDestroy {
  ...
  ngOnInit() {
    this.isComponentAlive = true;
    this.route.paramMap.pipe(
      takeWhile(() => !!this.isComponentAlive),
      flatMap(params => {
        this.user = null;
        this.similarUsers = null;
        const userId = params.get('uuid');
        return this.userService.getUser(userId)
          .pipe(
            flatMap((user: IUser) => {
              this.user = user;
              return this.userService.
              getSimilarUsers(userId);
            })
          );
      })
```

```
    ).subscribe((similarUsers: IUser[]) => {
        this.similarUsers = similarUsers;
    })
}
    ...
}
```

And yes! If you now refresh the app, you'll notice that the calls are sequential as we first get the main user, and then the similar users. To confirm, you can open Chrome DevTools and see the network log for the **application programming interface (API)** calls. You should see something like this:

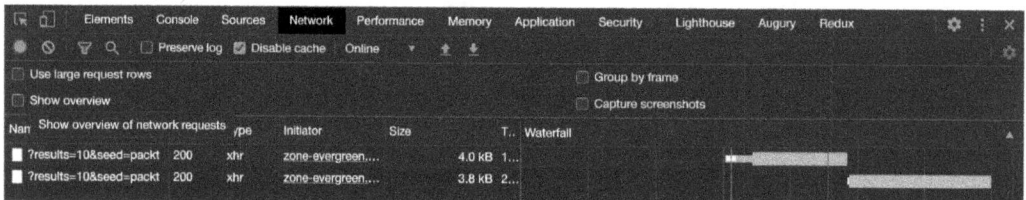

Figure 5.15 – API calls executing synchronously

Now that you've finished the recipe, see the next section on how this works.

How it works...

The `flatMap` operator takes the output from the previous Observable and is supposed to return a new Observable back. This helps us to sequentially execute our HTTP calls to be sure that the data is loaded according to its priority, or our business logic.

Since we wanted to execute the calls whenever a new user is selected, which can happen from the `UserDetailComponent` class itself, we put a `flatMap` operator on the `route.paramsMap` directly. Whenever that happens, we first set the `user` and `similarUsers` properties to `null`. "*But why?*" Well, because if we're on the `UserDetailsComponent` page and we click on any similar user, the page wouldn't change since we're already on it. This means the user and `similarUsers` variables will still contain their previous values. And since they'll have values already (that is, they're not `null`), the loader will not show in that case on tapping any similar user. Smart, right?

Anyways, after assigning the variables to `null`, we return the Observable back from the `this.userService.getUser(userId)` chunk, which results in executing the first HTTP call to get the main user. Then, we use a pipe and `flatMap` on the first call's Observable to get the main user, assign it to the `this.user` chunk, and then return the Observable from the second call—that is, the `this.userService.getSimilarUsers(userId)` code. Finally, we use the `.subscribe` method to receive the value from `getSimilarUsers(userId)` and once the value is received, we assign it to `this.similarUsers`.

See also

- `flatMap/mergeMap` documentation (`https://www.learnrxjs.io/learn-rxjs/operators/transformation/mergemap`)

Using the switchMap operator to switch the last subscription with a new one

For a lot of apps, we have features such as searching content as the user types. This is a really good **user experience** (**UX**) as the user doesn't have to press a button to do a search. However, if we send a call to the server on every keyboard press, that's going to result in a lot of HTTP calls being sent, and we can't know which HTTP call will complete first; thus, we can't be sure if we will have the correct data shown on the view or not. In this recipe, you'll learn to use the `switchMap` operator to cancel out the last subscription and create a new one instead. This would result in canceling previous calls and keeping only one call—the last one.

Getting ready

The project that we are going to work with resides in `chapter05/start_here/using-switchmap-operator`, inside the cloned repository.

1. Open the project in VS Code.

2. Open the Terminal and run `npm install` to install the dependencies of the project.

3. Once done, run ng serve -o.

    This should open the app in a new browser tab, and you should see something like this:

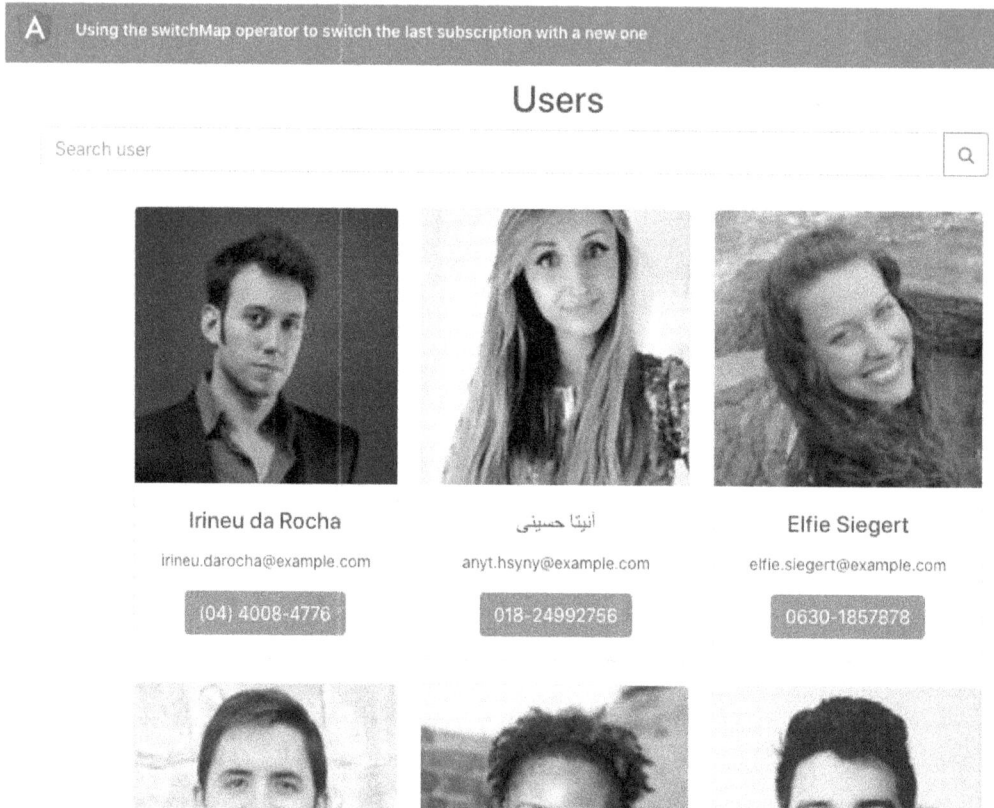

Figure 5.16 – The using-switchmap-operator app running on http://localhost:4200

Now that we have the app running locally, open Chrome DevTools and go to the **Network** tab. Type 'huds' in the search input, and you'll see four calls being sent to the API server, as follows:

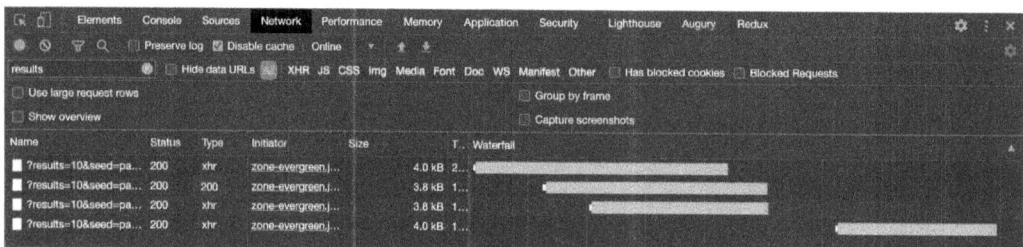

Figure 5.17 – A separate call sent for each input change

How to do it...

You can start typing into the search box on the home page to see the filtered users, and if you see the **Network** tab, you'll notice that whenever the input changes, we send a new HTTP call. Let's avoid sending a call on each keypress by using the `switchMap` operator.

1. First, import the `switchMap` operator from `rxjs/operators` in the home/home.component.ts file, as follows:

    ```
    . . .
    import { switchMap, takeWhile } from 'rxjs/operators';
    ```

2. We will now modify our subscription to the `username` form control— specifically, the `valueChanges` Observable to use the `switchMap` operator for the `this.userService.searchUsers(query)` method call. This returns an `Observable` containing the result of the HTTP call. The code should look like this:

    ```
    . . .
    ngOnInit() {
      this.componentAlive = true;
      this.searchForm = new FormGroup({
        username: new FormControl('', [])
      })
      this.searchUsers();
      this.searchForm.get('username').valueChanges
        .pipe(
          takeWhile(() => !!this.componentAlive),
          switchMap((query) => this.userService.
          searchUsers(query))
        )
        .subscribe((users) => {
          this.users = users;
        })
    }
    ```

If you refresh the app now, open Chrome DevTools, and check the network type while typing `'huds'`, you'll see that all the previous calls are canceled and we only have the latest HTTP call succeeding:

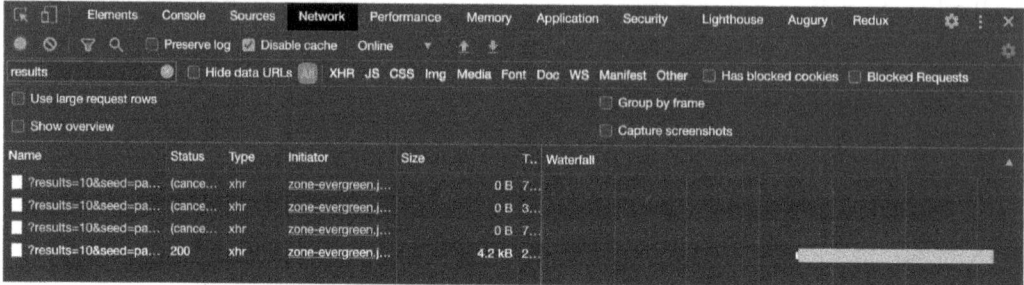

Figure 5.18 – switchMap canceling prior HTTP calls

Woot! We now have only one call that'll succeed, process the data, and end up in the view. See the next section on how it works.

How it works...

The `switchMap` operator cancels the previous (inner) subscription and subscribes to a new Observable instead. That's why it cancels all the HTTP calls sent before in our example and just subscribes to the last one. This was the intended behavior for our app.

See also

- `switchMap` operator documentation (`https://www.learnrxjs.io/learn-rxjs/operators/transformation/switchmap`)

Debouncing HTTP requests using RxJS

In the previous recipe, we learned how to use the `switchMap` operator to cancel previous HTTP calls if a new HTTP call comes. This is fine, but why even send multiple calls when we can use a technique to wait a while before we send an HTTP call? Ideally, we'll just keep listening to duplicate requests for a period of time and will then proceed with the latest request. In this recipe, we'll be using the `debounceTime` operator to make sure we're only sending the HTTP call when the user stops typing for a while.

Getting ready

The project that we are going to work with resides in `chapter05/start_here/`
`using-debouncetime-operator`, inside the cloned repository.

1. Open the project in VS Code.
2. Open the Terminal and run `npm install` to install the dependencies of
    the project.
3. Once done, run `ng serve -o`.

    This should open the app in a new browser tab, and you should see something
    like this:

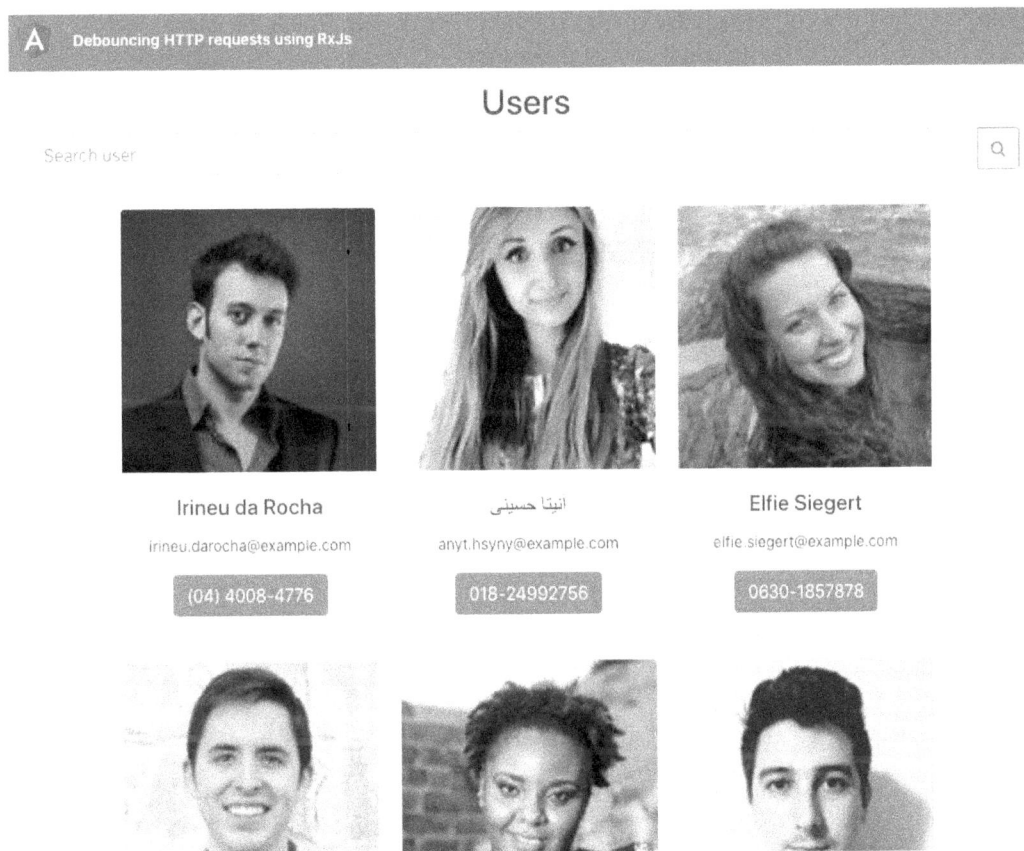

Figure 5.19 – The using-debouncetime-operator app running on http://localhost.4200

Now that we have the app running, open Chrome DevTools, go to the **Network** tab, and then type `'Irin'` in the user search bar. You should see something like this:

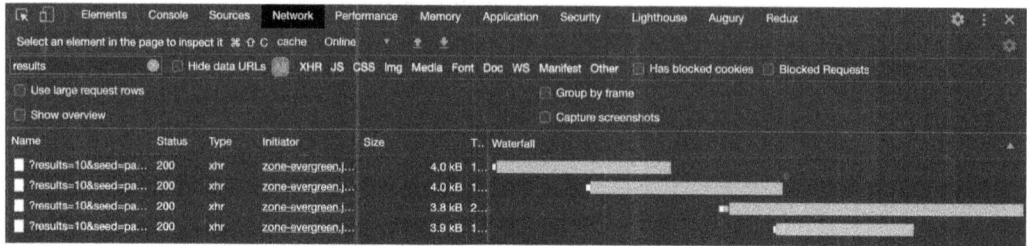

Figure 5.20 – A new call being sent to the server on each keyboard input

Notice how the third call's response comes after the fourth call? That's the issue we're trying to solve by using some sort of debounce.

Let's jump into the recipe steps in the next section.

How to do it...

As we see when we type into the search box on the home page (that is, whenever the input changes), we send a new HTTP call.

In order to make sure we only send one call when the search input is idle after typing, we'll put a `debounceTime` operator on the `this.searchForm.get('username').valueChanges` Observable. Update the `home/home.component.ts` file, as follows:

```
...
import { debounceTime, takeWhile } from 'rxjs/operators';
...
export class HomeComponent implements OnInit, OnDestroy {
  ...
  ngOnInit() {
    ...
    this.searchForm.get('username').valueChanges
      .pipe(
        takeWhile(() => !!this.componentAlive),
        debounceTime(300),
      )
      .subscribe(() => {
        this.searchUsers();
      })
```

```
    }

    searchUsers() {...}

    ngOnDestroy() {}
}
```

And that's it! If you type `'irin'` in the search input while inspecting the **Network** tab, you should see only one call being sent to the server, as follows:

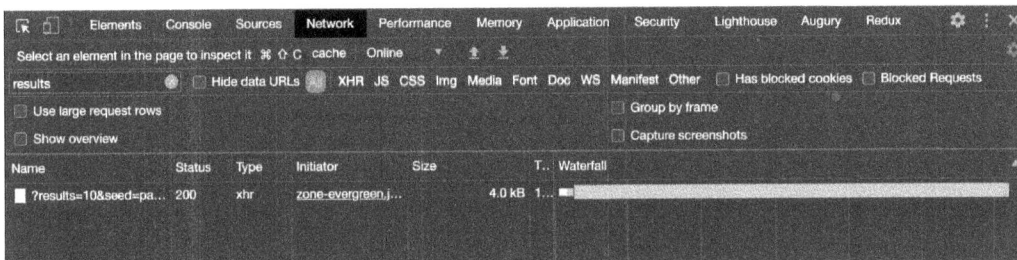

Figure 5.21 – debounceTime causing only one call to be sent to the server

See the next section to understand how it all works.

How it works...

The `debounceTime` operator waits for a particular time before emitting a value from the source Observable, and that too only when there's no more source emission at hand. This allows us to use the operator on the input's `valueChanges` Observable. When you type something in the input, the `debounceTime` operator waits for 300ms to see if you're still typing. And if you've not typed for those 300ms, it moves forward with the emission, causing the HTTP call at the end.

See also

- `debounceTime` operator documentation (https://rxjs-dev.
  firebaseapp.com/api/operators/debounceTime)
- `debounce` operator documentation (https://rxjs-dev.firebaseapp.
  com/api/operators/debounce)
- `delay` operator documentation (https://rxjs-dev.firebaseapp.com/
  api/operators/delay)

6
Reactive State Management with NgRx

Angular and Reactive programming are best buddies, and handling an app's state reactively is one of the best things you can do with your app. NgRx is a framework that provides a set of libraries as reactive extensions for Angular. In this chapter, you'll learn how to use the NgRx ecosystem to manage your app's state reactively, and you'll also learn a couple of cool things the NgRx ecosystem will help you with.

Here are the recipes we're going to cover in this chapter:

- Creating your first NgRx store with actions and reducer
- Using `@ngrx/store-devtools` to debug the state changes
- Creating an effect to fetch third-party **application programming interface (API)** data
- Using selectors to fetch data from stores in multiple components
- Using `@ngrx/component-store` for local state management within a component
- Using `@ngrx/router-store` to work with route changes reactively

Technical requirements

For the recipes in this chapter, make sure you have **Git** and **Node.js** installed on your machine. You also need to have the @angular/cli package installed, which you can do with npm install -g @angular/cli from your terminal. The code for this chapter can be found at https://github.com/PacktPublishing/Angular-Cookbook/tree/master/chapter06.

Creating your first NgRx store with actions and reducer

In this recipe, you'll work your way through understanding NgRx's basics by setting up your first NgRx store. You'll also create some actions along with a reducer, and to see the changes in the reducer, we'll be putting in appropriate console logs.

Getting ready

The project that we are going to work with resides in chapter06/start_here/ngrx-actions-reducer, inside the cloned repositor:

1. Open the project in **Visual Studio Code** (**VS Code**).

2. Open the terminal and run npm install to install the dependencies of the project.

3. Once done, run ng serve -o.

This should open the app in a new browser tab. Tap the **Login as Admin** button and you should see the following screen:

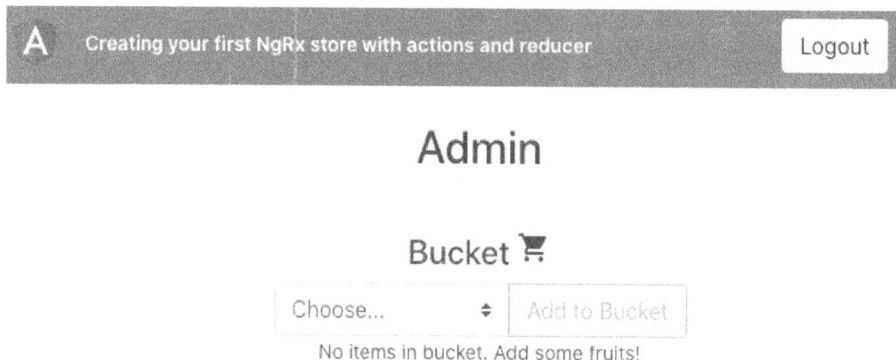

Figure 6.1 – ngrx-actions-reducers app running on http://localhost:4200

Now that we have the app running, we'll move on to the steps for the recipe.

How to do it...

We have an existing Angular app that we used in the prior recipes as well. If you log in as an Admin user, you can add and remove items from the bucket. However, if you log in as an Employee, you can only add items and not remove them. We'll now start integrating NgRx into the app and create a reducer and some actions:

1. Begin by installing the `@ngrx/store` package via **Node Package Manager** (**npm**) in your project. Open Terminal (Mac/Linux) or Command Prompt (Windows), navigate to the project root, and run the following command:

    ```
    npm install @ngrx/store@12.0.0 --save
    ```

    Make sure to rerun the `ng-serve` command if you already have it running.

2. Update the `app.module.ts` file to include `StoreModule`, as follows:

    ```
    . . .
    import { StoreModule } from '@ngrx/store';
    @NgModule({
      declarations: [
        AppComponent
      ],
      imports: [
        BrowserModule,
        AppRoutingModule,
        FormsModule,
        BrowserAnimationsModule,
        StoreModule.forRoot({})
      ],
      providers: [],
      bootstrap: [AppComponent]
    })
    export class AppModule { }
    ```

    Notice that we've passed an empty object { } to the `forRoot` method; we'll change that going forward.

3. Now, we'll create some actions. Create a folder named `store` inside the app folder. Then, create a file named `app.actions.ts` inside the `store` folder, and finally, add the following code to the newly created file:

```
import { createAction, props } from '@ngrx/store';
import { IFruit } from '../interfaces/fruit.interface';

export const addItemToBucket = createAction(
  '[Bucket] Add Item',
  props<IFruit>()
);

export const removeItemFromBucket = createAction(
  '[Bucket] Remove Item',
  props<IFruit>()
);
```

Since we have the actions in place now, we have to create a reducer.

4. Create a new file inside the `store` folder, name it `app.reducer.ts`, and add the following code to it to define the necessary imports:

```
import { Action, createReducer, on } from '@ngrx/store';
import { IFruit } from '../interfaces/fruit.interface';
import * as AppActions from './app.actions';
```

5. Now, define an `AppState` interface to reflect the app's state, and an `initialState` variable to reflect what the app's state will look like when the app starts. Add the following code to the `app.reducer.ts` file:

```
import { Action, createReducer, on } from '@ngrx/store';
import { IFruit } from '../interfaces/fruit.interface';
import * as AppActions from './app.actions';

export interface AppState {
  bucket: IFruit[];
}
```

```
const initialState: AppState = {
  bucket: []
}
```

6. It's time to actually create a reducer now. Add the following code to the app.
    reducer.ts file to create a reducer:

```
...
const initialState: AppState = {
  bucket: []
}
const appReducer = createReducer(
  initialState,
  on(AppActions.addItemToBucket, (state, fruit) =>
  ({ ...state, bucket: [fruit, ...state.bucket] })),
  on(AppActions.removeItemFromBucket, (state, fruit) => {
    return {
      ...state,
      bucket: state.bucket.filter(bucketItem => {
        return bucketItem.id !== fruit.id;
      }) }
  }),
);


export function reducer(state: AppState = initialState,
action: Action) {
  return appReducer(state, action);
}
```

7. We'll also add some sweet little console.logs calls into the reducer method
    to see all the actions firing up on our console. Add a log as follows to the
    app.reducer.ts file:

```
export function reducer(state: AppState = initialState,
action: Action) {
  console.log('state', state);
  console.log('action', action);
  return appReducer(state, action);
}
```

8. Finally, let's register this reducer in the `app.module.ts` file using the `StoreModule.forRoot()` method as follows so that we can see things working:

```
. . .
import { StoreModule } from '@ngrx/store';
import * as appStore from './store/app.reducer';
@NgModule({
  declarations: [
    AppComponent
  ],
  imports: [
    . . .
    StoreModule.forRoot({app: appStore.reducer})
  ],
  providers: [],
  bootstrap: [AppComponent]
})
export class AppModule { }
```

If you refresh the app now, you should see the following logs on the console as soon as the app starts:

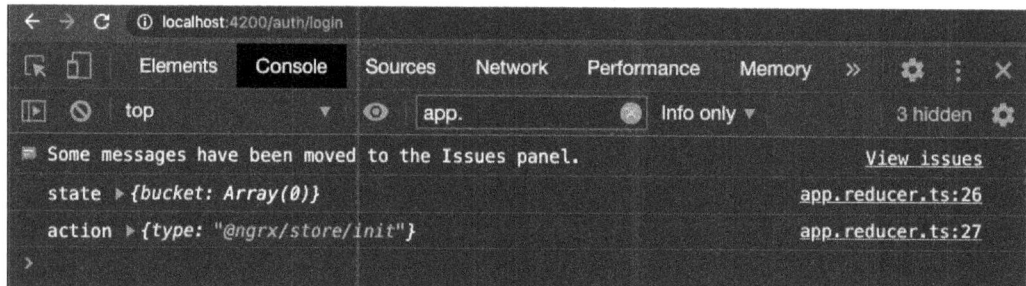

Figure 6.2 – Logs showing initial state and @ngrx/store/init action on app start

9. Now that we can see that the reducer works, let's dispatch our actions on adding and removing items from the basket. For that, dispatch the actions as follows in the `shared/components/bucket/bucket.component.ts` file:

```
...
import { Store } from '@ngrx/store';
import { AppState } from 'src/app/store/app.reducer';
import { addItemToBucket, removeItemFromBucket } from
'src/app/store/app.actions';

export class BucketComponent implements OnInit {
  ...
  constructor(
    private bucketService: BucketService,
    private store: Store<AppState>
  ) { }

  ngOnInit(): void {...}

  addSelectedFruitToBucket() {
const newItem: IFruit = {
    id: Date.now(),
    name: this.selectedFruit
  }
    this.bucketService.addItem(newItem);
    this.store.dispatch(addItemToBucket(newItem));
  }
  deleteFromBucket(fruit: IFruit) {
    this.bucketService.removeItem(fruit);
    this.store.dispatch(removeItemFromBucket(fruit));
  }
}
```

10. Log in to the app as Admin, add a few items to the bucket, and then remove some items. You'll see something like this on the console:

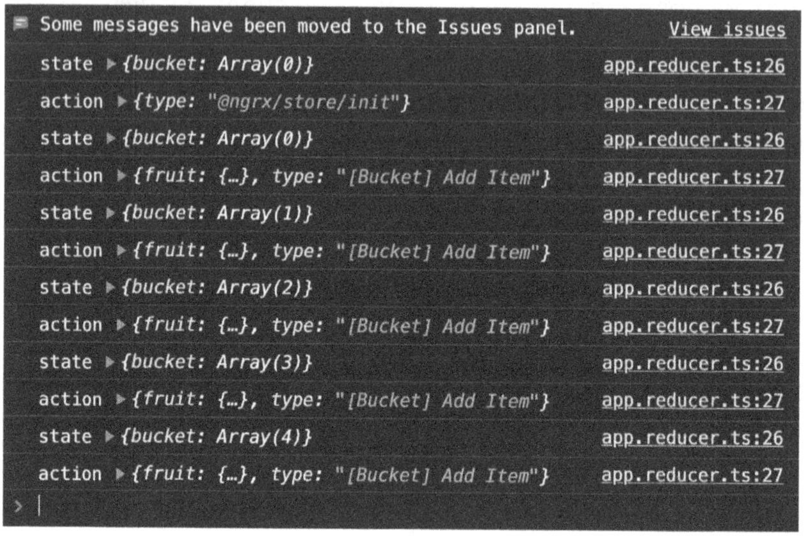

Figure 6.3 – Logs showing the actions for adding and removing items from a bucket

And that covers it all for this recipe! You now know how to integrate an NgRx store into an Angular app and how to create NgRx actions and dispatch them. You also know how to create a reducer, define its state, and listen to the actions to act on the ones dispatched.

See also

- NgRx reducers documentation (https://ngrx.io/guide/store/reducers)
- NgRx actions documentation (https://ngrx.io/guide/store/actions)
- RxJS merge operator documentation (https://www.learnrxjs.io/learn-rxjs/operators/combination/merge)

Using @ngrx/store-devtools to debug the state changes

In this recipe, you'll learn how to set up and use @ngrx/store-devtools to debug your app's state, the actions dispatch, and the difference in the state when the actions dispatch. We'll be using an existing app we're familiar with to learn about the process.

Getting ready

The project for this recipe resides in `chapter06/start_here/using-ngrx-store-devtool`:

1. Open the project in VS Code.

2. Open the terminal and run `npm install` to install the dependencies of the project.

3. Once done, run `ng serve -o`.

    This should open the app in a new browser tab.

4. Login as an Admin user, and you should see a screen like this:

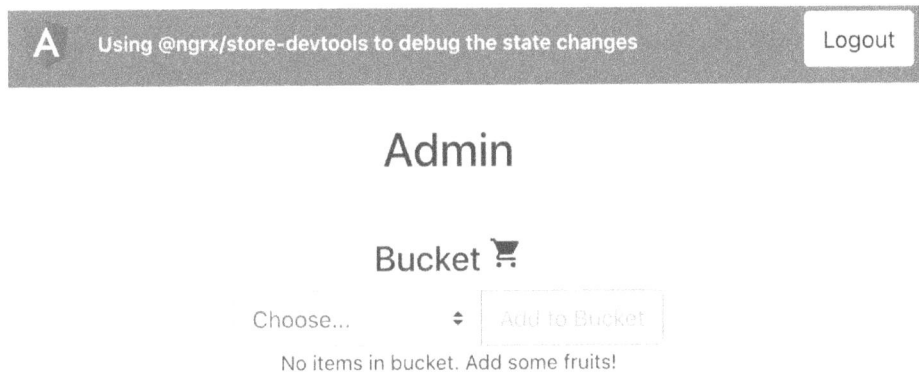

Figure 6.4 – Using ngrx-store-devtools app running on http://localhost:4200

Now that we have the app set up, let's see the steps of the recipe in the next section.

How to do it...

We have an Angular app that already has the `@ngrx/store` package integrated. We also have a reducer set up and some actions in place that are logged on the console as soon as you add or remove an item. Let's move toward configuring the store dev tools for our ap:

1. Begin with installing the `@ngrx/store-devtools` package in the project, as follows:

    ```
    npm install @ngrx/store-devtools@12.0.0 --save
    ```

2. Now, update your `app.module.ts` file to include a `StoreDevtoolsModule.instrument` entry, as follows:

```
...
import * as appStore from './store/app.reducer';
import { StoreDevtoolsModule } from '@ngrx/store-
devtools';

@NgModule({
  declarations: [
    AppComponent
  ],
  imports: [
    ...
    StoreModule.forRoot({app: appStore.reducer}),
    StoreDevtoolsModule.instrument({
      maxAge: 25, // Retains last 25 states
    }),
  ],
  providers: [],
  bootstrap: [AppComponent]
})
export class AppModule { }
```

3. And now, download the Redux DevTools extension from `https://github.com/zalmoxisus/redux-devtools-extension/` for your particular browser and install it. I'll be consistently using the Chrome browser in this book.

4. Open Chrome DevTools. There should be a new tab named **Redux**. Tap it and refresh the page. You'll see something like this:

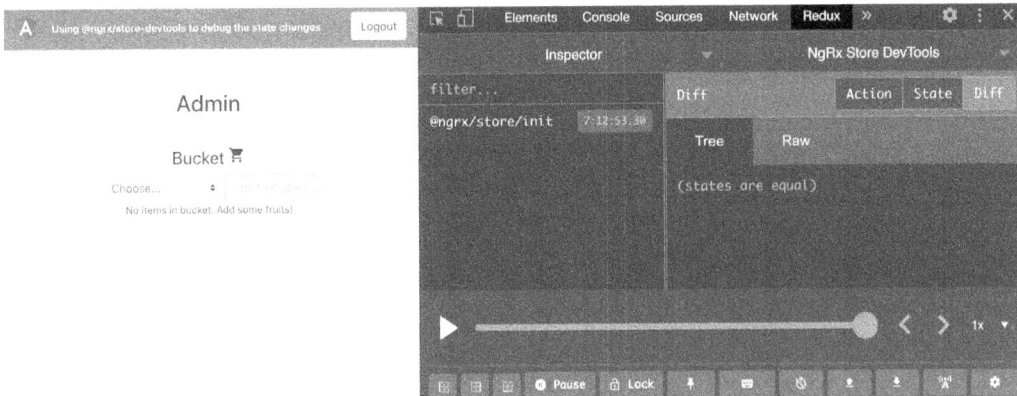

Figure 6.5 – Redux DevTools showing the initial Redux action dispatched

5. To see the state of the app right now, tap the **State** button, as shown in the following screenshot, and you should see that we have `app > bucket: []` as the current state:

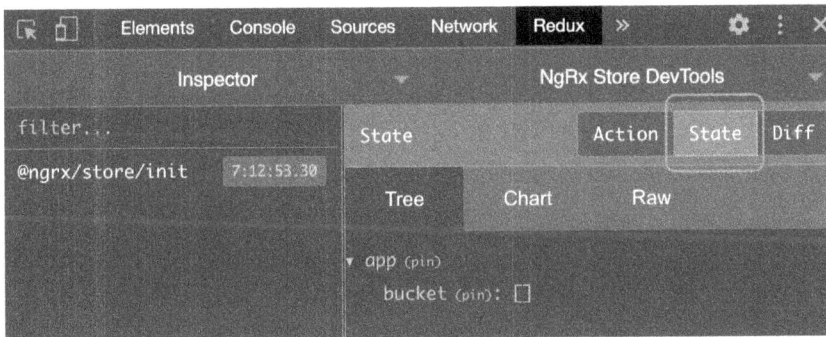

Figure 6.6 – Viewing current state in the Redux DevTools extension

6. Now, add a cherry 🍒 and a banana 🍌 to the bucket, and then remove the banana 🍌 from the bucket. You should see all the relevant actions being dispatched, as follows:

Figure 6.7 – Redux DevTools showing addItemToBucket and removeItemFromBucket actions

If you expand the bucket array from the state, you'll see that it reflects the current state of the bucket, as we can see in the following screenshot:

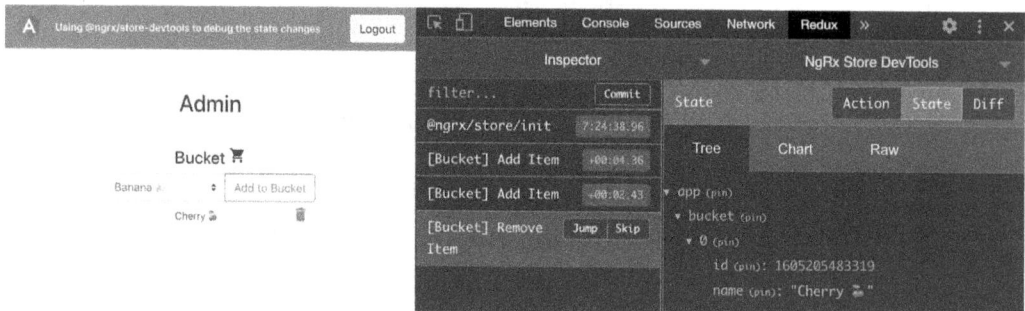

Figure 6.8 – Redux DevTools showing bucket's current state

Great! You've just learned how to use the Redux DevTools extension to see your NgRx state and the actions being dispatched.

How it works...

It is important to understand that NgRx is a combination of Angular and Redux (using RxJS). By using the Store Devtools package and the Redux DevTools extension, we're able to debug the app really easily, which helps us find potential bugs, predict state changes, and be more transparent about what's happening behind the scenes in the @ngrx/store package.

There's more...

You can also see the difference that an action caused within an app's state. That is, we have an addition of an item in the bucket when we dispatch the addItemToBucket action with the fruit, and we have an item removed from the bucket when we dispatch the removeItemFromBucket action. See the following screenshot and *Figure 6.10* for each cases:

Figure 6.9 – addItemToBucket action causing the addition of an item to the bucket

Notice the green background around the data {id:1605205728586,name:'Banana 🍌'} in *Figure 6.9*. This represents an addition to the state. You can see the removeItemFromBucket action depicted here:

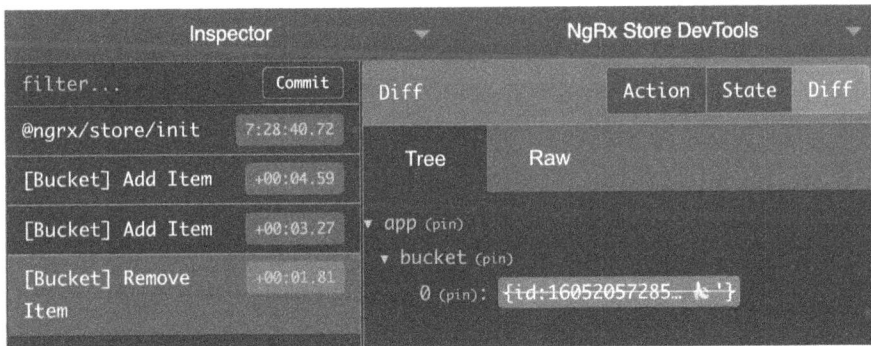

Figure 6.10 – removeItemFromBucket action causing the removal of an item from the bucket

Similarly, notice the red background and a strikethrough around the data {id:16052057285... 🍌'} in *Figure 6.10*. This represents removal from the state.

See also

- NgRx Store Devtools documentation (`https://ngrx.io/guide/store-devtools`)

Creating an effect to fetch third-party API data

In this recipe, you'll learn how to use NgRx effects using the `@ngrx/effects` package. You'll create and register an effect, and that effect will be listening for an event. Then, we'll react to that action to fetch third-party API data, and in response, we'll either dispatch a success or a failure action. This is gonna be fun.

Getting ready

The project for this recipe resides in `chapter06/start_here/using-ngrx-effect`:

1. Open the project in VS Code.

2. Open the terminal and run `npm install` to install the dependencies of the project.

3. Once done, run `ng serve -o`.

   This should open the app in a new browser tab, and you should see the app, as follows:

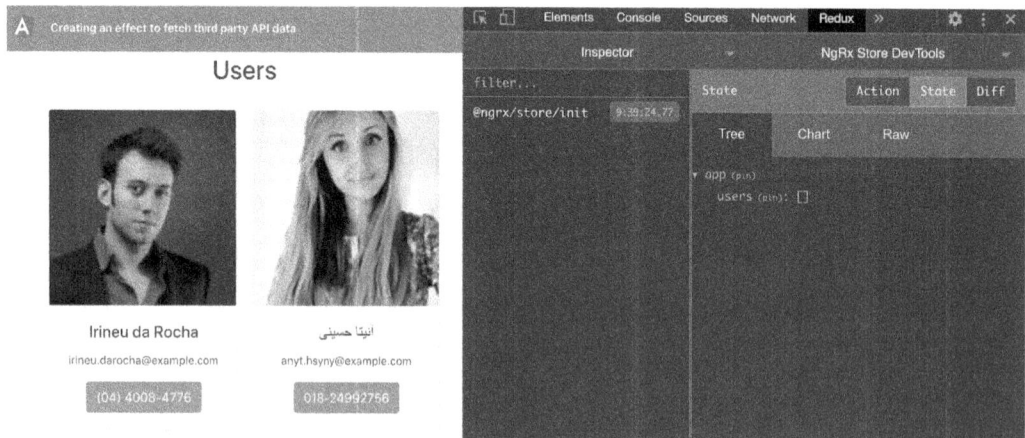

Figure 6.11 – Using ngrx-effects app running on http://localhost:4200

Now that we have the app running locally, let's see the steps of the recipe in the next section.

How to do it...

We have an app with a single route named **Home** page. In the HomeComponent class, we're using the UserService to send a **HyperText Transfer Protocol** (**HTTP**) call to get users and are then showing that on the browser. We already have the @ngrx/store and @ngrx/store-devtools packages integrated, as you can see in *Figure 6.1*:

1. Install the @ngrx/effects package in the project, as follows:

    ```
    npm install --save @ngrx/effects@12.0.0
    ```

2. We'll now create actions to get the users from the HTTP call. We'll have one action to get the users, one to dispatch on successfully getting the users, and one action to dispatch in case we get an error. Add the following code to the store/app. actions.ts file:

    ```
    import { createAction, props } from '@ngrx/store';
    import { IUser } from '../core/interfaces/user.
    interface';

    export const APP_ACTIONS = {
      GET_USERS: '[Users] Get Users',
      GET_USERS_SUCCESS: '[Users] Get Users Success',
      GET_USERS_FAILURE: '[Users] Get Users Failure',
    }

    export const getUsers = createAction(
      APP_ACTIONS.GET_USERS,
    );

    export const getUsersSuccess = createAction(
      APP_ACTIONS.GET_USERS_SUCCESS,
      props<{users: IUser[]}>()
    );

    export const getUsersFailure = createAction(
      APP_ACTIONS.GET_USERS_FAILURE,
      props<{error: string}>()
    );
    ```

Let's create an effect now so that we can listen to the GET_USERS action, perform the API call, and dispatch the success action in case of successful data fetch.

3. Create a file in the `store` folder named `app.effects.ts` and add the following code to it:

```
import { Injectable } from '@angular/core';
import { Actions, createEffect, ofType } from '@ngrx/
effects';
import { of } from 'rxjs';
import { map, mergeMap, catchError } from 'rxjs/
operators';
import { UserService } from '../core/services/user.
service';
import { APP_ACTIONS, getUsersFailure, getUsersSuccess }
from './app.actions';


@Injectable()
export class AppEffects {

  constructor(
    private actions$: Actions,
    private userService: UserService
  ) {}
}
```

4. We'll create a new effect in the `app.effects.ts` file now to register a listener for the GET_USERS action, as follows:

```
...
@Injectable()
export class AppEffects {
  getUsers$ = createEffect(() =>
    this.actions$.pipe(
      ofType(APP_ACTIONS.GET_USERS),
      mergeMap(() => this.userService.getUsers()
        .pipe(
          map(users => {
```

```
                    return getUsersSuccess({
                        users
                    })
                }),
                catchError((error) => of(getUsersFailure({
                    error
                }))))
            )
        )
    )
    );
    ...
}
```

5. We'll now register our effect as the root effects for the app in the app.module.ts file, as follows:

```
...
import { EffectsModule } from '@ngrx/effects';
import { AppEffects } from './store/app.effects';

@NgModule({
  declarations: [...],
  imports: [

    ...

    StoreDevtoolsModule.instrument({
      maxAge: 25, // Retains last 25 states
    }),
    EffectsModule.forRoot([AppEffects])
  ],
  providers: [],
  bootstrap: [AppComponent]
})
export class AppModule { }
```

As soon as we've registered the effects, you should see an additional action named `@ngrx/effects/init` firing in the Redux DevTools extension, as follows:

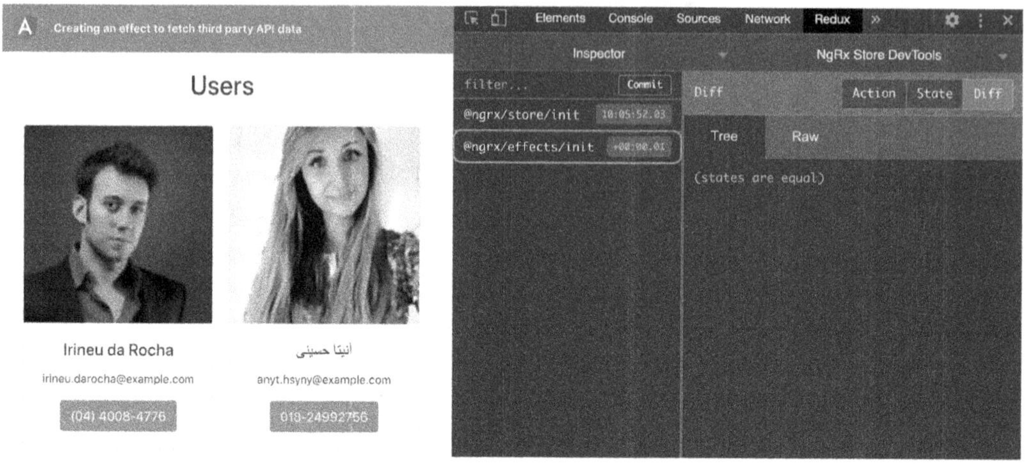

Figure 6.12 – @ngrx/effects/init action fired on app launch

6. Now that we have the effects listening to the actions, let's dispatch the GET_USERS action from the HomeComponent class, and we should see the GET_USERS_ SUCCESS action fired in return on the successful call fetch. Add the following code to dispatch the action from home/home.component.ts:

```
...
import { AppState } from '../store/app.reducer';
import { Store } from '@ngrx/store';
import { getUsers } from '../store/app.actions';

@Component({...})
export class HomeComponent implements OnInit, OnDestroy {
  users$: Observable<IUser[]>;
  constructor(
    private userService: UserService,
    private store: Store<AppState>
  ) {}

  ngOnInit() {
    this.store.dispatch(getUsers())
```

```
        this.users$ = this.userService.getUsers();
    }
    ngOnDestroy() {}
}
```

If you refresh the app now, you should see the `[Users] Get Users` action dispatched, and in return, the `[Users] Get Users Success` action dispatches on the successful HTTP call:

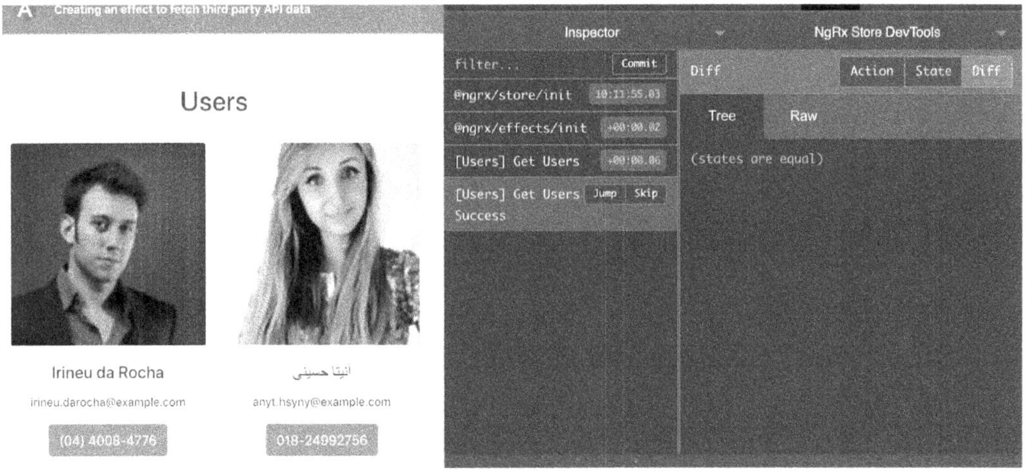

Figure 6.13 – GET_USERS and GET_USERS_SUCCESS actions being dispatched

Notice in *Figure 6.13* that the `Diff` is nothing after the `GET_USERS_SUCCESS` action is dispatched. This is because we haven't updated the state using the reducer so far.

7. Let's update the state in the `app.reducer.ts` file to listen to the `GET_USERS_SUCCESS` action and assign the users to the state accordingly. The code should look like this:

```
import { Action, createReducer, on } from '@ngrx/store';
import { IUser } from '../core/interfaces/user.
interface';
import { getUsersSuccess } from './app.actions';

export interface AppState {
  users: IUser[];
}
```

```
const initialState: AppState = {
  users: []
}


const appReducer = createReducer(
  initialState,
  on(getUsersSuccess, (state, action) => ({
    ...state,
    users: action.users
  }))
);


export function reducer(state: AppState = initialState,
action: Action) {
  return appReducer(state, action);
}
```

If you refresh the app now, you should see the users being assigned to the state, as follows:

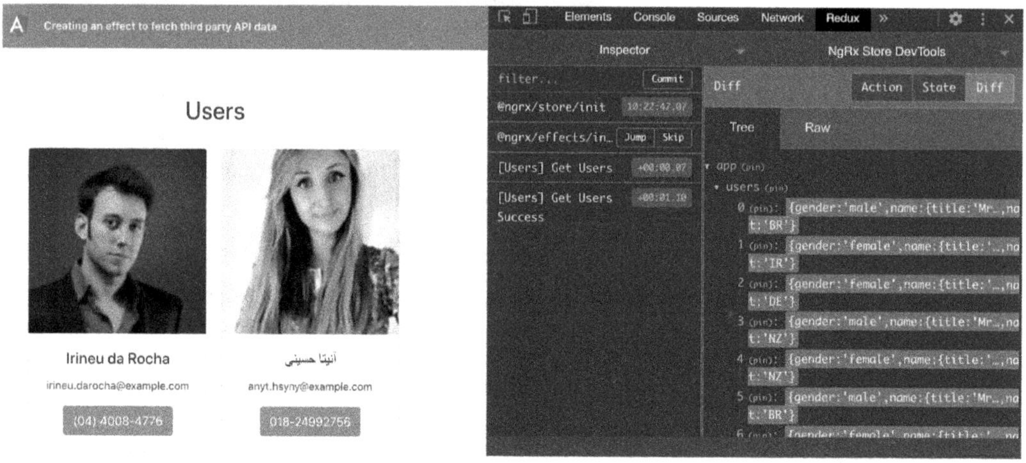

Figure 6.14 – GET_USERS_SUCCESS action adding users to the state

If you look at the app's state right now, you should see something like this:

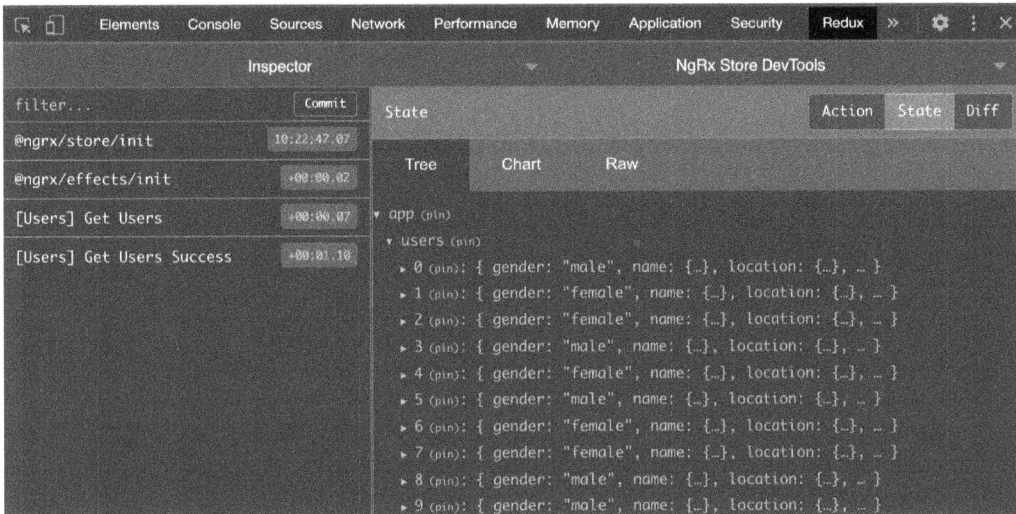

Figure 6.15 – App state containing users after the GET_USERS_SUCCESS action

Right now, we're sending two calls to the server—one through the effect, and one through the `ngOnInit` method of the `HomeComponent` class using the `UserService` instance directly. Let's remove the `UserService` from the `HomeComponent` class. We won't see any data right now, but that's what we're going to do in the next recipe.

8. Remove the `UserService` from the `HomeComponent` class and your `home.component.ts` file should now look like this:

```
    ...

    @Component({...})
    export class HomeComponent implements OnInit, OnDestroy {
      users$: Observable<IUser[]>;
      constructor(
      private userService: UserService, ← Remove this
        private store: Store<AppState>
      ) {}


      ngOnInit() {
        this.store.dispatch(getUsers());
```

```
        this.users$ = this.userService.getUsers();   ← Remove
        this
    }


    ngOnDestroy() {}
}
```

Great! You now know how to use NgRx effects in your Angular apps. See the next section to understand how NgRx effects work.

> **Important note**
>
> We now have an output, as shown in *Figure 6.15*—that is, we keep showing the loader even after the users' data has been set in the store. The recipe's main purpose is to use `@ngrx/effects`, and that has been done. We'll show the appropriate data in the next recipe, *Using selectors to fetch data from stores in multiple components*.

How it works...

In order for the NgRx effects to work, we needed to install the `@ngrx/effects` package, create an effect, and register it as an array of effects (root effects) in the `AppModule` class. When you create an effect, it has to listen to an action. When an action is dispatched to the store from any component or even from another effect, the registered effect triggers, does the job you want it to do, and is supposed to dispatch another action in return. For API calls, we usually have three actions—that is, the main action, and the following success and failure actions. Ideally, on the success action (and perhaps on the failure action too), you would want to update some of your state variables.

See also

- NgRx effects documentation (`https://ngrx.io/guide/effects`)

Using selectors to fetch data from stores in multiple components

In the previous recipe, we created an NgRx effect to fetch third-party API data as users, and we saved it in the Redux store. That's what we have as a starting point in this recipe. We have an effect that fetches the users from `api.randomuser.me` and stores it in the state, and we don't currently show anything on the **user interface** (**UI**). In this recipe, you'll create some NgRx selectors to show users on the **Home** page as well as on the **User Detail** page with similar users.

Getting ready

The project for this recipe resides in `chapter06/start_here/using-ngrx-selector`:

1. Open the project in VS Code.

2. Open the terminal and run `npm install` to install the dependencies of the project.

3. Once done, run `ng serve -o`.

    This should open the app in a new browser tab. As soon as the page is opened, you should see the app, as follows:

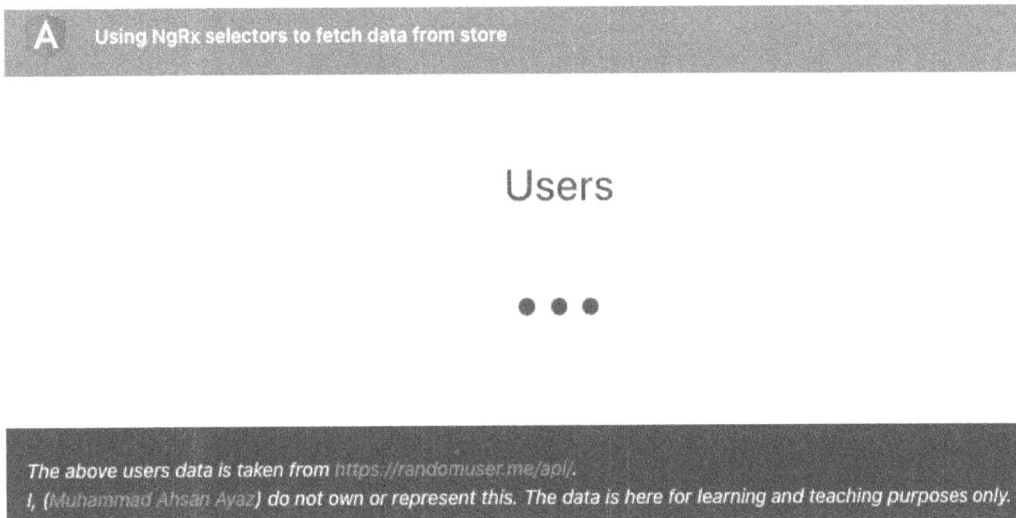

Figure 6.16 – Using ngrx-selectors app running on http://localhost:4200

Now that we have the app running locally, let's see the steps of the recipe in the next section.

How to do it...

All we have to do in this recipe is to work with NgRx selectors, the reducer we have, and the Redux state in general. Easy peasy. Let's get started!

We'll first show the users on the **Home** page and, in order to do that, we have to create our first NgRx selecto:

1. Create a new file inside the `store` folder. Name it `app.selectors.ts` and add the following code to it:

```
import { createSelector, createFeatureSelector } from '@
ngrx/store';
import { AppState } from './app.reducer';

export const selectApp =
createFeatureSelector<AppState>('app');

export const selectUsers = createSelector(
  selectApp,
  (state: AppState) => state.users
);
```

Now that we have the selector in place, let's use it in the `HomeComponent` class.

2. Modify the `ngOnInit` method in the `home.component.ts` file. It should look like this:

```
...
import { getUsers } from '../store/app.actions';
import { selectUsers } from '../store/app.selectors';

@Component({...})
export class HomeComponent implements OnInit, OnDestroy {
  ...

  ngOnInit() {
    this.users$ = this.store.select(selectUsers);
    this.store.dispatch(getUsers())
  }


  ngOnDestroy() {}
}
```

As soon as you refresh the app now, you should be able to see the users. And if you click on any one of the users, you'll navigate to the user details, but won't see any valuable date. The page should look like this:

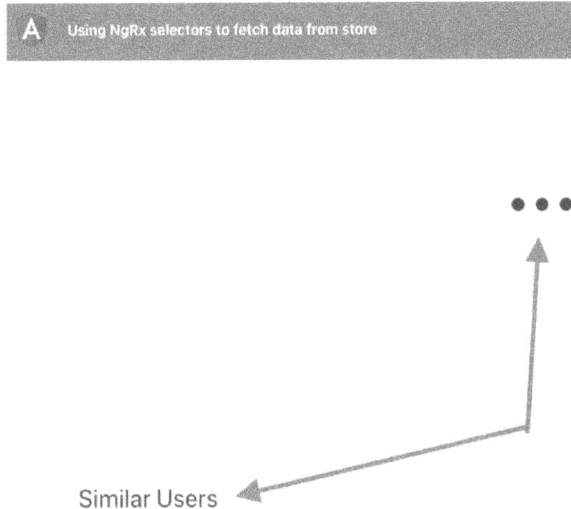

Figure 6.17 – Unable to display the current user and similar users

3. In order to see the current user and similar users, we'll first create two Observables in the UserDetailComponent class so that we can subscribe to their respective store selectors later on. Add the Observables to the user-detail.component.ts file, as follows:

```
...
import { ActivatedRoute } from '@angular/router';
import { Observable } from 'rxjs/internal/Observable';

@Component({...})
export class UserDetailComponent implements OnInit,
OnDestroy {
    user: IUser = null; ← Remove this
    similarUsers: IUser[] = []; ← Remove this
    user$: Observable<IUser> = null; ← Add this
    similarUsers$: Observable<IUser[]> = null; ← Add this
    isComponentAlive: boolean;
    constructor( ) {}
```

```
ngOnInit() {
    this.isComponentAlive = true;
}


ngOnDestroy() {
    this.isComponentAlive = false;
    }
}
```

4. Update the `user-detail.component.html` template to use the new Observable properties, as follows:

```html
<div class="user-detail">
  <div class="main-content user-card">
      <app-user-card *ngIf="user$ | async as user;
      else loader" [user]="user"></app-user-card>
  </div>


  <div class="secondary-container">
    <h4>Similar Users</h4>
    <div class="similar-users">
      <ng-container *ngIf="similarUsers$ | async
      as similarUsers; else loader">
         <app-user-card class="user-card" *ngFor="let user
         of similarUsers" [user]="user"></app-user-card>
      </ng-container>
    </div>
  </div>
</div>
...
```

5. Update the `app.selectors.ts` file to add both the selectors, as follows:

```typescript
...
import { IUser } from '../core/interfaces/user.
interface';
export const selectUsers = createSelector(...);
```

```
export const selectCurrentUser = (uuid) =>
createSelector(
  selectUsers,
  (users: IUser[]) => users ? users.find(user => {
    return user.login.uuid === uuid;
  }) : null
);


export const selectSimilarUsers = (uuid) =>
createSelector(
  selectUsers,
  (users: IUser[]) => users ? users.filter(user => {
    return user.login.uuid !== uuid;
  }): null
);
```

Since we navigated to the **User Detail** page with the user's **universally unique identifier (UUID)**, we will listen to the active route's `paramsMap` and assign the appropriate selectors.

6. First, add the correct imports to the `user-detail.component.ts` file, as follows:

```
. . .
import { takeWhile } from 'rxjs/operators';
import { Store } from '@ngrx/store';
import { AppState } from '../store/app.reducer';
import { selectCurrentUser, selectSimilarUsers } from
'../store/app.selectors';
import { ActivatedRoute } from '@angular/router';
```

7. Now, in the same `user-detail.component.ts` file, use the `Store` service and update the `ngOnInit` method, as follows:

```
@Component({...})
export class UserDetailComponent implements OnInit,
OnDestroy {
  . . .
  constructor(
```

```
      private route: ActivatedRoute,
      private store: Store<AppState>
  ) {}


  ngOnInit() {
    this.isComponentAlive = true;
    this.route.paramMap.pipe(
      takeWhile(() => !!this.isComponentAlive)
    )
    .subscribe(params => {
      const uuid = params.get('uuid');
      this.user$ = this.store.
      select(selectCurrentUser(uuid))
      this.similarUsers$ = this.store.
      select(selectSimilarUsers(uuid))
    });
  }
  ...
}
```

We'll add another method to the `UserDetailComponent` class that'll fetch the users if they haven't been fetched already in the app.

8. Add the `getUsersIfNecessary` method to the `user-detail.component.ts` file, as follows:

```
...
import { first, takeWhile } from 'rxjs/operators';
import { Store } from '@ngrx/store';
import { AppState } from '../store/app.reducer';
import { selectCurrentUser, selectSimilarUsers,
selectUsers } from '../store/app.selectors';
import { getUsers } from '../store/app.actions';


@Component({...})
export class UserDetailComponent implements OnInit,
OnDestroy {
  ...
```

```
ngOnInit() {
    ...
    this.getUsersIfNecessary();
}

getUsersIfNecessary() {
    this.store.select(selectUsers)
    .pipe(
        first ()
    )
    .subscribe((users) => {
        if (users === null) {
            this.store.dispatch(getUsers())
        }
    })
}
}
```

Refresh the app… and boom! You now see the current user and similar users as well. See the next section to understand how it all works.

How it works...

In this recipe, we already had a reducer and an effect that fetches the third-party API data as users. We started by creating a selector for the users for the home screen. That was easy—we just needed to create a simple selector. Note that the reducer's state is in the following form:

```
app: {
    users: []
}
```

That's why we first used `createFeatureSelector` to fetch the app state, and then we used `createSelector` to get the users state.

The hard part was getting the current users and similar users. For that, we created selectors that could take the uuid as input. Then, we listened to the paramMap in the UserDetailComponent class for the uuid, and as soon as it changed, we fetched it. We then used it with the selectors by passing the uuid into them so that the selectors could filter the current user and similar users.

Finally, we had the issue that if someone lands directly on the **User Detail** page with the uuid, they won't see anything because we wouldn't have fetched the users. This is due to the fact that we only fetch the users on the home page, so anyone landing directly on a user's detail page wouldn't cause the effect to be triggered. That's why we created a method named getUsersIfNecessary so that it can check the state and fetch the users if they're not already fetched.

See also

- NgRx selectors documentation (https://ngrx.io/guide/store/selectors)

Using @ngrx/component-store for local state management within a component

In this recipe, you'll learn how to use the NgRx Component Store and how to use it instead of the push-based Subject/BehaviorSubject pattern with services for maintaining a component's state locally.

Remember that @ngrx/component-store is a stand-alone library and doesn't correlate with Redux or @ngrx/store, and so on.

Getting ready

The project that we are going to work with resides in chapter06/start_here/ngrx-component-store, inside the cloned repositor:

1. Open the project in VS Code.

2. Open the terminal and run npm install to install the dependencies of the project.

3. Once done, run ng serve -o.

This should open the app in a new browser tab. Log in as Admin and you should see it, as follows:

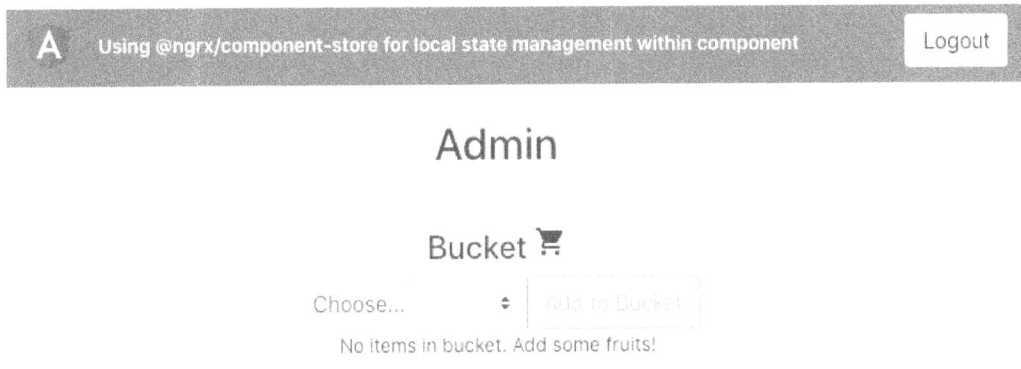

Figure 6.18 – ngrx-component-store app running on http://localhost:4200

Now that we have the app running locally, let's see the steps of the recipe in the next section.

How to do it...

We have our favorite bucket app that we've used in a lot of recipes so far. The state of the buckets right now is stored in the `BucketService`, which uses a `BehaviorSubject` pattern. We're going to replace it with the NgRx Component Store. Let's begin:

1. Add the `@ngrx/component-store` package to the project's dependencies by running the following command in the project root:

    ```
    npm install @ngrx/component-store@12.0.0 --save
    ```

2. We first have to make our `BucketService` compatible with a `ComponentStore`. In order to do that, we'll create an interface for the bucket state, extend the `BucketService` from `ComponentStore`, and initialize the service by calling the `super` method. Update the `file services/bucket.service.ts` file, as follows:

    ```
    ...
    import { IBucketService } from '../interfaces/bucket-service';
    import { ComponentStore } from '@ngrx/component-store';
    ```

```
export interface BucketState {
  bucket: IFruit[]
}


@Injectable({
  providedIn: 'root'
})
export class BucketService extends
ComponentStore<BucketState>  implements IBucketService {
  bucketSource = new BehaviorSubject([]);
  bucket$: Observable<IFruit[]> =
  this.bucketSource.asObservable();
  constructor() {
    super({
      bucket: []
    })
  }
  ...
}
```

None of this will make sense until we actually show the data from the ComponentStore. Let's work on that now.

3. Modify the bucket$ Observable to use the ComponentStore state rather than relying on the BehaviorSubject pattern, as follows:

```
...
export class BucketService extends
ComponentStore<BucketState>  implements IBucketService {
  bucketSource = new BehaviorSubject([]);
  readonly bucket$: Observable<IFruit[]> =
  this.select(state => state.bucket);
  constructor() {
    super({
      bucket: []
    })
  }
}
```

```
    . . .
  }
```

You should potentially see that none of the bucket items show anymore, or that even if you add an item, it won't show. That's because it still requires some work.

4. First, let's make sure that instead of initializing the `bucket` from the Component Store with an empty array, we initialize it with the values from `localStorage`. Just try adding a few items, even if they don't show up yet. Then, modify the `loadItems()` method to use the `setState` method on `BucketService`. The code should look like this:

```
loadItems() {
  const bucket = JSON.parse(window.localStorage.
  getItem('bucket') || '[]');
  this.bucketSource.next(bucket);   ← Remove this
  this.setState({   ← Add this
    bucket
  })
}
```

Notice that we've removed the `this.bucketSource.next(bucket);` line from the code. This is because we're not going to work with the `bucketSource` property anymore, which is a `BehaviorSubject` pattern. We'll do the same for the next set of functions.

Also, you should now see the items that you added previously and that weren't shown.

5. Let's replace the `addItem` method in the `BucketService` now so that it updates the state correctly and shows the new items on view, as intended. For this, we'll use the `updater` method of the `ComponentStore` and modify our `addItem` method to be an updater, as follows:

```
readonly addItem = this.updater((state, fruit: IFruit)
=> {
  const bucket = [fruit, ...state.bucket]
  window.localStorage.setItem('bucket',
  JSON.stringify(bucket));
  return ({
```

```
        bucket
    })
  });
```

If you add an item now, you should see it appearing on the view.

6. We can now replace the `removeItem` method as well to be an `updater` method in the `BucketService` as well. The code should look like this:

```
    readonly removeItem = this.updater((state, fruit:
    IFruit) => {
        const bucket = state.bucket.filter(item =>
        item.id !== fruit.id);
        window.localStorage.setItem('bucket',
        JSON.stringify(bucket));
        return ({
          bucket
        })
    });
```

With this change, you should see the app working. But we do have an issue to fix, and that is that the `EmployeeService` also needs to be updated to make the `removeItem` method an `updater` method.

7. Let's replace the `removeItem` method in the `EmployeeBucketService` to be an `updater` method as well. Modify the `employee/services/employee-bucket.service.ts` file, as follows:

```
import { Injectable } from '@angular/core';
import { IFruit } from 'src/app/interfaces/fruit.
interface';
import { BucketService } from 'src/app/services/bucket.
service';

...

export class EmployeeBucketService extends BucketService
{

  constructor() {
    super();
  }
```

```
    readonly removeItem = this.updater((state, _: IFruit)
    => {
      alert('Employees can not delete items');
      return state;
    });
}
```

And voilà! Everything should actually be fine right now, and you shouldn't see any errors.

8. Since we've got rid of all usages of the `BehaviorSubject` pattern in the `BucketService` property named `bucketSource`, we can remove the property itself from the `BucketService`. The final code should look like this:

```
import { Injectable } from '@angular/core';
import { BehaviorSubject ← Remove this, Observable }
from 'rxjs';

...

export class BucketService extends
ComponentStore<BucketState>  implements IBucketService {
  bucketSource = new BehaviorSubject([]); ← Remove
  readonly bucket$: Observable<IFruit[]> =
  this.select((state) => state.bucket);
  constructor() {
    super({
      bucket: []
    })
  }
  ...
}
```

Congratulations! You finished the recipe. See the next section to understand how it works.

How it works...

As mentioned earlier, `@ngrx/component-store` is a standalone package that can easily be installed in your Angular apps without having to use `@ngrx/store`, `@ngrx/effects`, and so on. It is supposed to replace the usage of `BehaviorSubject` in Angular services, and that's what we did in this recipe. We covered how to initialize a `ComponentStore` and how to set the initial state using the `setState` method when we already had the values without accessing the state, and we learned how to create `updater` methods that can be used to update the state, as they can access the state and allow us to even pass arguments for our own use cases.

See also

- `@ngrx/component-store` documentation (https://ngrx.io/guide/component-store)

- Effects in `@ngrx/component-store` documentation (https://ngrx.io/guide/component-store/effect)

Using @ngrx/router-store to work with route changes reactively

NgRx is awesome because it allows you to have your data stored in a centralized place. However, listening to route changes is still something that is out of the NgRx scope for what we've covered so far. We did rely on the `ActivatedRoute` service to watch for route changes, and when we want to test such components, the `ActivatedRoute` service becomes a dependency. In this recipe, you'll install the `@ngrx/router-store` package and will learn how to listen to the route changes using some actions built into the package.

Getting ready

The project that we are going to work with resides in `chapter06/start_here/ngrx-router-store`, inside the cloned repositor:

1. Open the project in VS Code.

2. Open the terminal and run `npm install` to install the dependencies of the project.

3. Once done, run `ng serve -o`.

    This should open the app in a new browser tab, and you should see something like this:

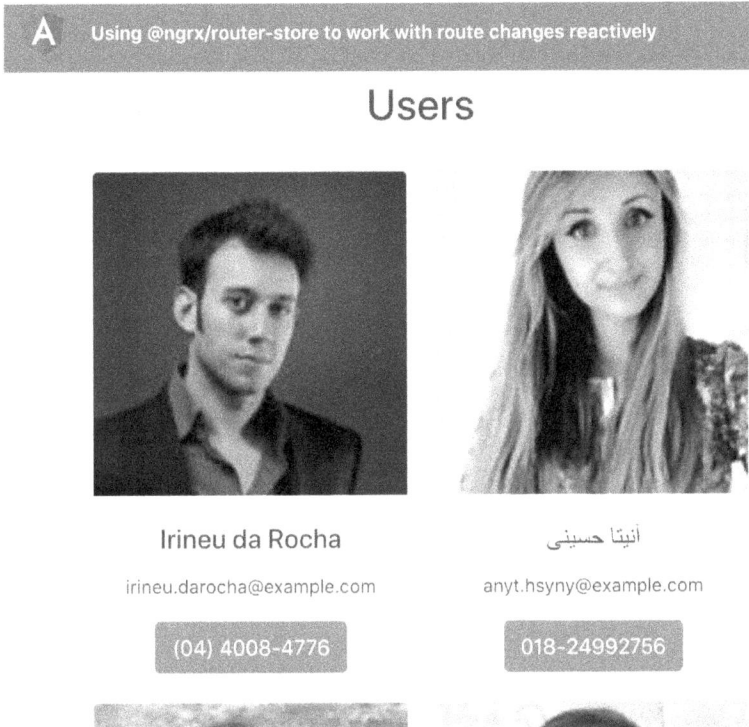

Figure 6.19 – ngrx-router-store app running on http://localhost:4200

Now that the app is running, see the next section for the steps of the recipe.

How to do it...

In order to utilize the power of NgRx even for route changes, we'll utilize the `@ngrx/router-store` package to listen to route changes. Let's begin!

1. First, install the `@ngrx/router-store` package by running the following command in your project root:

    ```
    npm install @ngrx/router-store@12.0.0 --save
    ```

2. Now, import `StoreRouterConnectingModule` and `routerReducer` from
    the `@ngrx/router-store` package in your `app.module.ts` file and set up
    the `imports`, as follows:

```
...

import { StoreRouterConnectingModule, routerReducer }
from '@ngrx/router-store';

@NgModule({
    declarations: [...],
    imports: [
        BrowserModule,
        AppRoutingModule,
        HttpClientModule,
        StoreModule.forRoot({
            app: appStore.reducer,
        router: routerReducer
        }),
    StoreRouterConnectingModule.forRoot(),
        StoreDevtoolsModule.instrument({
            maxAge: 25, // Retains last 25 states
        }),
        EffectsModule.forRoot([AppEffects])
    ],
    providers: [],
    bootstrap: [AppComponent]
})
export class AppModule { }
```

As soon as you refresh the app now and inspect it via the Redux DevTools
extension, you should see some additional actions named `@ngrx/router-store/*` being dispatched as well. You should also see that the `router` property
in the state has the current routes' information, as depicted in the following
screenshot:

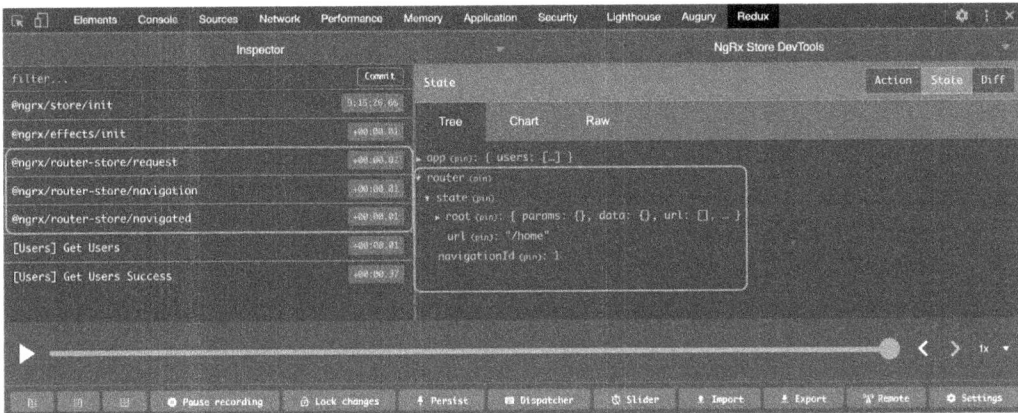

Figure 6.20 – @ngrx/router-store actions and the router state reflected in the NgRx store

3. We now have to modify our reducer—or, more precisely, the `AppState` interface—
    to reflect that we have the `router` property as well from the `@ngrx/router-store` package. To do so, modify the `store/app.reducer.ts` file, as follows:

```
...
import { getUsersSuccess } from './app.actions';
import { RouterReducerState } from '@ngrx/router-store'

export interface AppState {
  users: IUser[];
  router: RouterReducerState<any>;
}

const initialState: AppState = {
  users: null,
  router: null
}
...
```

4. Essentially, we have to get rid of the `ActivatedRoute` service's usage from our `UserDetailComponent` class. In order to do so, we'll first modify our selectors to get the params from the router state directly. Modify the `app.selectors.ts` file, as follows:

```
. . .
import { getSelectors, RouterReducerState } from '@ngrx/
router-store';

export const selectApp =
createFeatureSelector<AppState>('app');
export const selectUsers = createSelector(
  selectApp,
  (state: AppState) => state.users
);


. . .
export const selectRouter = createFeatureSelector<
  AppState,
  RouterReducerState<any>
>('router');


const { selectRouteParam } = getSelectors(selectRouter);


export const selectUserUUID = selectRouteParam('uuid');

export const selectCurrentUser = createSelector(
  selectUserUUID,
  selectUsers,
  (uuid, users: IUser[]) => users ? users.find(user => {
    return user.login.uuid === uuid;
  }) : null
);


export const selectSimilarUsers = createSelector(
  selectUserUUID,
```

```
  selectUsers,
  (uuid, users: IUser[]) => users ? users.filter(user =>
  {
    return user.login.uuid !== uuid;
  }): null
);
```

You should see some errors on the console right now. That's because we changed the signature of the `selectSimilarUsers` and `selectCurrentUser` selectors, but it'll be fixed in the next step.

5. Modify the `user-detail/user-detail.component.ts` file to use the updated selectors correctly, as follows:

```
...
export class UserDetailComponent implements OnInit,
OnDestroy {
  ...
  ngOnInit() {
    ...
    this.route.paramMap.pipe(
      takeWhile(() => !!this.isComponentAlive)
    )
    .subscribe(params => {
      const uuid = params.get('uuid');
      this.user$ = this.store.select(selectCurrentUser)
      this.similarUsers$ = this.store.
      select(selectSimilarUsers)
    })
  }
  ...
}
```

This change should have resolved the errors on the console, and you should actually see the app working perfectly fine, even though we're not passing any `uuid` from the `UserDetailComponent` class anymore.

6. With the changes from the previous step, we can now safely remove the usage of the `ActivatedRoute` service from the `UserDetailComponent` class, and the code should now look like this:

```
. . .
import { Observable } from 'rxjs/internal/Observable';
import { first } from 'rxjs/operators';
import { Store } from '@ngrx/store';
. . .
export class UserDetailComponent implements OnInit,
OnDestroy {
  . . .
  constructor(
    private store: Store<AppState>
  ) {}

  ngOnInit() {
    this.isComponentAlive = true;
    this.getUsersIfNecessary();
    this.user$ = this.store.select(selectCurrentUser)
    this.similarUsers$ = this.store.
    select(selectSimilarUsers)
  }
  . . .
}
```

Woohoo! You've finished the recipe now. See the next section to find out how this works.

How it works...

`@ngrx/router-store` is an amazing package that does a lot of magic to make our development a lot easier with NgRx. You saw how we could remove the `ActivatedRoute` service completely from the `UserDetailComponent` class by using the selectors from the package. Essentially, this helped us get the **route params** right in the selectors, and we could use it in our selectors to get and filter out the appropriate data. Behind the scenes, the package listens to the route changes in the entire Angular app and fetches from the route itself. It then stores the respective information in the NgRx Store so that it remains in the Redux state and can be selected via the package-provided selectors easily. In my opinion, it's freaking awesome! I say this because the package is doing all the heavy lifting that we would have to do otherwise. As a result, our `UserDetailComponent` class now relies only on the `Store` service, which makes it even easier to test because of fewer dependencies.

See also

- `@ngrx/router-store` documentation (`https://ngrx.io/guide/router-store/`)

7
Understanding Angular Navigation and Routing

One of the most amazing things about Angular is that it is an entire ecosystem (a framework) rather than a library. In this ecosystem, the Angular router is one of the most critical blocks to learn and understand. In this chapter, you'll learn some really cool techniques about routing and navigation in Angular. You'll learn about how to guard your routes, listen to route changes, and configure global actions on route changes.

The following are the recipes we're going to cover in this chapter:

- Creating an Angular app and modules with routes using the CLI
- Feature modules and lazily loaded routes
- Authorized access to routes using route guards
- Working with route parameters
- Showing a global loader between route changes
- Preloading route strategies

Technical requirements

For the recipes in this chapter, make sure you have **Git** and **Node.js** installed on your machine. You also need to have the `@angular/cli` package installed, which you can do with `npm install -g @angular/cli` from your terminal. The code for this chapter can be found at `https://github.com/PacktPublishing/Angular-Cookbook/tree/master/chapter07`.

Creating an Angular app with routes using the CLI

If you ask me about how we used to create projects for web applications 7-8 years ago, you'll be astonished to learn how difficult it was. Luckily, the tools and standards have evolved in the software development industry and when it comes to Angular, starting a project is super easy. You can even configure different things out of the box. In this recipe, you'll create a fresh Angular project using the Angular CLI and will also enable the routing configuration as you create the project.

Getting ready

The project that we are going to work on does not have a starter file. So, you can open the `chapter07/start_here` folder from the cloned repository directly into the Visual Studio Code app.

How to do it...

We'll be creating the app using the Angular CLI first. It'll have routing enabled out of the box. Similarly, going forward, we'll create some feature modules with components as well, but they'll have eagerly loaded routes. So, let's get started:

1. First, open the terminal and make sure you're inside the `chapter07/start_here` folder. Once inside, run the following command:

    ```
    ng new basic-routing-app --routing --style scss
    ```

    The command should create a new Angular app for you with routing enabled and SCSS selected as your styling choice.

2. Run the following commands to open up the app in the browser:

    ```
    cd basic-routing app
    ng serve -o
    ```

3. Now, let's create a top-level component named `landing` by running the following command:

```
ng g c landing
```

4. Remove all the content from `app.component.html` and keep only `router-outlet`, as follows:

```
<router-outlet></router-outlet>
```

5. We'll now make `LandingComponent` the default route by adding it to the `app-routing.module.ts` file, as follows:

```
import { NgModule } from '@angular/core';
import { Routes, RouterModule } from '@angular/router';
import { LandingComponent } from './landing/landing.
component';

const routes: Routes = [{
  path: '',
  redirectTo: 'landing',
  pathMatch: 'full'
}, {
  path: 'landing',
  component: LandingComponent
}];

...
```

6. Refresh the page and you should see the URL automatically changed to `http://localhost:4200/landing` as the app redirected to the default route.

7. Replace the contents of `landing.component.html` with the following code:

```
<div class="landing">
  <div class="landing__header">
    <div class="landing__header__main">
      Creating an Angular app with routes using CLI
    </div>
    <div class="landing__header__links">
      <div class="landing__header__links__link">
        Home
```

```
      </div>
      <div class="landing__header__links__link">
        About
      </div>
    </div>
  </div>
  <div class="landing__body">
    Landing Works
  </div>
</div>
```

8. Now, add some styles for the header in the `landing.component.scss` file, as follows:

```
.landing {
  display: flex;
  flex-direction: column;
  height: 100%;
  &__header {
    height: 60px;
    padding: 0 20px;
    background-color: #333;
    color: white;
    display: flex;
    align-items: center;
    justify-content: flex-end;
    &__main {
      flex: 1;
    }
  }
}
```

9. Add the styles for the links in the header as follows:

```
.landing {
  ...
  &__header {
    ...
    &__links {
```

```scss
      padding: 0 20px;
      display: flex;
      &__link {
        margin-left: 16px;
        &:hover {
          color: #ececec;
          cursor: pointer;
        }
      }
    }
  }
}
```

10. Furthermore, add the styles for the body of the landing page after the `&__header` selector, as follows:

```scss
.landing {
  ...
  &__header {
    ...
  }
  &__body {
    padding: 30px;
    flex: 1;
    display: flex;
    justify-content: center;
    background-color: #ececec;
  }
}
```

11. Finally, to make it all look good, add the following styles to the `styles.scss` file:

```scss
html, body {
  width: 100%;
  height: 100%;
  margin: 0;
  padding: 0;
}
```

12. Now, add a feature module for both the `home` and `about` routes by running the following commands in the project root:

```
ng g m home
ng g c home

ng g m about
ng g c about
```

13. Next, import both `HomeModule` and `AboutModule` in your `app.module.ts` file as follows:

```
...
import { LandingComponent } from './landing/landing.
component';
import { HomeModule } from './home/home.module';
import { AboutModule } from './about/about.module';

@NgModule({
  declarations: [...],
  imports: [
    BrowserModule,
    AppRoutingModule,
    HomeModule,
    AboutModule
  ],
  providers: [],
  bootstrap: [AppComponent]
})
export class AppModule { }
```

14. Now, we can configure the routes. Modify the `app-routing.module.ts` file to add the appropriate routes as follows:

```
import { NgModule } from '@angular/core';
import { Routes, RouterModule } from '@angular/router';
import { AboutComponent } from './about/about.component';
import { HomeComponent } from './home/home.component';
import { LandingComponent } from './landing/landing.
```

```
component';

const routes: Routes = [{
  path: '',
  redirectTo: 'landing',
  pathMatch: 'full'
}, {
  path: 'landing',
  component: LandingComponent
}, {
  path: 'home',
  component: HomeComponent
}, {
  path: 'about',
  component: AboutComponent
}];
...
```

15. We can style our Home and About components in just a bit. Add the following CSS to both the home.component.scss file and the about.component.scss file:

```
:host {
  display: flex;
  width: 100%;
  height: 100%;
  justify-content: center;
  align-items: center;
  background-color: #ececec;
  font-size: 24px;
}
```

16. Now, we can bind our links to the appropriate routes in the landing page. Modify landing.component.html as follows:

```
<div class="landing">
  <div class="landing__header">
    <div class="landing__header__links">
      <div class="landing__header__links__link"
```

```
      routerLink="/home">
        Home
      </div>
      <div class="landing__header__links__link"
      routerLink="/about">
        About
      </div>
    </div>
  </div>
  <div class="landing__body">
    Landing Works
  </div>
</div>
```

Awesome! Within a few minutes, and with the help of the amazing Angular CLI and Angular router, we were able to create a landing page, two feature modules, and feature routes (although eagerly loaded) and we styled some stuff as well. The wonders of the modern web!

Now that you know how basic routing is implemented, see the next section to understand how it works.

How it works...

When we use the `--routing` argument while creating the app, or when creating a module, the Angular CLI automatically creates a module file named `<your module>-routing.module.ts`. This file basically contains a routing module. In this recipe, we just created the feature modules without routing to keep the implementation simpler and faster. In the next recipe, you'll learn about routes within modules as well. Anyway, since we've created the eagerly loaded feature modules, this means that all the JavaScript of all the feature modules loads as soon as the app is loaded. You can inspect the **Network** tab in Chrome DevTools and see the content of the `main.js` file since it contains all our components and modules. See the following screenshot, which shows both the `AboutComponent` and `HomeComponent` code in the `main.js` file:

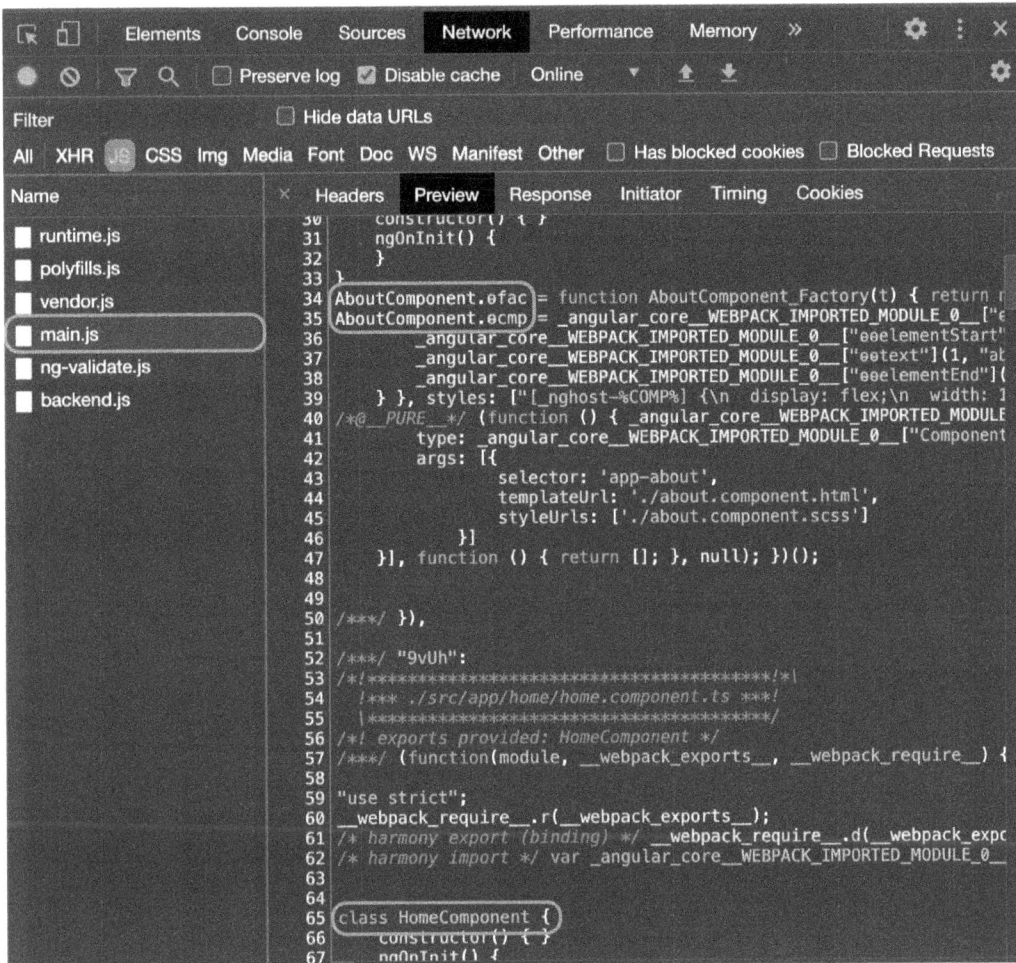

Figure 7.1 – main.js containing AboutComponent and HomeComponent code

Since we've established that all our components in the recipe are loaded eagerly on the app start, it is necessary to understand that it happens because we import `HomeModule` and `AboutModule` in the `imports` array of `AppModule`.

See also

- Angular router docs (`https://angular.io/guide/router`)

Feature modules and lazily loaded routes

In the previous recipe, we learned how to create a basic routing app with eagerly loaded routes. In this recipe, you'll learn how to work with feature modules to lazily load them instead of loading them when the app loads. For this recipe, we'll assume that we already have the routes in place and we just need to load them lazily.

Getting ready

The project for this recipe resides in `chapter07/start_here/lazy-loading-modules`:

1. Open the project in Visual Studio Code.

2. Open the terminal and run `npm install` to install the dependencies of the project.

3. Once done, run `ng serve -o`.

    This should open the app in a new browser tab and you should see the app as follows:

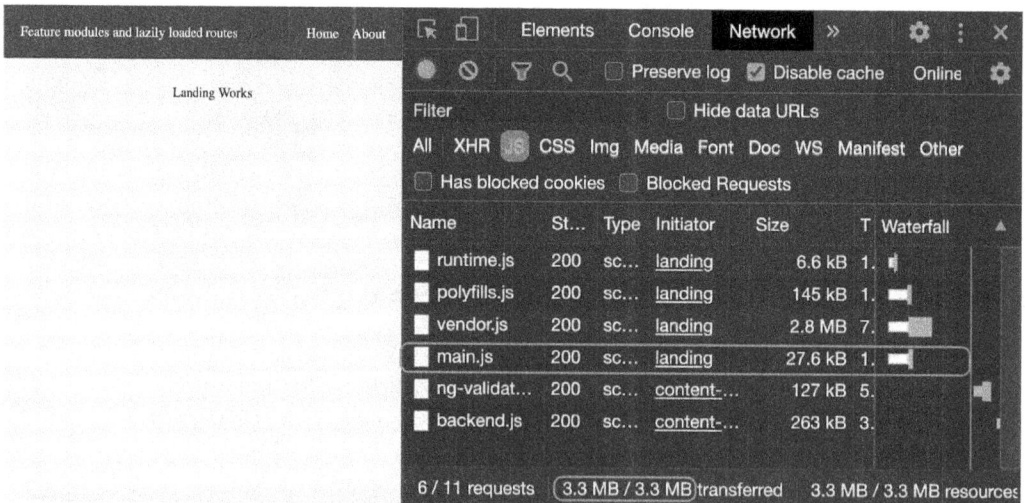

Figure 7.2 – lazy-loading-modules app running on http://localhost:4200

Now that we have the app running locally, let's see the steps of the recipe in the next section.

How to do it...

As shown in *Figure 7.2*, we have all the components and modules inside the `main.js` file. Therefore, we have about 23.4 KB in size for the `main.js` file. We'll modify the code and the routing structure to achieve lazy loading. As a result, we'll have the particular files of the routes loading when we actually navigate to them:

1. First, we have to make our target modules capable of being lazily loaded. For that, we'll have to create a `<module>-routing.module.ts` file for both `AboutModule` and `HomeModule`. So, let's create a new file in both the `about` and `home` folders:

   a) Name the first file `about-routing.module.ts` and add the following code to it:

   ```
   // about-routing.module.ts
   import { NgModule } from '@angular/core';
   import { Routes, RouterModule } from '@angular/router';
   import { AboutComponent } from './about.component';

   const routes: Routes = [{
     path: '',
     component: AboutComponent
   }];

   @NgModule({
     imports: [RouterModule.forChild(routes)],
     exports: [RouterModule]
   })
   export class AboutRoutingModule { }
   ```

   b) Name the second file `home-routing.module.ts` and add the following code to it:

   ```
   // home-routing.module.ts
   import { NgModule } from '@angular/core';
   import { Routes, RouterModule } from '@angular/router';
   import { HomeComponent } from './home.component';

   const routes: Routes = [{
     path: '',
     component: HomeComponent
   ```

```
}];

@NgModule({
  imports: [RouterModule.forChild(routes)],
  exports: [RouterModule]
})
export class HomeRoutingModule { }
```

2. Now, we'll add these routing modules to the appropriate modules, that is, we'll import `HomeRoutingModule` in `HomeModule` as follows:

```
// home.module.ts
import { NgModule } from '@angular/core';
import { CommonModule } from '@angular/common';
import { HomeComponent } from './home.component';
import { HomeRoutingModule } from './home-routing.
module';

@NgModule({
  declarations: [HomeComponent],
  imports: [
    CommonModule,
    HomeRoutingModule
  ]
})
export class HomeModule { }
```

Add `AboutRoutingModule` in `AboutModule` as follows:

```
// about.module.ts
import { NgModule } from '@angular/core';
import { CommonModule } from '@angular/common';
import { AboutComponent } from './about.component';
import { AboutRoutingModule } from './about-routing.
module';

@NgModule({
  declarations: [AboutComponent],
  imports: [
```

```
      CommonModule,
      AboutRoutingModule
    ]
  })
  export class AboutModule { }
```

3. Our modules are now capable of being lazily loaded. We just need to lazy load
    them now. In order to do so, we need to modify `app-routing.module.ts` and
    change our configurations to use the ES6 imports for the `about` and `home` routes,
    as follows:

```
import { NgModule } from '@angular/core';
import { Routes, RouterModule } from '@angular/router';
import { LandingComponent } from './landing/landing.
component';

const routes: Routes = [{
  path: '',
  redirectTo: 'landing',
  pathMatch: 'full'
}, {
  path: 'landing',
  component: LandingComponent
}, {
  path: 'home',
  loadChildren: () => import('./home/home.module').then
  (m => m.HomeModule)
}, {
  path: 'about',
  loadChildren: () => import('./about/about.module').
  then(m => m.AboutModule)
}];

@NgModule({
  imports: [RouterModule.forRoot(routes)],
  exports: [RouterModule]
})
export class AppRoutingModule { }
```

4. Finally, we will remove the AboutModule and HomeModule imports from the
    imports array of AppModule so that we get the desired code-splitting out of the
    box. The content of app.module.ts should look as follows:

```
import { BrowserModule } from '@angular/platform-
browser';
import { NgModule } from '@angular/core';
import { AppRoutingModule } from './app-routing.module';
import { AppComponent } from './app.component';
import { LandingComponent } from './landing/landing.
component';
import { HomeModule } from './home/home.module'; ← Remove
import { AboutModule } from './about/about.module'; ←
Remove

@NgModule({
  declarations: [
    AppComponent,
    LandingComponent
  ],
  imports: [
    BrowserModule,
    AppRoutingModule,
    HomeModule, ← Remove
    AboutModule ← Remove
  ],
  providers: [],
  bootstrap: [AppComponent]
})
export class AppModule { }
```

Refresh the app and you'll see that the bundle size for the main.js file is down
to 18.1 KB, which was about 23.4 KB before. See the following screenshot:

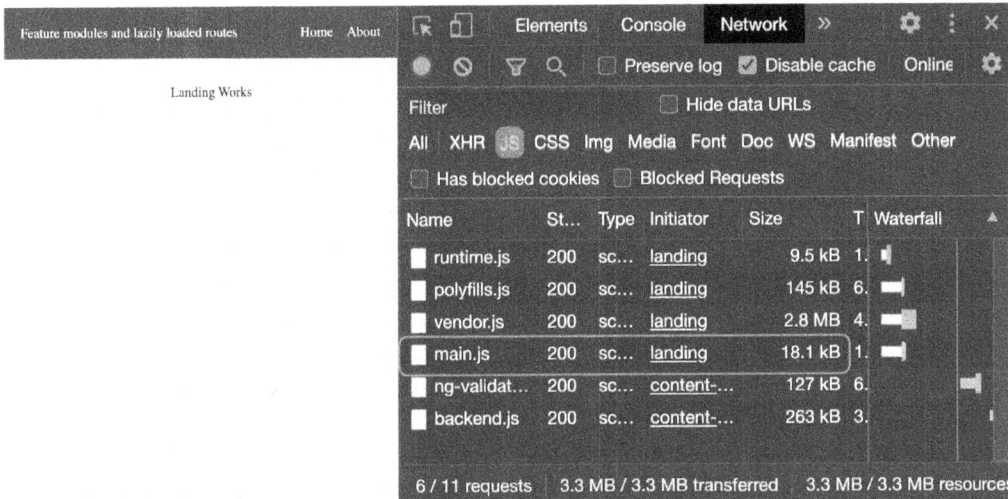

Figure 7.3 – Reduced size of main.js on app load

But what about the Home and About routes? And what about lazy loading? Well, tap the **Home** route from the header and you'll see a new JavaScript file being downloaded in the **Network** tab specifically for the route. That's lazy loading in action! See the following screenshot:

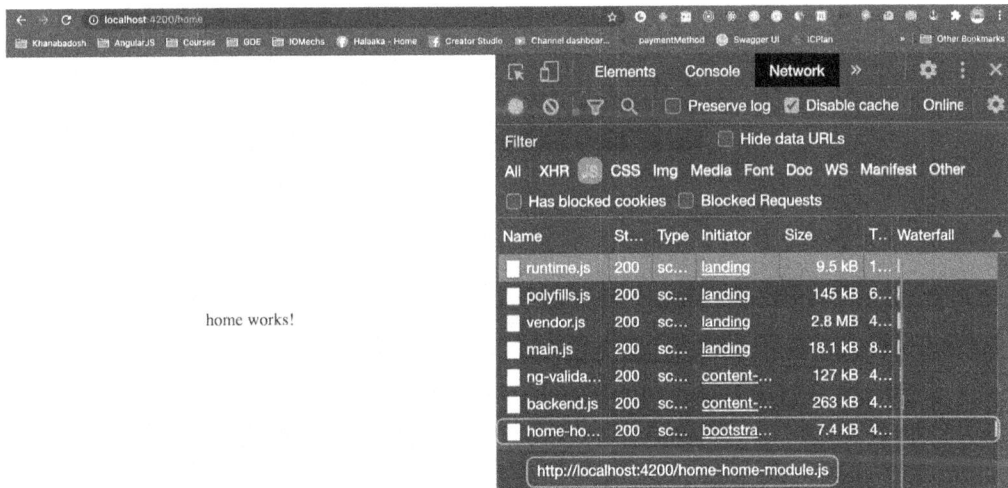

Figure 7.4 – home route being lazily loaded

Awesomesauce! You just became lazy! Just kidding. You just learned the art of lazily loading routes and feature modules in your Angular app. You can now show this off to your friends as well.

How it works...

Angular works with modules and usually the features are broken down into modules. As we know, `AppModule` serves as the entry point for the Angular app and Angular will import and bundle anything that is imported in `AppModule` during the build process, resulting in the `main.js` file. However, if we want to lazy load our routes/feature modules, we need to avoid importing feature modules in `AppModule` directly and use the `loadChildren` method for our routes to load the feature modules instead, on-demand. That's what we did in this recipe. It is important to note that the routes stayed the same in `AppRoutingModule`. However, we had to put `path: ''` in our feature routing modules since that'll combine the route in `AppRoutingModule` and then in the feature routing module to become what's defined in `AppRoutingModule`. That's why our routes were still `'about'` and `'home'`.

See also

- Lazy loading modules in Angular (`https://angular.io/guide/lazy-loading-ngmodules`)

Authorized access to routes using route guards

Not all routes in your Angular app should be accessible by everyone in the world. In this recipe, we'll learn how to create route guards in Angular to prevent unauthorized access to routes.

Getting ready

The project for this recipe resides in `chapter07/start_here/using-route-guards`:

1. Open the project in Visual Studio Code.

2. Open the terminal and run `npm install` to install the dependencies of the project.

3. Once done, run `ng serve -o`.

This should open the app in a new browser tab, and you should see the app as follows:

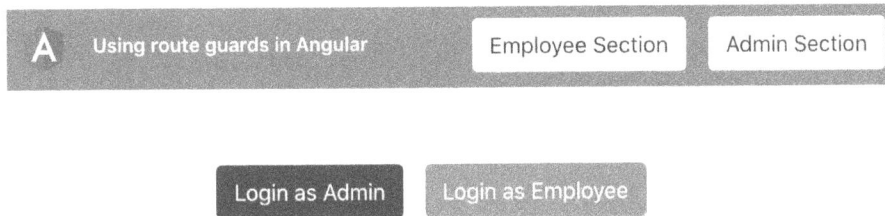

Figure 7.5 – using-route-guards app running on http://localhost:4200

Now that we have the app running locally, let's see the steps of the recipe in the next section.

How to do it...

We have an app with a couple of routes already set up. You can log in as either an employee or an admin to get to the bucket list of the app. However, if you tap any of the two buttons in the header, you'll see that you can navigate to the Admin and Employee sections even without being logged in. This is what we want to prevent from happening. Notice in the `auth.service.ts` file that we already have a way for the user to do a login, and we can check whether the user is logged in or not using the `isLoggedIn()` method:

1. First, let's create a route guard that will only allow the user to go to the particular routes if the user is logged in. We'll name it `AuthGuard`. Let's create it by running the following command in the project root:

    ```
    ng g guard guards/Auth
    ```

    Once the command is run, you should be able to see some options to select which interfaces we'd like to implement.

2. Select the `CanActivate` interface and press *Enter*.

3. Now, add the following logic to the `auth.guard.ts` file to check whether the user is logged in, and if the user is not logged in, we'll redirect the user to the login page, which is the `'/auth'` route:

```
import { Injectable } from '@angular/core';
import { CanActivate, ActivatedRouteSnapshot,
RouterStateSnapshot, UrlTree, Router } from '@angular/
router';
import { Observable } from 'rxjs';
import { AuthService } from '../services/auth.service';

@Injectable({
  providedIn: 'root'
})
export class AuthGuard implements CanActivate {
  constructor(private auth: AuthService, private router:
  Router) {  }
  canActivate(
    route: ActivatedRouteSnapshot,
    state: RouterStateSnapshot): Observable<boolean |
    UrlTree> | Promise<boolean | UrlTree> | boolean |
    UrlTree {
      const loggedIn = !!this.auth.isLoggedIn();
      if (!loggedIn) {
        this.router.navigate(['/auth']);
        return false;
      }
    return true;
  }
}
```

4. Now, let's apply `AuthGuard` to our Admin and Employee routes in the `app-routing.module.ts` file, as follows:

```
...
import { AuthGuard } from './guards/auth.guard';

const routes: Routes = [{...}, {
  path: 'auth',
```

```
    loadChildren: () => import('./auth/auth.module').then
    (m => m.AuthModule)
  }, {
    path: 'admin',
    loadChildren: () => import('./admin/admin.module').
    then(m => m.AdminModule),
    canActivate: [AuthGuard]
  }, {
    path: 'employee',
    loadChildren: () => import('./employee/employee.
    module').then(m => m.EmployeeModule),
    canActivate: [AuthGuard]
  }];
  ...
export class AppRoutingModule { }
```

If you now log out and try to tap either the **Employee Section** or **Admin Section** buttons in the header, you'll notice that you're now not able to go to the routes until you log in. The same is the case if you try to enter the URL directly for the routes in the address bar and hit *Enter*.

5. Now we'll try to create a guard, one for the Employee route and one for the Admin route. Run the following commands one by one and select the CanActivate interface for both the guards:

```
ng g guard guards/Employee


ng g guard guards/Admin
```

6. Since we have the guards created, let's put the logic for AdminGuard first. We'll try to see what type of user has logged in. If it is an admin, then we allow the navigation, else we prevent it. Add the following code to admin.guard.ts:

```
...
import { UserType } from '../constants/user-type';
import { AuthService } from '../services/auth.service';
...
export class AdminGuard implements CanActivate {
  constructor(private auth: AuthService) {}
  canActivate(
```

```
    route: ActivatedRouteSnapshot,

    state: RouterStateSnapshot): Observable<boolean |
    UrlTree> | Promise<boolean | UrlTree> | boolean |
    UrlTree {

    return this.auth.loggedInUserType === UserType.Admin;
  }
}
```

7. Add AdminGuard to the Admin route in app-routing.module.ts as follows:

```
...
import { AdminGuard } from './guards/admin.guard';
import { AuthGuard } from './guards/auth.guard';

const routes: Routes = [{
  path: '',
  ...
}, {
  path: 'auth',
  ...
}, {
  path: 'admin',
  loadChildren: () => import('./admin/admin.module').
  then(m => m.AdminModule),
  canActivate: [AuthGuard, AdminGuard]
}, {
  path: 'employee',
  ...
}];
...
```

Try to log out and log in as an employee now. Then try tapping the **Admin Section**
button in the header. You'll notice that you can't go to the Admin section of the
bucket list anymore. This is because we have AdminGuard in place and you're not
logged in as an admin right now. Logging in as an admin should work just fine.

8. Similarly, we'll add the following code to `employee.guard.ts`:

```
...
import { UserType } from '../constants/user-type';
import { AuthService } from '../services/auth.service';

@Injectable({
  providedIn: 'root'
})
export class EmployeeGuard implements CanActivate {
  constructor(private auth: AuthService) {}
  canActivate(
    route: ActivatedRouteSnapshot,
    state: RouterStateSnapshot): Observable<boolean |
    UrlTree> | Promise<boolean | UrlTree> | boolean |
    UrlTree {
    return this.auth.loggedInUserType === UserType.
    Employee;
  }
}
```

9. Now, add `EmployeeGuard` to the Employee route in `app-routing.module.ts` as follows:

```
...
import { EmployeeGuard } from './guards/employee.guard';
const routes: Routes = [
  ...
  , {
    path: 'employee',
    loadChildren: () => import('./employee/employee.
    module').then(m => m.EmployeeModule),
    canActivate: [AuthGuard, EmployeeGuard]
  }];
...
```

Now, only the appropriate routes should be accessible by checking which type of user is logged in.

Great! You now are an authorization expert when it comes to guarding routes. With great power comes great responsibility. Use it wisely.

How it works...

The `CanActivate` interface of the route guards is the heart of our recipe because it corresponds to the fact that each route in Angular can have an array of guards for the `CanActivate` property of the route definition. When a guard is applied, it is supposed to return a Boolean value or a `UrlTree`. We've focused on the Boolean value's usage in our recipe. We can return the Boolean value directly using a promise or even using an Observable. This makes guards really flexible for use even with remote data. Anyway, for our recipe, we've kept it easy to understand by checking whether the user is logged in (for `AuthGuard`) and by checking whether the expected type of user is logged in for the particular routes (`AdminGuard` and `EmployeeGuard`).

See also

- Preventing unauthorized access in Angular routes (`https://angular.io/guide/router#preventing-unauthorized-access`)

Working with route parameters

Whether it is about building a REST API using Node.js or configuring routes in Angular, setting up routes is an absolute art, especially when it comes to working with parameters. In this recipe, you'll create some routes with parameters and will learn how to get those parameters in your components once the route is active.

Getting ready

The project for this recipe resides in `chapter07/start_here/working-with-route-params`:

1. Open the project in Visual Studio Code.

2. Open the terminal and run `npm install` to install the dependencies of the project.

3. Once done, run `ng serve -o`.

   This should open the app in a new browser tab. As soon as the page is opened, you should see a list of users.

4. Tap the first user, and you should see the following view:

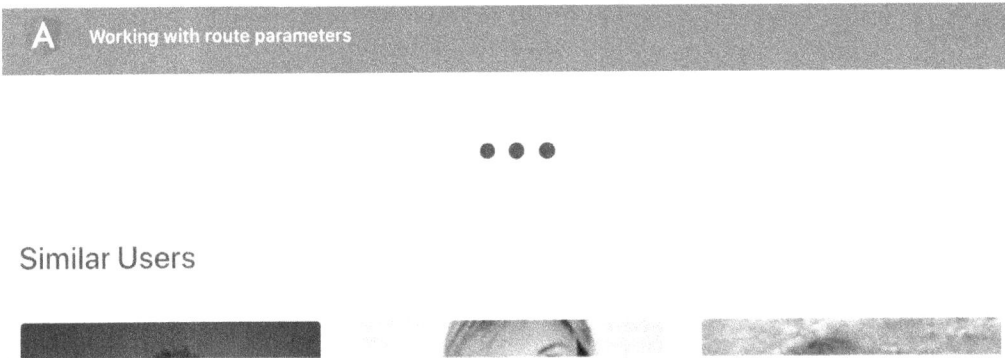

Figure 7.6 – user-details not bringing the correct user

Now that we have the app running locally, let's see the steps of the recipe in the next section.

How to do it...

The problem right now is that we have a route for opening the user details, but we don't have any idea in `UserDetailComponent` about which user was clicked, that is, which user to fetch from the service. Therefore, we'll implement the route parameters to pass the user's ID (`uuid`) from the home page to the user details page:

1. First, we have to make our user route capable of accepting the route parameter named `uuid`. This is going to be a **required** parameter, which means the route will not work without passing this. Let's modify `app-routing.module.ts` to add this required parameter to the route definition, as follows:

```
...
import { UserDetailComponent } from './user-detail/user-detail.component';

const routes: Routes = [
  ...
, {
  path: 'user/:uuid',
  component: UserDetailComponent
}];
...
```

With this change, clicking on a user on the home page will not work anymore. If you try it, you'll see an error as follows and that's because `uuid` is a required parameter:

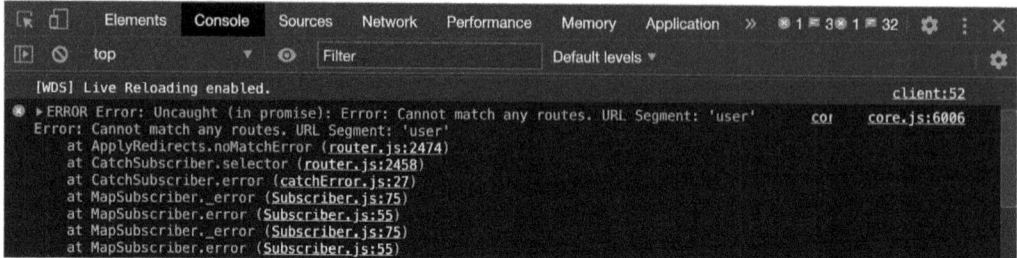

Figure 7.7 – Angular complaining about not being able to match the requested route

2. The fix is easy for the error; we need to pass the `uuid` when navigating to the user route. Let's do this by modifying the `user-card.component.ts` file as follows:

```
import { Component, Input, OnInit } from '@angular/core';
import { Router } from '@angular/router';
import { IUser } from '../../interfaces/user.interface';


@Component({
  selector: 'app-user-card',
  templateUrl: './user-card.component.html',
  styleUrls: ['./user-card.component.scss']
})
export class UserCardComponent implements OnInit {
  @Input('user') user: IUser;
  constructor(private router: Router) { }

  ngOnInit(): void {
  }

  cardClicked() {
    this.router.navigate(['
    /user/${this.user.login.uuid}'])
  }

}
```

We're now able to navigate to a particular user's route, and you should also be able to see the UUID in the address bar as follows:

Figure 7.8 – The UUID being shown in the address bar

3. To get the current user from `UserService`, we need to get the `uuid` value in `UserDetailComponent`. Right now, we're sending `null` when calling the `getUser` method of `UserService` from `UserDetailComponent`. In order to use the user's ID, we can fetch the `uuid` value from the route parameters by importing the `ActivatedRoute` service. Update `user-detail.component.ts` as follows:

```
...
import { ActivatedRoute } from '@angular/router';
...
export class UserDetailComponent implements OnInit,
OnDestroy {
  user: IUser;
  similarUsers: IUser[];
  constructor(
    private userService: UserService,
    private route: ActivatedRoute
  ) {}

  ngOnInit() {
    ...
```

```
    }

    ngOnDestroy() {
    }
}
```

4. We'll create a new method named `getUserAndSimilarUsers` in `UserDetailComponent` and move the code from the `ngOnInit` method into the new method as follows:

```
...
export class UserDetailComponent implements OnInit,
OnDestroy {
    ...
  ngOnInit() {
    const userId = null;
    this.getUserAndSimilarUsers(userId);
  }

  getUserAndSimilarUsers(userId) {
    this.userService.getUser(userId)
      .pipe(
        mergeMap((user: IUser) => {
          this.user = user;
          return this.userService.
          getSimilarUsers(userId);
        })
      ).subscribe((similarUsers: IUser[]) => {
        this.similarUsers = similarUsers;
      })
  }
    ...
}
```

5. Now that we have the code refactored a bit, let's try to access the `uuid` from the route parameters using the `ActivatedRoute` service, and pass it into our `getUserAndSimilarUsers` method as follows:

```
...
import { mergeMap, takeWhile } from 'rxjs/operators';
import { ActivatedRoute } from '@angular/router';
...
export class UserDetailComponent implements OnInit,
OnDestroy {
  componentIsAlive = false;
  constructor(private userService: UserService, private
  route: ActivatedRoute ) {}

  ngOnInit() {
    this.componentIsAlive = true;
    this.route.paramMap
      .pipe(
        takeWhile (() => this.componentIsAlive)
      )
      .subscribe((params) => {
        const userId = params.get('uuid');
        this.getUserAndSimilarUsers(userId);
      })
  }

  getUserAndSimilarUsers(userId) {...}

  ngOnDestroy() {
    this.componentIsAlive = false;
  }
}
```

Grrreat!! With this change, you can try refreshing the app on the home page and then click any user. You should see the current user as well as similar users being loaded. To understand all the magic behind the recipe, see the next section.

How it works...

It all begins when we change our route's path to `user/:userId`. This makes `userId` a required parameter for our route. The other piece of the puzzle is to retrieve this parameter in `UserDetailComponent` and then use it to get the target user, as well as similar users. For that, we use the `ActivatedRoute` service. The `ActivatedRoute` service holds a lot of necessary information about the current route and, therefore, we were able to fetch the current route's `uuid` parameter by subscribing to the `paramMap` Observable, so even if the parameter changes while staying on a user's page, we still execute the necessary operations. Notice that we also create a property named `componentIsAlive`. As you might have seen in our prior recipes, we use it in conjunction with the `takeWhile` operator to automatically unsubscribe from the Observable streams as soon as the user navigates away from the page, or essentially when the component is destroyed.

See also

- Tour of Heroes tutorial – sample usage of the `ActivatedRoute` service (`https://angular.io/guide/router-tutorial-toh#route-parameters-in-the-activatedroute-service`)

- Link parameters array – Angular docs (`https://angular.io/guide/router#link-parameters-array`)

Showing a global loader between route changes

Building user interfaces that are snappy and fast is key to winning users. The apps become much more enjoyable for the end users and it could bring a lot of value to the owners/creators of the apps. One of the core experiences on the modern web is to show a loader when something is happening in the background. In this recipe, you'll learn how to create a global user interface loader in your Angular app that shows whenever there is a route transition in the app.

Getting ready

The project that we are going to work with resides in `chapter07/start_here/` `routing-global-loader` inside the cloned repository:

1. Open the project in Visual Studio Code.

2. Open the terminal and run `npm install` to install the dependencies of the project.

3. Once done, run `ng serve -o`.

   This should open the app in a new browser tab and you should see it as follows:

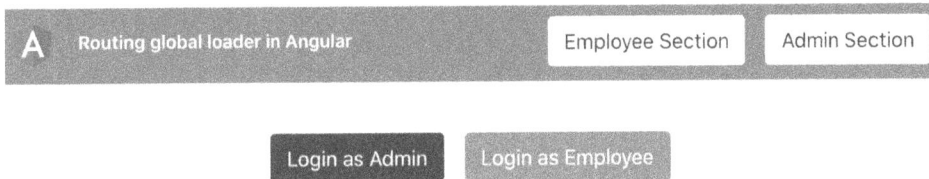

Figure 7.9 – routing-global-loader app running on http://localhost:4200

Now that we have the app running locally, let's see the steps of the recipe in the next section.

How to do it...

For this recipe, we have the bucket app with a couple of routes in it. We also have `LoaderComponent` already created, which we have to use during the route changes:

1. We'll begin showing `LoaderComponent` by default in the entire app. To do that, add the `<app-loader>` selector in the `app.component.html` file right before the `div` with the `content` class as follows:

```
<div class="toolbar" role="banner" id="toolbar">
  . . .
</div>
<app-loader></app-loader>
<div class="content" role="main">
  <div class="page-section">
    <router-outlet></router-outlet>
  </div>
</div>
```

2. Now we'll create a property in the `AppComponent` class to show the loader conditionally. We'll mark this property as `true` during the routing and will mark it as `false` when the routing is finished. Create the property as follows in the `app.component.ts` file:

```
. . .
export class AppComponent {
  isLoadingRoute = false;
  // DO NOT USE THE CODE BELOW IN PRODUCTION
  // IT WILL CAUSE PERFORMANCE ISSUES
  constructor(private auth: AuthService, private router:
  Router) {

  }

  get isLoggedIn() {
    return this.auth.isLoggedIn();
  }

  logout() {
    this.auth.logout();
    this.router.navigate(['/auth']);
  }
}
```

3. We'll now make sure that `<app-loader>` is shown only when the `isLoadingRoute` property is `true`. To do that, update the `app.component.html` template file to include an `*ngIf` statement as follows:

```
. . .
<app-loader *ngIf="isLoadingRoute"></app-loader>
<div class="content" role="main">
  <div class="page-section">
    <router-outlet></router-outlet>
  </div>
</div>
```

4. Now that the *ngIf statement is in place, we need to set the isLoadingRoute property to true somehow. To do that, we'll listen to the router service's events property, and take an action upon the NavigationStart event. Modify the code in the app.component.ts file as follows:

```
import { Component } from '@angular/core';
import { NavigationStart, Router } from '@angular/
router';
import { AuthService } from './services/auth.service';

...

export class AppComponent {
  isLoadingRoute = false;
  // DO NOT USE THE CODE BELOW IN PRODUCTION
  // IT WILL CAUSE PERFORMANCE ISSUES
  constructor(private auth: AuthService, private router:
  Router) {
    this.router.events.subscribe((event) => {
      if (event instanceof NavigationStart) {
        this.isLoadingRoute = true;
      }
    })
  }

  get isLoggedIn() {...}

  logout() {...}
}
```

If you refresh the app, you'll notice that <app-loader> never goes away. It is now being shown forever. That's because we're not marking the isLoadingRoute property as false anywhere.

5. To mark `isLoadingRoute` as `false`, we need to check for three different events: `NavigationEnd`, `NavigationError`, and `NavigationCancel`. Let's add some more logic to handle these three events and mark the property as `false`:

```
import { Component } from '@angular/core';
import { NavigationCancel, NavigationEnd,
NavigationError, NavigationStart, Router } from '@
angular/router';

. . .

export class AppComponent {

    . . .

    constructor(private auth: AuthService, private router:
    Router) {
      this.router.events.subscribe((event) => {
        if (event instanceof NavigationStart) {
          this.isLoadingRoute = true;
        }

        if (
          event instanceof NavigationEnd ||
          event instanceof NavigationError ||
          event instanceof NavigationCancel
        ) {
          this.isLoadingRoute = false;
        }
      })
    }

    get isLoggedIn() {...}

    logout() {...}
}
```

And boom! We now have a global loader that shows during the route navigation among different pages.

> **Important note**
>
> When running the app locally, you experience the best internet conditions possible (especially if you're not fetching remote data). Therefore, you might not see the loader at all or might see it for only a fraction of a second. In order to see it for a longer period, open Chrome DevTools, go to the **Network** tab, simulate slow 3G, refresh the app, and then navigate between routes.
>
> If the routes have static data, then you'll only see the loader the first time you navigate to that route. The next time you navigate to the same route, it would already have been cached, so the global loader might not show.

Congrats on finishing the recipe. You now can implement a global loader in Angular apps, which will show from the navigation start to the navigation end.

How it works...

The router service is a very powerful service in Angular. It has a lot of methods as well as Observables that we can use for different tasks in our apps. For this recipe, we used the `events` Observable. By subscribing to the `events` Observable, we can listen to all the events that the `Router` service emits through the Observable. For this recipe, we were only interested in the `NavigationStart`, `NavigationEnd`, `NavigationError`, and `NavigationCancel` events. The `NavigationStart` event is emitted when the router starts navigation. The `NavigationEnd` event is emitted when the navigation ends successfully. The `NavigationCancel` event is emitted when the navigation is canceled due to a **route guard** returning `false`, or redirects by using `UrlTree` due to some reason. The `NavigationError` event is emitted when there's an error due to any reason during the navigation. All of these events are of the `Event` type and we can identify the type of the event by checking whether it is an instance of the target event, using the `instanceof` keyword. Notice that since we had the subscription to the `Router.events` property in `AppComponent`, we didn't have to worry about unsubscribing the subscription because there's only one subscription in the app, and `AppComponent` will not be destroyed throughout the life cycle of the app.

See also

- Router events docs (`https://angular.io/guide/router#router-events`)
- Router service docs (`https://angular.io/api/router/Router`)

Preloading route strategies

We're already familiar with how to lazy load different feature modules upon navigation. Although sometimes, you might want to preload subsequent routes to make the next route navigation instantaneous or might even want to use a custom preloading strategy based on your application's business logic. In this recipe, you'll learn about the `PreloadAllModules` strategy and will also implement a custom strategy to cherry-pick which modules should be preloaded.

Getting ready

The project that we are going to work with resides in `chapter07/start_here/ route-preloading-strategies` inside the cloned repository:

1. Open the project in Visual Studio Code.

2. Open the terminal and run `npm install` to install the dependencies of the project.

3. Once done, run `ng serve -o`.

    This should open the app in a new browser tab and you should see something like the following:

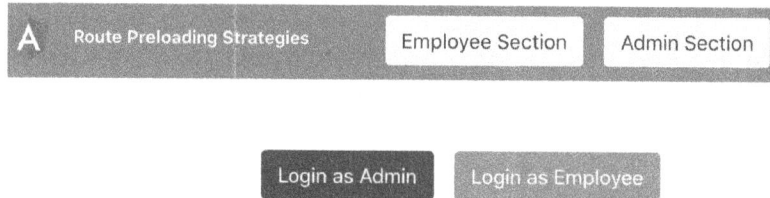

Figure 7.10 – route-preloading-strategies app running on http://localhost:4200

4. Open Chrome DevTools by pressing *Ctrl* + *Shift* + *C* on Windows or *Cmd* + *Shift* + *C* on Mac.

5. Navigate to the **Network** tab and filter on JavaScript files only. You should see something like this:

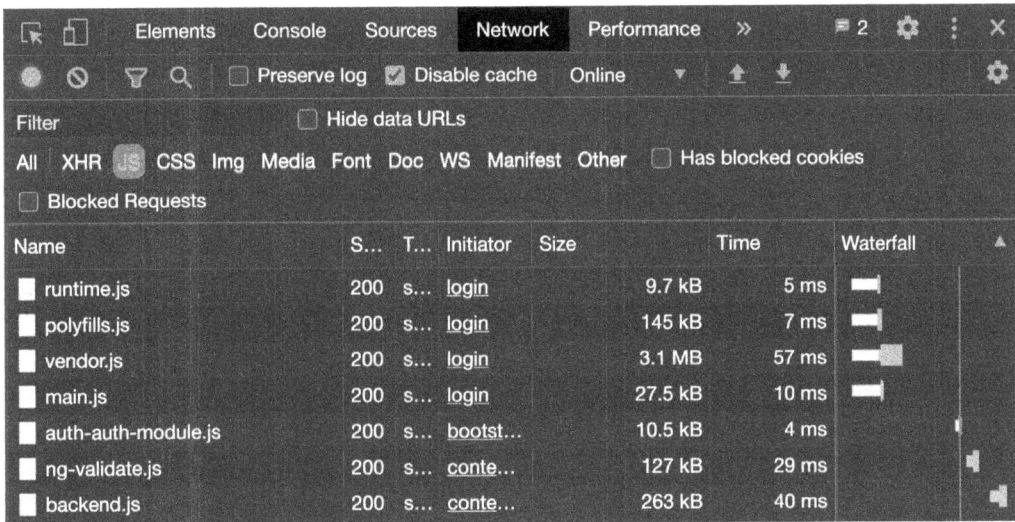

Figure 7.11 – JavaScript files loaded on app load

Now that we have the app running locally, let's see the next section for this recipe

How to do it...

Notice in *Figure 7.11* how we're automatically loading the auth-auth-module.js file since we're logged out. Although the routes in AuthModule are all configured to be lazily loaded, we can still look into what happens if we use the PreloadAllModules strategy, and then a custom preload strategy:

1. We're going to try out the PreloadAllModules strategy first. To use it, let's modify the app-routing.module.ts file as follows:

```
import { NgModule } from '@angular/core';
import { Routes, RouterModule, PreloadAllModules } from
'@angular/router';

const routes: Routes = [...];

@NgModule({
  imports: [RouterModule.forRoot(routes, {
    preloadingStrategy: PreloadAllModules
  })],
  exports: [RouterModule]
})
export class AppRoutingModule { }
```

If you refresh the app, you should see not only the `auth-auth-module.js` file but also the module files for Admin and Employee, as follows:

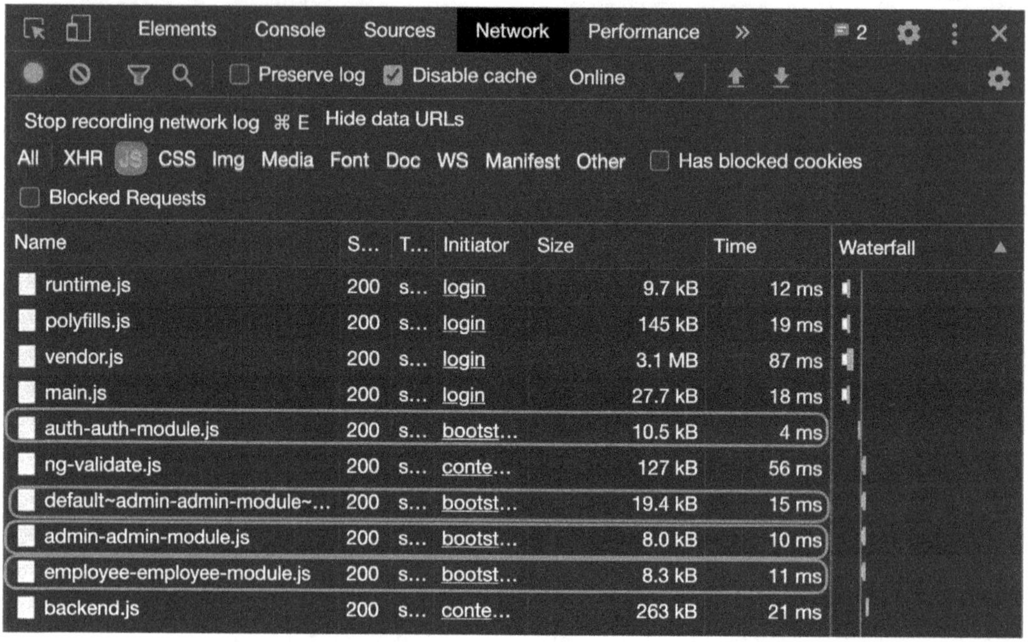

Figure 7.12 – JavaScript files loaded with the PreloadAllModules strategy

So far so good. But what if we wanted to preload only the Admin module, supposing our app is intended for admins mostly? We'll create a custom preload strategy for that.

2. Let's create a service named `CustomPreloadStrategy` by running the following command in our project:

```
ng g s services/custom-preload-strategy
```

3. In order to use our preload strategy service with Angular, our service needs to implement the `PreloadingStrategy` interface from the `@angular/router` package. Modify the newly created service as follows:

```
import { Injectable } from '@angular/core';
import { PreloadingStrategy } from '@angular/router';

@Injectable({
  providedIn: 'root'
})
```

```
export class CustomPreloadStrategyService implements
PreloadingStrategy {

    constructor() { }
}
```

4. Next, we need to implement the `preload` method from the
   `PreloadingStrategy` interface for our service to work properly. Let's modify
   `CustomPreloadStrategyService` to implement the `preload` method,
   as follows:

```
import { Injectable } from '@angular/core';
import { PreloadingStrategy, Route } from '@angular/
router';
import { Observable, of } from 'rxjs';

@Injectable({
  providedIn: 'root'
})
export class CustomPreloadStrategyService implements
PreloadingStrategy {

    constructor() { }

    preload(route: Route, load: () => Observable<any>):
    Observable<any> {
      return of(null)
    }
}
```

5. Right now, our `preload` method returns `of(null)`. Instead, in order to decide
   which routes to preload, we're going to add an object to our route definitions as the
   `data` object having a Boolean named `shouldPreload`. Let's quickly do that by
   modifying `app-routing.module.ts` as follows:

```
...
const routes: Routes = [{...}, {
    path: 'auth',
    loadChildren: () => import('./auth/auth.module').then(m
```

```
=> m.AuthModule),
   data: { shouldPreload: true }
}, {
   path: 'admin',
   loadChildren: () => import('./admin/admin.module').
   then(m => m.AdminModule),
   data: { shouldPreload: true }
}, {
   path: 'employee',
   loadChildren: () => import('./employee/employee.
   module').then(m => m.EmployeeModule),
   data: { shouldPreload: false }
}];
...
```

6. All the routes with `shouldPreload` set to `true` should be preloaded and if they are set to `false`, then they should not be preloaded. We'll create two methods. One for the case where we want to preload a route and one for the route which we don't want to preload a route. Let's modify `custom-preload-strategy.service.ts` to add the methods as follows:

```
export class CustomPreloadStrategyService implements
PreloadingStrategy {
   ...
   loadRoute(route: Route, loadFn: () => Observable<any>):
   Observable<any> {
      console.log('Preloading done for route: ${route.
      path}')
      return loadFn();
   }

   noPreload(route: Route): Observable<any> {
      console.log('No preloading set for: ${route.path}');
      return of(null);
   }
   ...
}
```

7. Awesome! Now we have to use the methods created in *Step 6* inside the `preload` method. Let's modify the method to use the `shouldPreload` property of the `data` object from the route definitions. The code should look as follows:

```
...
export class CustomPreloadStrategyService implements
PreloadingStrategy {
...
  preload(route: Route, load: () => Observable<any>):
  Observable<any> {
    try {
      const { shouldPreload } = route.data;
      return shouldPreload ? this.loadRoute(route, load)
      : this.noPreload(route);
    }
    catch (e) {
      console.error(e);
      return this.noPreload(route);
    }
  }
}
```

8. The final step is to use our custom preload strategy. In order to do so, modify the `app-routing-module.ts` file as follows:

```
import { NgModule } from '@angular/core';
import { Routes, RouterModule, PreloadAllModules ← Remove
} from '@angular/router';
import { CustomPreloadStrategyService } from './services/
custom-preload-strategy.service';

const routes: Routes = [...];

@NgModule({
  imports: [RouterModule.forRoot(routes, {
    preloadingStrategy: CustomPreloadStrategyService
  })],
  exports: [RouterModule]
})
export class AppRoutingModule { }
```

Voilà! If you refresh the app now and monitor the **Network** tab, you'll notice that only the JavaScript files for Auth and Admin are preloaded, and there's no preloading of the Employee module, as follows:

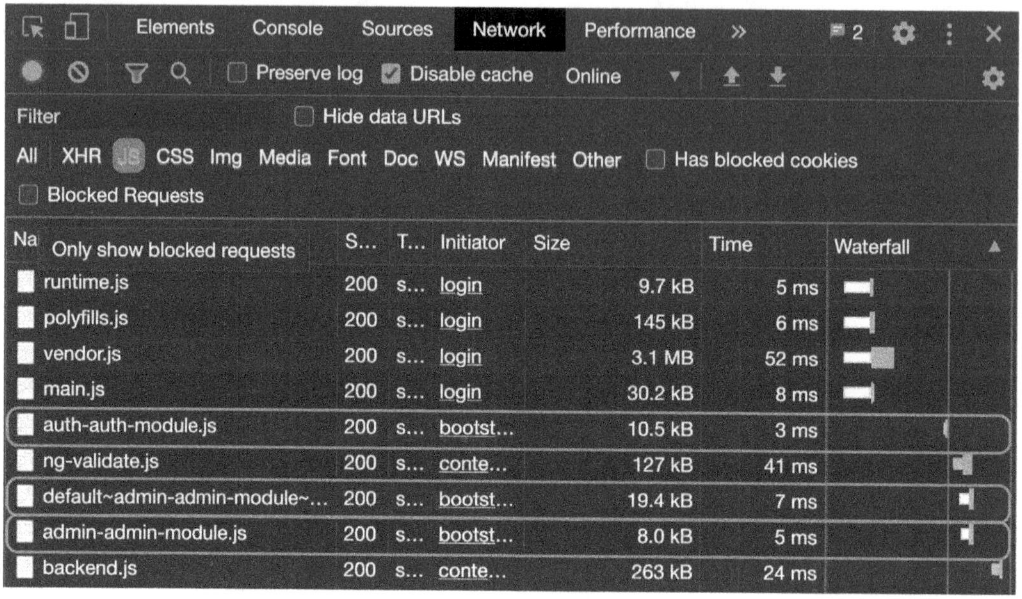

Figure 7.13 – Preloading only the Auth and Admin modules using a custom preload strategy

You can also have a look at the console logs to see which routes were preloaded. You should see the logs as follows:

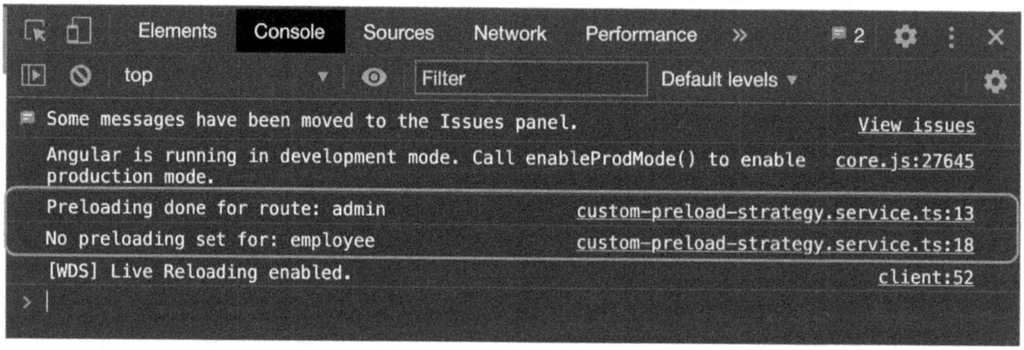

Figure 7.14 – Logs for preloading only the Auth and Admin modules

Now that you've finished the recipe, see the next section on how this works.

How it works...

Angular provides a great way to implement our own custom preloading strategy for our feature modules. We can decide easily which modules should be preloaded and which should not. In the recipe, we learned a very simple way to configure the preloading using the `data` object of the routes configuration by adding a property named `shouldPreload`. We created our own custom preload strategy service named `CustomPreloadStrategyService`, which implements the `PreloadingStrategy` interface from the `@angular/router` package. The idea is to use the `preload` method from the `PreloadingStrategy` interface, which allows us to decide whether a route should be preloaded. That's because Angular goes through each route using our custom preload strategy and decides which routes to preload. And that's it. We can now assign the `shouldPreload` property in the `data` object to any route we want to preload on app start.

See also

- Route preloading strategies article on `web.dev` (`https://web.dev/route-preloading-in-angular/`)

8
Mastering Angular Forms

Getting user inputs is an integral part of almost any modern app that we use. Whether it is authenticating users, asking for feedback, or filling out business-critical forms, knowing how to implement and present forms to end users is always an interesting challenge. In this chapter, you'll learn about Angular forms and how you can create great user experiences using them.

Here are the recipes that we're going to cover in this chapter:

- Creating your first template-driven Angular form
- Form validation with template-driven forms
- Testing template-driven forms
- Creating your first Reactive form
- Form validation with Reactive forms
- Creating an asynchronous validator function
- Testing Reactive forms
- Using debounce with Reactive form control
- Writing your own custom form control using `ControlValueAccessor`

Technical requirements

For the recipes in this chapter, make sure you have **Git** and **NodeJS** installed on your machine. You also need to have the `@angular/cli` package installed, which you can do with `npm install -g @angular/cli` from your terminal. The code for this chapter can be found at `https://github.com/PacktPublishing/Angular-Cookbook/tree/master/chapter08`.

Creating your first template-driven Angular form

Let's start getting familiar with Angular forms in this recipe. In this one, you'll learn about the basic concepts of template-driven forms and will create a basic Angular form using the template-driven forms API.

Getting ready

The project for this recipe resides in `chapter08/start_here/template-driven-forms`:

1. Open the project in Visual Studio Code.

2. Open the terminal and run `npm install` to install the dependencies of the project.

3. Once done, run `ng serve -o`.

    This should open the app in a new browser tab and you should see the following view:

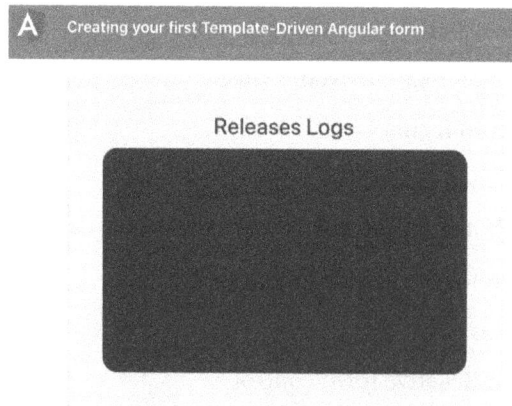

Figure 8.1 – Template-driven forms app running on http://localhost:4200

How to do it...

We have an Angular app that already has a release log component and a bunch of things set up, such as the `ReleaseLog` class under the `src/app/classes` folder. So, in this recipe, we'll create a template-driven form to allow the user to pick an app and submit a release version. Let's get started:

1. First, open the terminal in the project's root and create a component for the release form as follows:

   ```
   ng g c components/release-form
   ```

   The command should create a new component in the `src/app/components` folder named `ReleaseFormComponent`.

2. Add the newly created component to the template of `VersionControlComponent` and modify the `version-control.component.html` file as follows:

   ```html
   <div class="version-control">
     <app-release-form></app-release-form>
     <app-release-logs [logs]="releaseLogs"></app-release-logs>
   </div>
   ```

   Next, let's adjust some styles for the release form to be used within `VersionControlComponent`.

3. Modify the `version-control.component.scss` file as follows:

   ```scss
   :host {
     ...
     min-width: 400px;
     .version-control {
       display: flex;
       justify-content: center;
     }

     app-release-logs,
     app-release-form {
       flex: 1;
     }
   ```

```
app-release-form {
    margin-right: 20px;
  }
}
```

We'll have two inputs in the `ReleaseFormComponent` template. One to select the app we want to release, and the second for the version we want to release.

4. Let's modify the `release-form.component.ts` file to add the `Apps` enum as a local property that we can later use in the template:

```
import { Component, OnInit } from '@angular/core';
import { IReleaseLog } from 'src/app/classes/release-log';
import { Apps } from 'src/app/constants/apps';
...
export class ReleaseFormComponent implements OnInit {
  apps = Object.values(Apps);
  newLog: IReleaseLog = {
    app: Apps.CALENDAR,
    version: '0.0.0'
  };
  constructor() { }

  ngOnInit(): void {
  }
}
```

5. Let's now add the template for our form. Modify the `release-form.component.html` file and add the following code:

```
<form>
  <div class="form-group">
    <label for="appName">Select App</label>
    <select class="form-control" id="appName" required>
      <option value="">--Choose--</option>
      <option *ngFor="let app of apps"
      [value]="app">{{app}}</option>
    </select>
  </div>
```

```
<div class="form-group">
  <label for="versionNumber">Version Number</label>
  <input type="text" class="form-control"
  id="versionNumber" aria-describedby="versionHelp"
  placeholder="Enter version number">
  <small id="versionHelp" class="form-text
  text-muted">Use semantic versioning (x.x.x)</small>
</div>
<button type="submit" class="btn btn-primary">
Submit</button>
</form>
```

6. We now need to integrate the template-driven form. Let's add `FormsModule` to the `app.module.ts` file as follows:

```
...
import { ReleaseFormComponent } from './components/
release-form/release-form.component';
import { FormsModule } from '@angular/forms';

@NgModule({
  declarations: [...],
  imports: [
    BrowserModule,
    AppRoutingModule,
    FormsModule
  ],
  ...
})
export class AppModule { }
```

7. We can now make our form work in the template. Let's modify the `release-form.component.html` file to create a template variable for the form, named `#releaseForm`. We will also use the `[(ngModel)]` binding for both the inputs against appropriate values for the `newLog` property as follows:

```
<form #releaseForm="ngForm">
  <div class="form-group">
    <label for="appName">Select App</label>
```

```html
<select name="app" [(ngModel)]="newLog.app"
class="form-control" id="appName" required>
    <option value="">--Choose--</option>
    <option *ngFor="let app of apps"
    [value]="app">{{app}}</option>
</select>
</div>
<div class="form-group">
    <label for="versionNumber">Version Number</label>
    <input name="version" [(ngModel)]="newLog.version"
    type="text" class="form-control" id="versionNumber"
    aria-describedby="versionHelp" placeholder="Enter
    version number">
    <small id="versionHelp" class="form-text text-
    muted">Use semantic versioning (x.x.x)</small>
</div>
<button type="submit" class="btn btn-primary">
Submit</button>
</form>
```

8. Create a method for when the form will be submitted. Modify the `release-form.component.ts` file to add a new method named `formSubmit`. When this method is called, we'll emit a new instance of `ReleaseLog` using an Angular `@Output` emitter as follows:

```typescript
import { Component, EventEmitter, OnInit, Output } from
'@angular/core';
import { NgForm } from '@angular/forms';
import { IReleaseLog, ReleaseLog } from 'src/app/classes/
release-log';

...

export class ReleaseFormComponent implements OnInit {
  @Output() newReleaseLog = new
  EventEmitter<ReleaseLog>();
  apps = Object.values(Apps);
  ...


  ngOnInit(): void {
```

```
    }

    formSubmit(form: NgForm): void {
      const { app, version } = form.value;
      const newLog: ReleaseLog = new ReleaseLog(app,
      version)
      this.newReleaseLog.emit(newLog);
    }

  }
```

9. Update the template now to use the `formSubmit` method on the form's submission and modify the `release-form.component.html` file as follows:

```
<form  #releaseForm="ngForm"
(ngSubmit)="formSubmit(releaseForm)">

...

</form>
```

10. We now need to modify `VersionControlComponent` to be able to act on the new release log emitted. In order to do so, modify the `version-control.component.html` file to listen to the `newReleaseLog` output event from `ReleaseFormComponent` as follows:

```
<div class="version-control">
  <app-release-form (newReleaseLog)="addNewReleaseLog
  ($event)"></app-release-form>
  <app-release-logs [logs]="releaseLogs"></app-release-
  logs>
</div>
```

11. Cool! Let's create the `addNewReleaseLog` method in the `version-control.component.ts` file and add the `ReleaseLog` received to the `releaseLogs` array. Your code should look as follows:

```
...
export class VersionControlComponent implements OnInit { -
  releaseLogs: ReleaseLog[] = [];
  ...
```

```
addNewReleaseLog(log: ReleaseLog) {
    this.releaseLogs.unshift(log);
}

}
```

Awesome! Within a few minutes, we were able to create our first template-driven form in Angular. If you refresh the app now and try creating some releases, you should see something similar to the following:

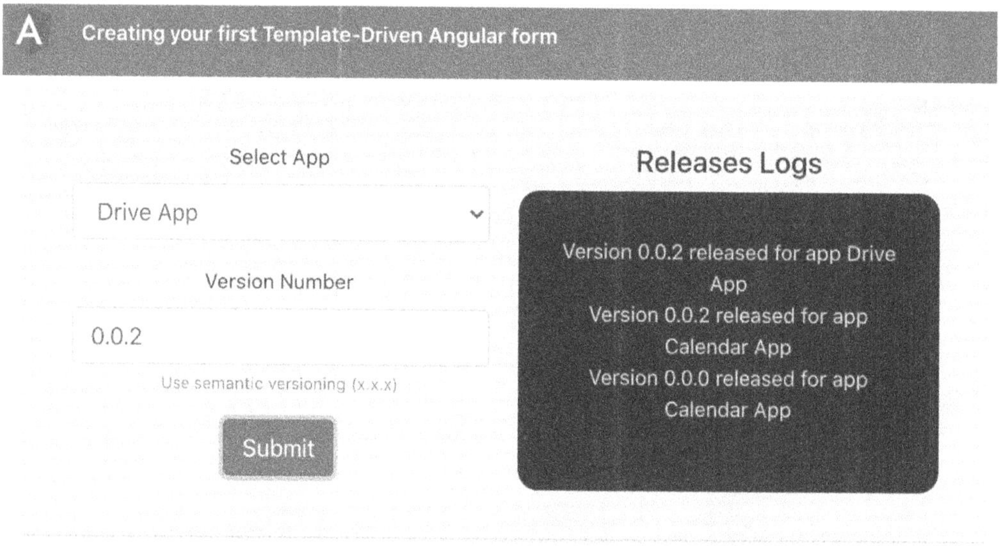

Figure 8.2 – Template-driven forms app final output

Now that you know how the template-driven forms are created, let's see the next section to understand how it works.

How it works...

The key to using template-driven forms in Angular resides in `FormsModule`, the `ngForm` directive, by creating a **template variable** using the `ngForm` directive and using the `[(ngModel)]` two-way data binding along with the `name` attributes for inputs in the template. We began by creating a simple form with some inputs. Then, we added the `FormsModule`, which is necessary for using the `ngForm` directive and the `[(ngModel)]` two-way data binding. Once we added the module, we could use both the directive and the data binding with our newly created local property named `newLog` in the `ReleaseFormComponent`. Notice that it could be an instance of the `ReleaseLog` class, but we kept it as an object of the `IReleaseLog` type instead because we don't want the `ReleaseLog` class's `message` property as we don't use it. With the `[(ngModel)]` usages and the `#releaseForm` template variable in place, we could submit the form using the `ngSubmit` emitter of Angular's `<form>` directive. Notice that we pass the `releaseForm` variable to the `formSubmit` method, which makes it easier to test the functionality for us. Upon submitting the form, we use the form's value to create a new `ReleaseLog` item and we emit it using the `newReleaseLog` output emitter. Notice that if you provide an invalid `version` for the new release log, the app will throw an error and will not create a release log. This is because we validate the version in the `constructor` of the `ReleaseLog` class. Finally, when this `newReleaseLog` event is captured by `VersionControlComponent`, it calls the `addNewReleaseLog` method, which adds our newly created release log to the `releaseLogs` array. And since the `releaseLogs` array is passed as an `@Input()` to `ReleaseLogsComponent`, it immediately shows it right away.

See also

- Building a template-driven form in Angular: `https://angular.io/guide/forms#building-a-template-driven-form`

Form validation with template-driven forms

A great user experience is key to acquiring more users that love to use your applications. And using forms is one of those things that users don't really enjoy. To make sure that users spend the least amount of time filling in forms and are done with them faster, we can implement form validation to make sure that users enter the appropriate data a.s.a.p. In this recipe, we're going to look at how we can implement form validation in template-driven forms.

Getting ready

The project for this recipe resides in `chapter08/start_here/tdf-form-validation`:

1. Open the project in Visual Studio Code.

2. Open the terminal and run `npm install` to install the dependencies of the project.

3. Once done, run `ng serve -o`.

    This should open the app in a new browser tab and you should see the app as follows:

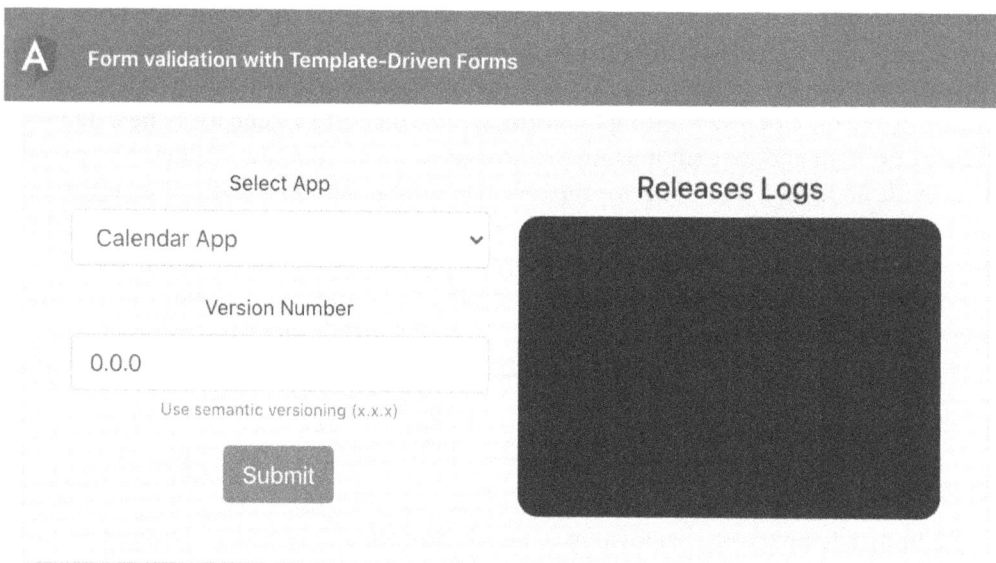

Figure 8.3 – TDF form validation app running on http://localhost:4200

Now that we have the app running locally, let's see the steps involved in this recipe in the next section.

How to do it...

We now have the app from the previous recipe, a simple Angular app with one template-driven form using the ngForm and ngModel directives. The form is used to create release logs. In this recipe, we're going to make this form better in terms of validating the input as the user types the input. Let's get started:

1. First of all, we'll add some validators from the @angular/forms package that are part of the Reactive forms API. We'll apply the **required** validation to both inputs and a **regex** validation on the version input. We need to create template variables for both our inputs. We will name them nameInput and versionInput, respectively. Modify the code in the release-form.component.html file so that it looks as follows:

```
<form  #releaseForm="ngForm"
(ngSubmit)="formSubmit(releaseForm)">
  <div class="form-group">
    <label for="appName">Select App</label>
    <select #nameInput="ngModel" name="app"
    [(ngModel)]="newLog.app" class="form-control"
    id="appName" required>
      <option value="">--Choose--</option>
      <option *ngFor="let app of apps"
      [value]="app">{{app}}</option>
    </select>
  </div>
  <div class="form-group">
    <label for="versionNumber">Version Number</label>
    <input #versionInput="ngModel" name="version"
    [(ngModel)]="newLog.version" type="text"
    class="form-control" id="versionNumber" aria-
    describedby="versionHelp" placeholder="Enter
    version number" required>
    <small id="versionHelp" class="form-text
    text-muted">Use semantic versioning (x.x.x)</small>
  </div>
  <button type="submit" class="btn btn-primary">
  Submit</button>
</form>
```

2. We can now use the template variables to apply validations. Let's start with the name input. In terms of validation, the name input shouldn't be empty and an app should be selected from the select box. Let's show a default Bootstrap alert when the input is invalid. Modify the code in the `release-form.component.html` file. It should look as follows:

```
<form  #releaseForm="ngForm"
(ngSubmit)="formSubmit(releaseForm)">
  <div class="form-group">
    <label for="appName">Select App</label>
    <select #nameInput="ngModel" name="app"
    [(ngModel)]="newLog.app" class="form-control"
    id="appName" required>
      <option value="">--Choose--</option>
      <option *ngFor="let app of apps"
      [value]="app">{{app}}</option>
    </select>
    <div [hidden]="nameInput.valid || nameInput.pristine"
    class="alert alert-danger">
      Please choose an app
    </div>
  </div>
  <div class="form-group">
    . . .
  </div>
  <button type="submit" class="btn btn-primary">Submit
  </button>
</form>
```

3. To validate the version name input, we need to apply the SEMANTIC_VERSION regex from our `src/app/constants/regexes.ts` file. Add the constant as a local property in the `ReleaseFormComponent` class to the `release-form.component.ts` file as follows:

```
. . .
import { Apps } from 'src/app/constants/apps';
import { REGEXES } from 'src/app/constants/regexes';
. . .
export class ReleaseFormComponent implements OnInit {
```

```
      @Output() newReleaseLog = new
      EventEmitter<ReleaseLog>();

      apps = Object.values(Apps);

      versionInputRegex = REGEXES.SEMANTIC_VERSION;

      ...

    }
```

4. Now, use `versionInputRegex` in the template to apply the validation and show the related error as well. Modify the `release-form.component.html` file so that the code looks as follows:

```
<form  #releaseForm="ngForm"
(ngSubmit)="formSubmit(releaseForm)">
  <div class="form-group">
    ...
  </div>
  <div class="form-group">
    <label for="versionNumber">Version Number</label>
    <input #versionInput="ngModel"
    [pattern]="versionInputRegex" name="version"
    [(ngModel)]="newLog.version" type="text"
    class="form-control" id="versionNumber" aria-
    describedby="versionHelp" placeholder="Enter
    version number" required>
    <small id="versionHelp" class="form-text
    text-muted">Use semantic versioning (x.x.x)</small>
    <div
      [hidden]="versionInput.value &&
      (versionInput.valid || versionInput.pristine)"
      class="alert alert-danger"
    >
      Please write an appropriate version number
    </div>
  </div>
  <button type="submit" class="btn btn-primary">
  Submit</button>
</form>
```

5. Refresh the app and try to invalidate both inputs by selecting the first option named
    --Choose-- from the **Select App** drop-down menu and by emptying the version
    input field. You should see the following errors:

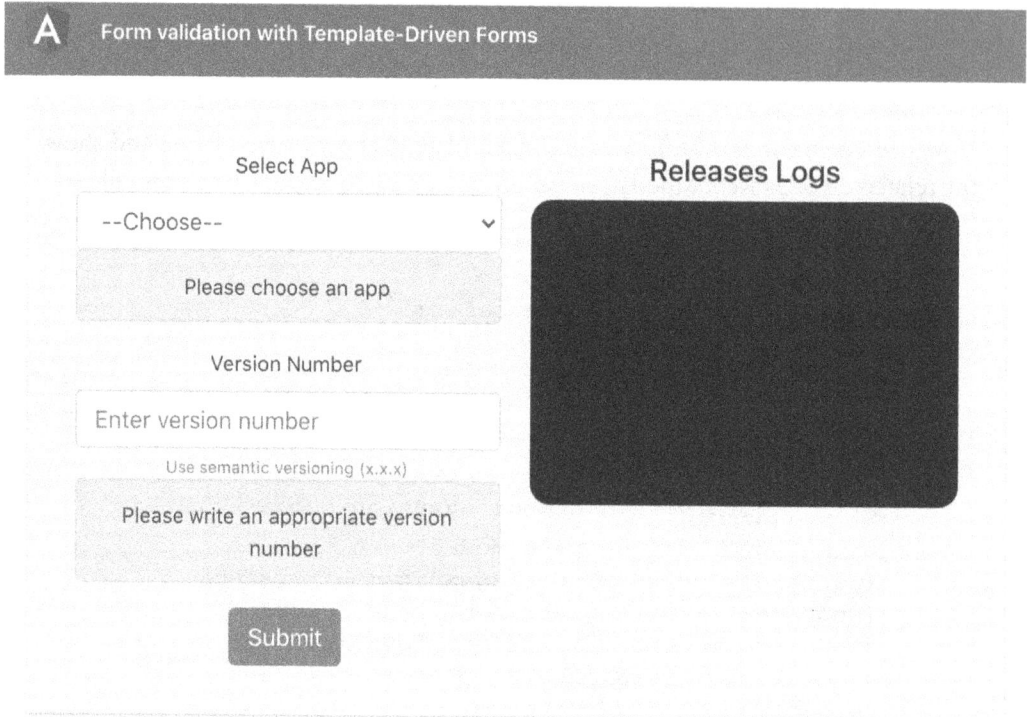

Figure 8.4 – Displaying input errors using ngModel and validation

6. Next, we're going to add some styles to make our inputs more visual when it comes
    to validation. Let's add some styles to the release-form.component.scss file
    as follows:

```
:host {
  /* Error messages */
  .alert {
    margin-top: 16px;
  }

  /* Valid form input */
  .ng-valid[required], .ng-valid.required  {
    border-bottom: 3px solid #259f2b;
```

```
    }

    /* Invalid form input */
    .ng-invalid:not(form)   {
        border-bottom: 3px solid #c92421;
    }
}
```

7. Finally, let's make the validation around the form submission. We'll disable the **Submit** button if the inputs do not have valid values. Let's modify the template in `release-form.component.html` as follows:

```
<form #releaseForm="ngForm"
(ngSubmit)="formSubmit(releaseForm)">
    <div class="form-group">

        . . .

    </div>
    <div class="form-group">

        . . .

    </div>
    <button type="submit" [disabled]="releaseForm.invalid"
    class="btn btn-primary">Submit</button>
</form>
```

If you refresh the app now, you'll see that the submit button is disabled whenever one or more inputs are invalid.

Great! You just learned how to validate template-driven forms and to make the overall user experience with template-driven forms slightly better.

How it works...

The core components of this recipe were the ngForm and ngModel directives. We could easily identify whether the submit button should be clickable (not disabled) or not based on whether the form is valid, that is, if all the inputs in the form have valid values. Note that we used the template variable created using the #releaseForm="ngForm" syntax on the <form> element. This is possible due to the ability of the ngForm directive to be exported into a template variable. Therefore, we were able to use the releaseForm. invalid property in the [disabled] binding of the submit button to conditionally disable it. We also showed the errors on individual inputs based on the condition that the input might be invalid. In this case, we show the Bootstrap alert element (a <div> with the CSS class alert). We also use Angular's provided classes, ng-valid and ng-invalid, on the form inputs to highlight the input in a certain way depending on the validity of the input's value. What's interesting about this recipe is that we validated the app name's input by making sure it contains a non-falsy value where the first <option> of the <select> box has the value "". And what's even more fun is that we also validated the version name right when the user types it using the [pattern] binding on the input to a regex. Otherwise, we'd have to wait for the user to submit the form, and then it would have been validated. Thus, we're providing a great user experience by providing the errors as the user types the version.

See also

- Show and hide validation error messages (Angular Docs): https://angular.io/guide/forms#show-and-hide-validation-error-messages
- NgForm docs: https://angular.io/api/forms/NgForm

Testing template-driven forms

To make sure we build robust and bug-free forms for end users, it is a really good idea to have tests relating to your forms. It makes the code more resilient and less prone to errors. In this recipe, you'll learn how to test your template-driven forms using unit tests.

Getting ready

The project for this recipe resides in chapter08/start_here/testing-td-forms:

1. Open the project in Visual Studio Code.
2. Open the terminal and run npm install to install the dependencies of the project.

3. Once done, run `ng serve -o`.

    This should open the app in a new browser tab and you should see the app as follows:

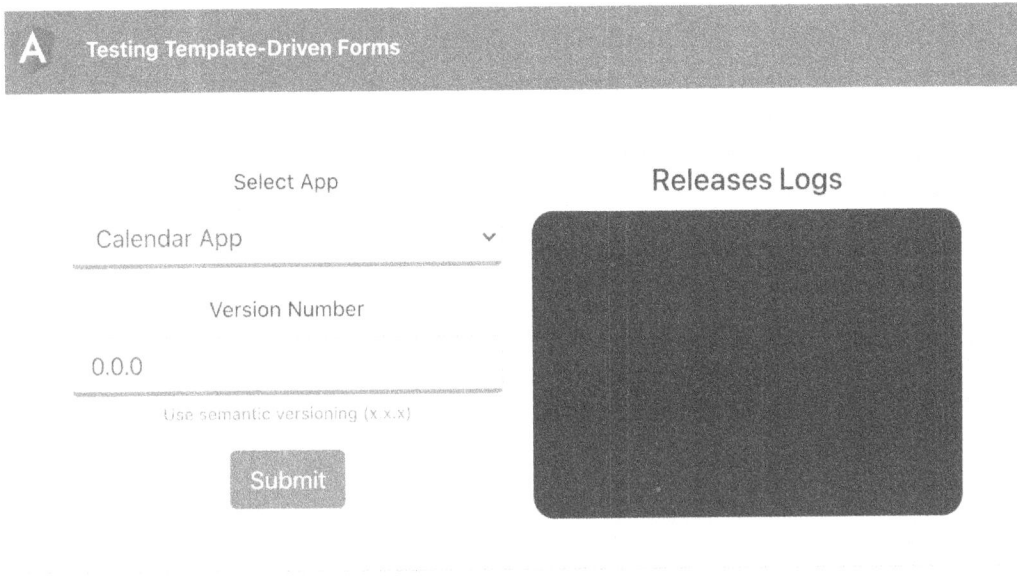

Testing Template-Driven Forms

Select App

Calendar App

Version Number

0.0.0

Use semantic versioning (x.x.x)

Submit

Releases Logs

Figure 8.5 – The Testing Template-Driven Forms app running on http://localhost:4200

Now that we have the app running locally, let's see the steps involved in this recipe in the next section.

How to do it...

We have the app from the previous recipe that contains a template-driven form used to create release logs. The form also has validations applied to the inputs. Let's start looking into how we can test this form:

1. First of all, run the following command to run the unit tests:

```
npm run test
```

Once the command is run, you should see a new instance of the Chrome window being opened that runs the unit tests. One test out of the six we have has failed. You will probably see something like the following in the automated Chrome window:

Karma v 6.3.2 - connected; test: complete;

Chrome 90.0.4430.212 (Mac OS 10.15.7) is idle

✳ Jasmine   3.6.0

● ● ● ● ● ✕

6 specs, 1 failure, randomized with seed 81739

Spec List | Failures

```
ReleaseFormComponent > should create

Error: NG0301: Export of name 'ngForm' not found!. Find more at https://angular.io/errors/NG0301
error properties: Object({ code: '301' })
Error: NG0301: Export of name 'ngForm' not found!. Find more at https://angular.io/errors/NG0301
    at cacheMatchingLocalNames (http://localhost:9876/_karma_webpack_/webpack:/node_modules/@angular/core
    at resolveDirectives (http://localhost:9876/_karma_webpack_/webpack:/node_modules/@angular/core/__ivy
    at elementStartFirstCreatePass (http://localhost:9876/_karma_webpack_/webpack:/node_modules/@angular/
    at ɵɵelementStart (http://localhost:9876/_karma_webpack_/webpack:/node_modules/@angular/core/__ivy_ng
    at ReleaseFormComponent_Template (ng:///ReleaseFormComponent.js:29:9)
    at executeTemplate (http://localhost:9876/_karma_webpack_/webpack:/node_modules/@angular/core/__ivy_n
    at renderView (http://localhost:9876/_karma_webpack_/webpack:/node_modules/@angular/core/__ivy_ngcc__
    at renderComponent (http://localhost:9876/_karma_webpack_/webpack:/node_modules/@angular/core/__ivy_n
    at renderChildComponents (http://localhost:9876/_karma_webpack_/webpack:/node_modules/@angular/core/_
    at renderView (http://localhost:9876/_karma_webpack_/webpack:/node_modules/@angular/core/__ivy_ngcc__
```

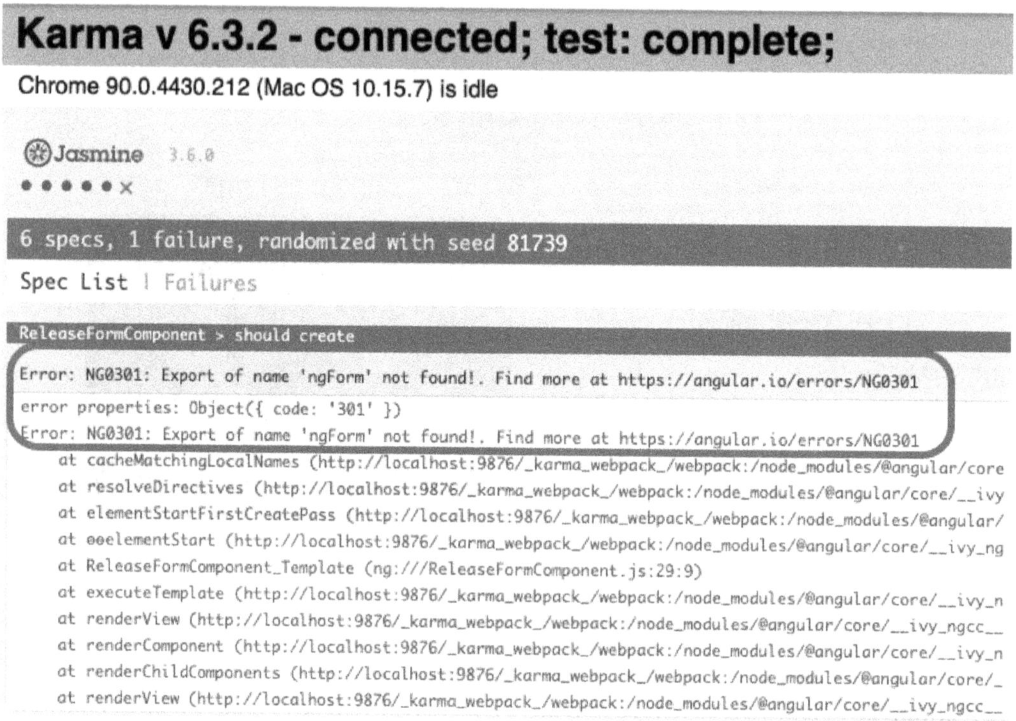

Figure 8.6 – Unit tests with Karma and Jasmine running in an automated Chrome window

2. The `ReleaseFormComponent > should create` test is failing because we don't have `FormsModule` added to the tests. Notice the `Export of name 'ngForm' not found` error. Let's import `FormsModule` into the testing module's configuration in `release-form.component.spec.ts` as follows:

```
import { ComponentFixture, TestBed } from '@angular/core/
testing';
import { FormsModule } from '@angular/forms';
import { ReleaseFormComponent } from './release-form.
component';

describe('ReleaseFormComponent', () => {
    ...
    beforeEach(async () => {
```

```
    await TestBed.configureTestingModule({
      declarations: [ ReleaseFormComponent ],
      imports: [ FormsModule ]
    })
      .compileComponents();
  });
  ...
  it('should create', () => {
    expect(component).toBeTruthy();
  });
});
```

If you look at the tests now, you should see all the tests passing as follows:

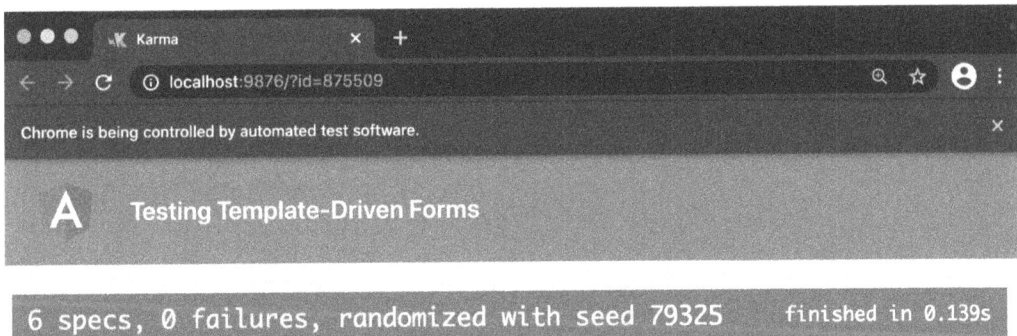

Figure 8.7 – All tests pass after importing FormsModule into the appropriate test

To test the form correctly, we'll add a couple of tests, one for successful input and one for each invalid input. For that, we need to access the form in our component since we're writing unit tests.

3. Let's access the `#releaseForm` in our component class using the `@ViewChild()` decorator in the `release-form.component.ts` file as follows:

```
import { Component, EventEmitter, OnInit, Output,
ViewChild } from '@angular/core';

. . .


@Component({
  selector: 'app-release-form',
  templateUrl: './release-form.component.html',
  styleUrls: ['./release-form.component.scss']
})
export class ReleaseFormComponent implements OnInit {
  @Output() newReleaseLog = new
  EventEmitter<ReleaseLog>();
  @ViewChild('releaseForm') releaseForm: NgForm;
  apps = Object.values(Apps);
  versionInputRegex = REGEXES.SEMANTIC_VERSION;
  . . .
}
```

4. Let's add a new test now. We'll write a test that should validate the case for when both the inputs have valid values. Add the test to the `release-form.component.spec.ts` file as follows:

```
import { ComponentFixture, TestBed, fakeAsync } from '@
angular/core/testing';
import { ReleaseFormComponent } from './release-form.
component';

describe('ReleaseFormComponent', () => {
  . . .
  it('should create', () => {
    expect(component).toBeTruthy();
  });


  it('should submit a new release log with the correct
  input values', fakeAsync( () => {
```

```
        expect(true).toBeFalsy();
    }));
});
```

5. The new test is failing so far. Let's try to fill the values in the form, submit the button, and make sure that our @Output emitter named newReleaseLog emits the correct value from releaseForm. The content of the test should look as follows:

```
...
import { ReleaseLog } from 'src/app/classes/release-log';
...
it('should submit a new release log with the correct
input values', fakeAsync(async () => {
    const submitButton = fixture.nativeElement.
    querySelector('button[type="submit"]');
    const CALENDAR_APP = component.apps[2];
    spyOn(component.newReleaseLog, 'emit');
    await fixture.whenStable(); // wait for Angular
    to configure the form
    component.releaseForm.controls[
    'version'].setValue('2.2.2');
    component.releaseForm.controls[
    'app'].setValue(CALENDAR_APP);
    submitButton.click();
    const expectedReleaseLog = new ReleaseLog(CALENDAR_
    APP, '2.2.2');
    expect(component.newReleaseLog.emit)
    .toHaveBeenCalledWith(expectedReleaseLog);
}));
```

When you save the file, you should see the new test passing with the expected values. It should appear as follows in the Chrome tab:

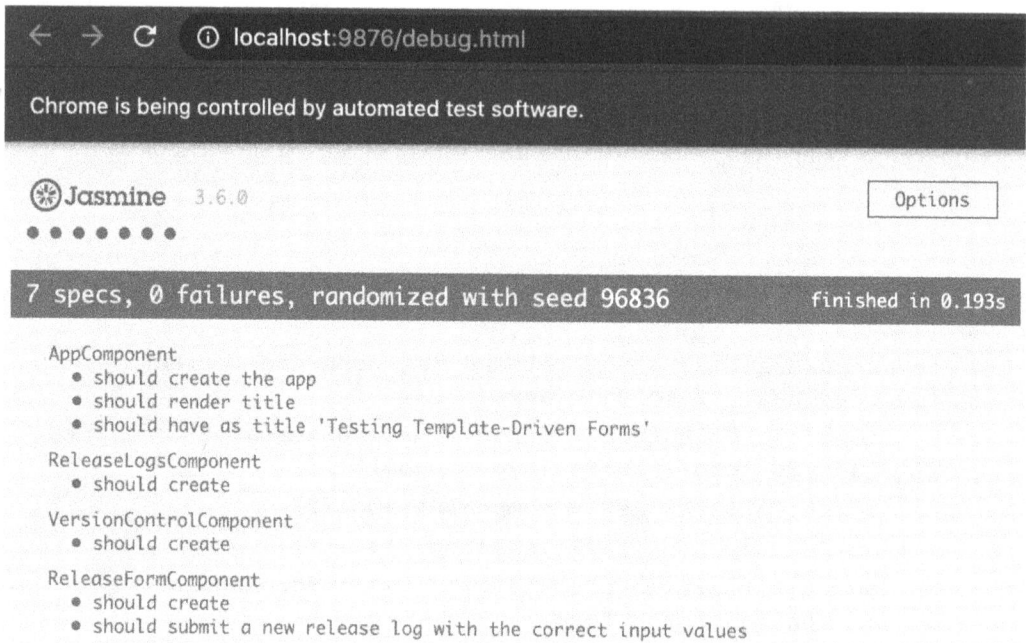

Figure 8.8 – New test for successful form submission passing

6. Let's add a test for the case when we have an incorrect version provided in the form. The submit button should be disabled and we should have an error thrown by the `formSubmit` method. Add a new test to your `release-form.component.spec.ts` file as follows:

```
...
describe('ReleaseFormComponent', () => {
    ...
    it('should submit a new release log with the correct
    input values', fakeAsync(async () => {
        const submitButton = fixture.nativeElement.
        querySelector('button[type="submit"]');

        const CALENDAR_APP = component.apps[2];

        spyOn(component.newReleaseLog, 'emit');

        await fixture.whenStable(); // wait for Angular
        to configure the form

        const expectedError = 'Invalid version provided.
```

```
          Please provide a valid version as
          (major.minor.patch)';
          component.releaseForm.controls[
          'version'].setValue('x.x.x');
          component.releaseForm.controls[
          'app'].setValue(CALENDAR_APP);
          expect(() => component.formSubmit(component.
          releaseForm))
            .toThrowError(expectedError);
          fixture.detectChanges();
          expect(submitButton.hasAttribute(
          'disabled')).toBe(true);
          expect(component.newReleaseLog.emit)
          .not.toHaveBeenCalled();
        }));
    });
```

7. Let's add our final test, which makes sure that the submit button is disabled when
    we have not selected an app for the release log. Add a new test to the `release-`
    `form.component.spec.ts` file as follows:

```
    ...
    describe('ReleaseFormComponent', () => {

      ...

      it('should disable the submit button when we
      don\'t have an app selected', fakeAsync(async () => {
        const submitButton = fixture.nativeElement.
        querySelector('button[type="submit"]');
        spyOn(component.newReleaseLog, 'emit');
        await fixture.whenStable(); // wait for Angular
        to configure the form
        component.releaseForm.controls[
        'version'].setValue('2.2.2');
        component.releaseForm.controls[
        'app'].setValue(null);
        fixture.detectChanges();
        expect(submitButton.hasAttribute(
        'disabled')).toBe(true);
```

```
      expect(component.newReleaseLog.emit
      ).not.toHaveBeenCalled();
   }));
});
```

If you look at the Karma tests window, you should see all the new tests passing as follows:

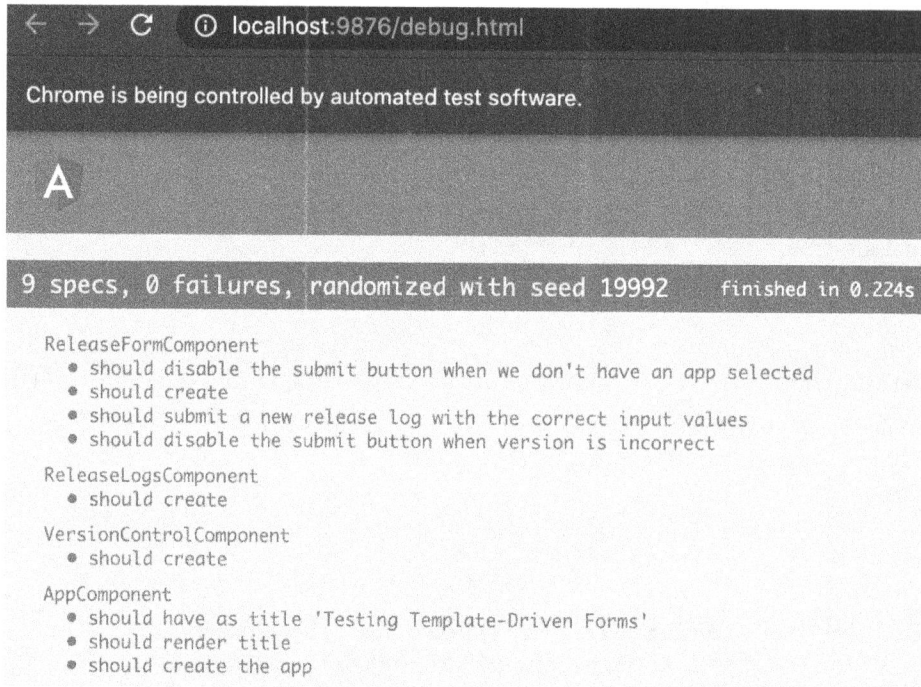

Figure 8.9 – All tests passing for the recipe

Awesome! You now know a bunch of techniques for testing your template-driven forms. Some of these techniques might still require some explanation. See the next section to understand how it all works.

How it works...

Testing template-driven forms can be a bit of a challenge as it depends on how complex the form is, what use cases you want to test, and how complex those use cases are. In our recipe, the first thing we did was to include `FormsModule` in the imports of the test file for `ReleaseFormComponent`. This makes sure that the tests know the `ngForm` directive and do not throw relevant errors. For the test with all the successful inputs, we spied on the `newReleaseLog` emitter's `emit` event defined in the `ReleaseFormComponent` class. This is because we know that when the inputs are correct, the user should be able to click the submit button, and as a result, inside the `formSubmit` method, the `emit` method of the `newReleaseLog` emitter will be called. Note that we're using `fixture.whenStable()` in each of our tests. This is to make sure that Angular has done the compilation and our `ngForm`, named `#releaseForm`, is ready. For the `should disable the submit button when version is incorrect` test, we rely on `formSubmit` to throw an error. This is because we know that an invalid version will cause an error in the `constructor` of the `ReleaseLog` class when creating a new release log. One interesting thing in this test is that we use the following code:

```
expect(() => component.formSubmit(component.releaseForm))
        .toThrowError(expectedError);
```

The interesting thing here is that we needed to call the `formSubmit` method ourselves with `releaseForm`. We couldn't just do it by writing `expect(component.formSubmit(component.releaseForm)).toThrowError(expectedError);` because that would rather call the function directly there and would result in the error. So, we need to pass an anonymous function here that Jasmine will call and would expect this anonymous function to throw an error. And finally, we make sure that our submit button is enabled or disabled by first getting the button using a `querySelector` on `fixture.nativeElement`. We then check the `disabled` attribute on the submit button using `submitButton.hasAttribute('disabled')`.

See also

- Testing template-driven forms: `https://angular.io/guide/forms-overview#testing-template-driven-forms`

Creating your first Reactive form

You've learned about template-driven forms in the previous recipes and are now confident in building Angular apps with them. Now guess what? Reactive forms are even better. Many known engineers and businesses in the Angular community recommend using Reactive forms. The reason is their ease of use when it comes to building complex forms. In this recipe, you'll build your first Reactive form and will learn its basic usage.

Getting ready

The project for this recipe resides in `chapter08/start_here/reactive-forms`:

1. Open the project in Visual Studio Code.

2. Open the terminal and run `npm install` to install the dependencies of the project.

3. Once done, run `ng serve -o`.

4. Click on the name of the first user and you should see the following view:

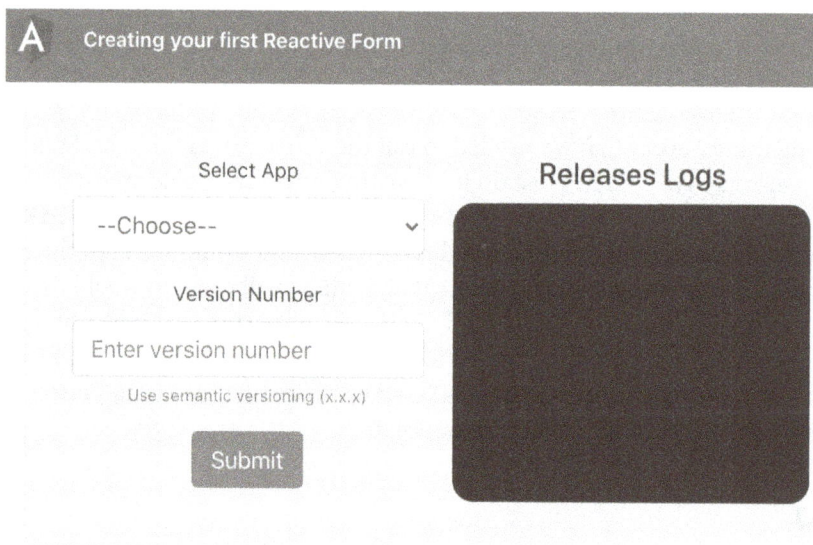

Figure 8.10 – The Reactive Form app running on http://localhost:4200

Now that we have the app running locally, let's see the steps involved in this recipe in the next section.

How to do it...

So far, we have an app that has `ReleaseLogsComponent`, which shows a bunch of release logs that we create. We also have `ReleaseFormComponent`, which has a form by means of which the release logs will be created. We now have to make our current form a Reactive form using the Reactive forms API. Let's get started:

1. First of all, we need to import `ReactiveFormsModule` into the imports of our `AppModule`. Let's do it by modifying the `app.module.ts` file as follows:

```
...
import { ReleaseFormComponent } from './components/
release-form/release-form.component';
import { ReactiveFormsModule } from '@angular/forms';

@NgModule({
  declarations: [...],
  imports: [
    BrowserModule,
    AppRoutingModule,
    ReactiveFormsModule
  ],
  providers: [],
  bootstrap: [AppComponent]
})
export class AppModule { }
```

2. Let's create the Reactive form now. We'll create a `FormGroup` in our `ReleaseFormComponent` class with the required controls. Modify the `release-form.component.ts` file as follows:

```
...
import { FormControl, FormGroup, Validators } from '@
angular/forms';
import { REGEXES } from 'src/app/constants/regexes';

@Component(...)
export class ReleaseFormComponent implements OnInit {
  apps = Object.values(Apps);
  versionInputRegex = REGEXES.SEMANTIC_VERSION;
```

```
releaseForm = new FormGroup({
  app: new FormControl('', [Validators.required]),
  version: new FormControl('', [
    Validators.required,
    Validators.pattern(REGEXES.SEMANTIC_VERSION)
  ]),
})
...
}
```

3. Now that we have the form named `releaseForm` in place, let's bind it to the form by using it in the template. Modify the `release-form.component.html` file as follows:

```
<form [formGroup]="releaseForm">
  ...
</form>
```

4. Great! Now that we have the form group bound, we can also bind the individual form controls so that when we finally submit the form, we can get the value out for each individual form control. Modify the `release-form.component.html` file further as follows:

```
<form [formGroup]="releaseForm">
  <div class="form-group">
    ...
    <select formControlName="app" class="form-control"
    id="appName" required>
      ...
    </select>
  </div>
  <div class="form-group">
    ...
    <input formControlName="version" type="text"
    class="form-control" id="versionNumber" aria-
    describedby="versionHelp" placeholder="Enter
    version number">
    <small id="versionHelp" class="form-text
    text-muted">Use semantic versioning (x.x.x)</small>
```

```
    </div>
    . . .
  </form>
```

5. Let's decide what will happen when we submit this form. We'll call a method named `formSubmit` in the template and pass `releaseForm` in it when the form is submitted. Modify the `release-form.component.html` file as follows:

```
<form [formGroup]="releaseForm"
(ngSubmit)="formSubmit(releaseForm)">

    . . .
</form>
```

6. The `formSubmit` method doesn't yet exist. Let's create it now in the `ReleaseFormComponent` class. We'll also log the value on the console and emit the value using an `@Output` emitter. Modify the `release-form.component.ts` file as follows:

```
import { Component, OnInit, Output, EventEmitter } from
'@angular/core';

. . .

import { ReleaseLog } from 'src/app/classes/release-log';

. . .


@Component(...)
export class ReleaseFormComponent implements OnInit {
  @Output() newReleaseLog = new
  EventEmitter<ReleaseLog>();
  apps = Object.values(Apps);
  . . .


  formSubmit(form: FormGroup): void {
    const { app, version } = form.value;
    console.log({app, version});
    const newLog: ReleaseLog = new ReleaseLog(app,
    version)
    this.newReleaseLog.emit(newLog);
  }
}
```

If you refresh the app now, complete the form, and hit **Submit**, you should see a log on the console as follows:

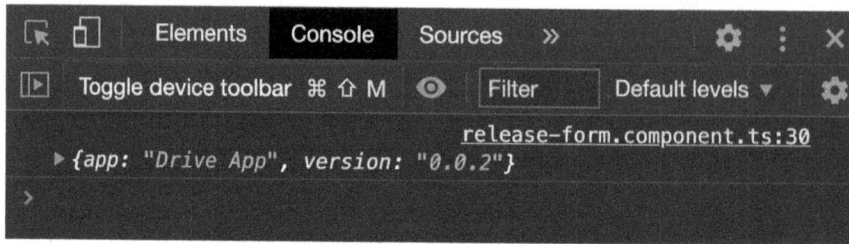

Figure 8.11 – Log displaying the values submitted using the Reactive form

7. Since we've emitted the value of the newly created release log via the `newReleaseLog` output emitter, we can listen to this event in the `version-control.component.html` file and add the new log accordingly. Let's modify the file as follows:

```
<div class="version-control">
    <app-release-form (newReleaseLog)="addNewReleaseLog
    ($event)"></app-release-form>
    <app-release-logs [logs]="releaseLogs">
    </app-release-logs>
</div>
```

8. Refresh the app and you should see the new release log being added to the release logs view. You should see the logs on the console as well, as shown in the following screenshot:

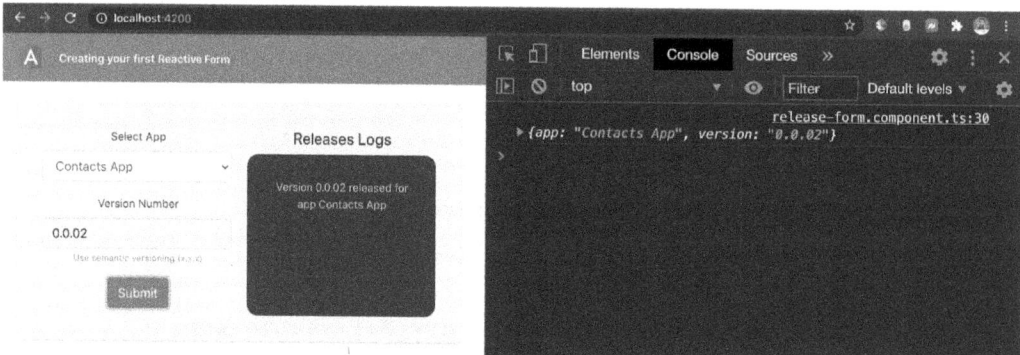

Figure 8.12 – New logs being added to the logs view on form submission

Awesome! So, now you know how to create a basic Reactive form using the Reactive forms API. Refer to the next section to understand how it all works.

How it works...

The recipe begins with having a basic HTML form in our Angular app with no Angular magic bound to it. We first started importing `ReactiveFormsModule` in the `AppModule`. If you're using the Angular Language Service with the editor of your choice, you might see an error as you import `ReactiveFormsModule` into the app and don't bind it with a Reactive form, in other words, with a `FormGroup`. Well, that's what we did. We created a reactive form using the `FormGroup` constructor and created the relevant form controls using the `FormControl` constructor. We then listened to the `ngSubmit` event on the `<form>` element to extract the value of `releaseForm`. Once done, we emitted this value using the `@Ouput()` named `newReleaseLog`. Notice that we also defined the type of the value that this emitter will emit as `IReleaseLog`; it is good practice to define those. This emitter was required because `ReleaseLogsComponent` is a sibling of `ReleaseFormComponent` in the component's hierarchy. Therefore, we're communicating through the parent component, `VersionControlComponent`. Finally, we listen to the `newReleaseLog` event's emission in the `VersionControlComponent` template and add a new log to the `releaseLogs` array via the `addNewReleaseLog` method. And this `releaseLogs` array is being passed to `ReleaseLogsComponent`, which displays all the logs as they're added.

See also

- Angular's guide to Reactive forms: `https://angular.io/guide/reactive-forms`

Form validation with Reactive forms

In the previous recipe, you learned how to create a Reactive form. Now, we're going to learn how to test them. In this recipe, you'll learn some basic principles of testing Reactive forms. We're going to use the same example from the previous recipe (the release logs app) and will implement a number of test cases.

Getting ready

The project that we are going to work with resides in `chapter08/start_here/validating-reactive-forms` inside the cloned repository:

1. Open the project in Visual Studio Code.

2. Open the terminal and run `npm install` to install the dependencies of the project.

3. Once done, run `ng serve -o`.

   This should open the app in a new browser tab and you should see it as follows:

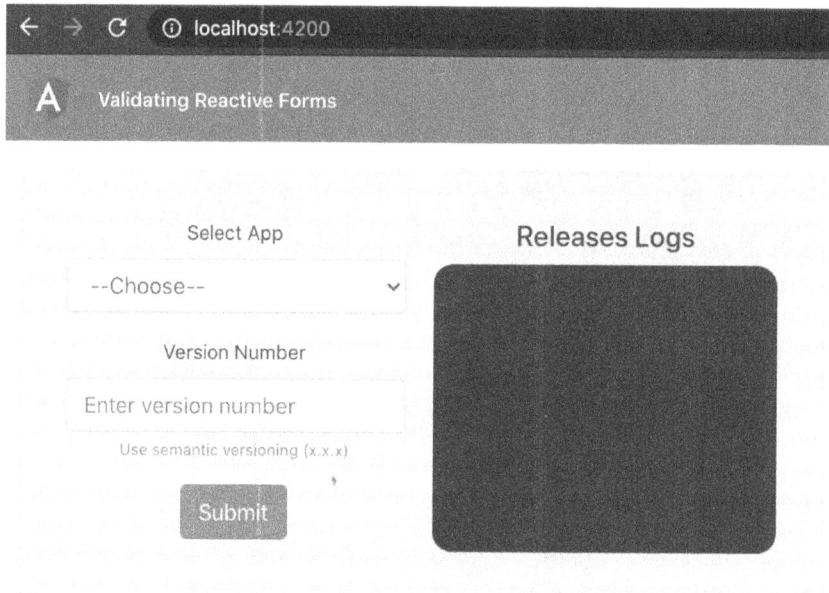

Figure 8.13 – The Validating Reactive Forms app running on http://localhost:4200

Now that we have the app running locally, let's see the steps involved in this recipe in the next section.

How to do it...

For this recipe, we're using the Release Logs application that has the Reactive form implemented already, although we don't have any sort of validation on the inputs so far. If you just select an app and submit the form, you'll see an error on the console as follows:

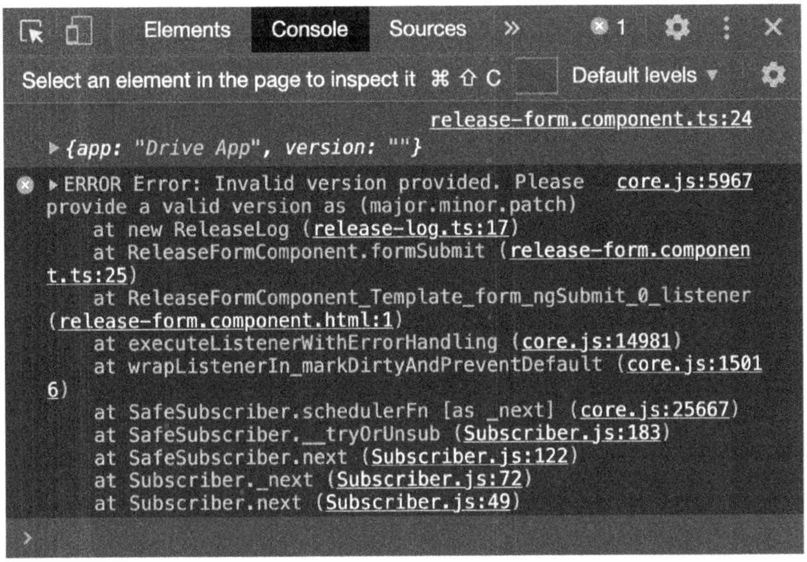

Figure 8.14 – Error when submitting the Reactive form app without form validations

We're going to incorporate some form validations to enhance the user experience and to make sure that the form can't be submitted with invalid input. Let's begin:

1. We'll first add some validations from the `@angular/forms` package, which are part of the Reactive Forms API. We'll apply the `required` validator on both inputs and the `pattern` validator on the `version` form control. Update the `release-form.component.ts` file as follows:

    ```
    import { Component, OnInit, Output, EventEmitter } from
    '@angular/core';
    import { FormControl, FormGroup, Validators } from '@
    angular/forms';
    ...
    import { REGEXES } from 'src/app/constants/regexes';
    ```

```
@Component({...})
export class ReleaseFormComponent implements OnInit {
  ...
  versionInputRegex = REGEXES.SEMANTIC_VERSION;
  releaseForm = new FormGroup({
    app: new FormControl('', Validators.required),
    version: new FormControl('', [
      Validators.required,
      Validators.pattern(this.versionInputRegex)
    ]),
  })
  ...
}
```

2. Now we'll add the hints to the view to show the user errors when an invalid input is selected. Modify the `release-form.component.html` file as follows:

```html
<form [formGroup]="releaseForm"
(ngSubmit)="formSubmit(releaseForm)">
  <div class="form-group">
    <label for="appName">Select App</label>
    <select formControlName="app" class="form-control"
    id="appName">

      ...
    </select>
    <div
      [hidden]="releaseForm.get('app').valid ||
      releaseForm.get('app').pristine"
      class="alert alert-danger">
      Please choose an app
    </div>
  </div>
  <div class="form-group">
    ...
    <small id="versionHelp" class="form-text
    text-muted">Use semantic versioning (x.x.x)</small>
```

```
        <div [hidden]="releaseForm.get('version').valid ||
        releaseForm.get('version').pristine"
           class="alert alert-danger">
           Please write an appropriate version number
        </div>
      </div>
      <button type="submit" class="btn btn-primary">Submit
      </button>
    </form>
```

3. We'll also add some styles to show the errors with a better UI. Add the following styles to the release-form.component.scss file:

```scss
:host {
  /* Error messages */
  .alert {
    margin-top: 16px;
  }

  /* Valid form input */
  .ng-valid:not(form),
  .ng-valid.required {
    border-bottom: 3px solid #259f2b;
  }

  /* Invalid form input */
  .ng-invalid:not(form) {
    border-bottom: 3px solid #c92421;
  }
}
```

Refresh the app and you should see the inputs with red borders when the input values are wrong. The errors once you enter or select an invalid input will look as follows:

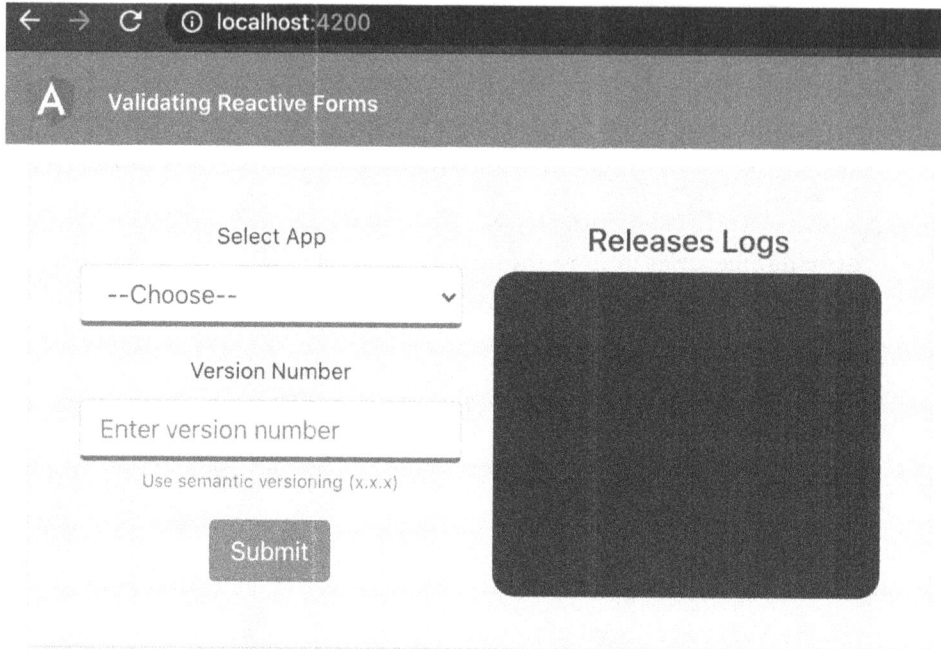

Figure 8.15 – Red borders shown on invalid input values

4. Finally, let's make the validation around the form submission. We'll disable the **Submit** button if the inputs do not have valid values. Let's modify the template in `release-form.component.html` as follows:

```html
<form [formGroup]="releaseForm"
(ngSubmit)="formSubmit(releaseForm)">
  <div class="form-group">
    ...
  </div>
  <div class="form-group">
    ...
  </div>
  <button type="submit" [disabled]="releaseForm.invalid"
  class="btn btn-primary">Submit</button>
</form>
```

If you refresh the app now, you'll see that the submit button is disabled whenever one or more inputs are invalid.

And that concludes the recipe. Let's look at the next section to see how it works.

How it works...

We started the recipe by adding the validators, and Angular has got a bunch of validators out of the box, including `Validators.email`, `Validators.pattern`, and `Validators.required`. We used the `required` validator with the `pattern` validator in our recipe for the inputs for the app name and the version, respectively. After that, to show the hints/errors for invalid inputs, we added some conditional styles to show a border-bottom on the inputs. We also added some `<div>` elements with `class="alert alert-danger"`, which are basically Bootstrap alerts to show the errors on invalid values for the form controls. Notice that we're using the following pattern to hide the error elements:

```
[hidden]="releaseForm.get(CONTROL_NAME).valid || releaseForm.
get(CONTROL_NAME).pristine"
```

We're using the condition with `.pristine` to make sure that as soon as the user selects the correct input and the input is modified, we hide the error again so that it doesn't show while the user is typing in the input or making another selection. Finally, we made sure that the form cannot even be submitted if the values of the form controls are invalid. We disabled the submit button using `[disabled]="releaseForm.invalid"`.

See also

* Angular docs for validating Reactive forms: `https://angular.io/guide/reactive-forms#validating-form-input`

Creating an asynchronous validator function

Form validations are pretty straightforward in Angular, the reason being the super-awesome validators that Angular provides out of the box. These validators are synchronous, meaning that as soon as you change the input, the validators kick in and provide you with information about the validity of the values right away. But sometimes, you might rely on some validations from a backend API, for instance. These situations would require something called asynchronous validators. In this recipe, you're going to create your first asynchronous validator.

Getting ready

The project that we are going to work with resides in `chapter08/start_here/` `asynchronous-validator` inside the cloned repository:

1. Open the project in Visual Studio Code.

2. Open the terminal and run `npm install` to install the dependencies of the project.

3. Once done, run `ng serve -o`.

   This should open the app in a new browser tab and you should see something like the following:

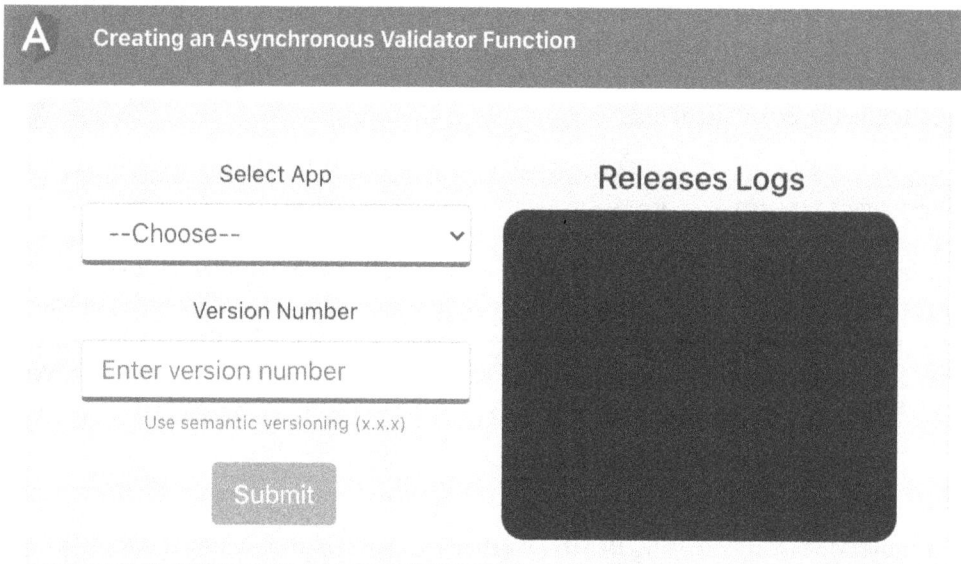

Figure 8.16 – Asynchronous validators app running on http://localhost:4200

Now that we have the app running, let's see the steps involved in this recipe in the next section.

How to do it...

We already have some things set up in the release logs app. We have a `data.json` file residing in the `src/assets` folder that holds the versions for each of our target apps for the release logs. We'll create an async validator to make sure that the new releases for each of the applications have a greater version than that specified in the `data.json` file. Let's begin:

1. First, we're going to create the async validator function for the recipe. Let's create a method named `versionValidator` in the `VersionService` class in the `version.service.ts` file as follows:

```
...
import { compareVersion } from 'src/app/utils';
import { AbstractControl, AsyncValidatorFn,
ValidationErrors } from '@angular/forms';
import { Observable, of } from 'rxjs';


@Injectable({...})
export class VersionService {
  ...
  versionValidator(appNameControl: AbstractControl):
  AsyncValidatorFn {
    // code here
  }
  ...
}
```

2. We'll now define the content of the validator function. Let's modify the `versionValidator` method as follows:

```
versionValidator(appNameControl: AbstractControl):
AsyncValidatorFn {
  return (control: AbstractControl):
  Observable<ValidationErrors> => {
  // if we don't have an app selected, do not validate
  if (!appNameControl.value) {
    return of(null);
  }
  return this.getVersionLog().pipe(
```

```
map(vLog => {
    const newVersion = control.value;
    const previousVersion = vLog[appNameControl.value];
    // check if the new version is greater than
        previous version
    return compareVersion(newVersion, previousVersion)
    === 1 ? null : {
        newVersionRequired: previousVersion
    };
    }))
}
}
```

3. Now that we have the validator function in place, let's add that to the form control for the version number. Let's modify the `release-form.component.ts` file as follows:

```
import { Component, OnInit, Output, EventEmitter } from
'@angular/core';
import { FormControl, FormGroup, Validators } from '@
angular/forms';
import { IReleaseLog, ReleaseLog } from 'src/app/classes/
release-log';
import { Apps } from 'src/app/constants/apps';
import { REGEXES } from 'src/app/constants/regexes';
import { VersionService } from 'src/app/core/services/
version.service';

@Component({...})
export class ReleaseFormComponent implements OnInit {
    ...
    constructor(private versionService: VersionService) { }
    ngOnInit(): void {
        this.releaseForm.get('version')
        .setAsyncValidators(
            this.versionService.versionValidator(
                this.releaseForm.get('app')
            )
        )
```

```
        }
      ...
    }
```

4. We will now use the validator to enhance the user's experience of the form by modifying the `release-form.component.html` file. For ease of usage, let's wrap the content inside an `<ng-container>` element using the `*ngIf` directive, and create a variable within the template for the version form control as follows:

```
<form [formGroup]="releaseForm"
(ngSubmit)="formSubmit(releaseForm)">
  <ng-container *ngIf="releaseForm.get('version')
  as versionControl">
    <div class="form-group">
      ...
    </div>
    <div class="form-group">
      ...
    </div>
    <button type="submit" [disabled]="releaseForm.
    invalid" class="btn btn-primary">Submit</button>
  </ng-container>
</form>
```

5. Let's now add the error message. We'll use our custom error, `newVersionRequired`, from the validator function to show the error when the specified version isn't newer than the previous version. Modify the `release-form.component.html` file as follows:

```
<form [formGroup]="releaseForm"
(ngSubmit)="formSubmit(releaseForm)">
  <ng-container *ngIf="releaseForm.get('version')
  as versionControl">
    <div class="form-group">
      ...
    </div>
    <div class="form-group">
      <label for="versionNumber">Version Number</label>
      <input formControlName="version" type="text"
      class="form-control" id="versionNumber"
```

```
        aria-describedby="versionHelp" placeholder="Enter
        version number">

        ...
        <div *ngIf="(versionControl.
        getError('newVersionRequired') &&
        !versionControl.pristine)"
          class="alert alert-danger">
          The version number should be greater
          than the last version '{{versionControl.
          errors['newVersionRequired']}}'
        </div>
      </div>
      <button [disabled]="releaseForm.invalid"
      class="btn btn-primary">Submit</button>
    </ng-container>
  </form>
```

Try to select an app and add a lower version number and you should now see the error as follows:

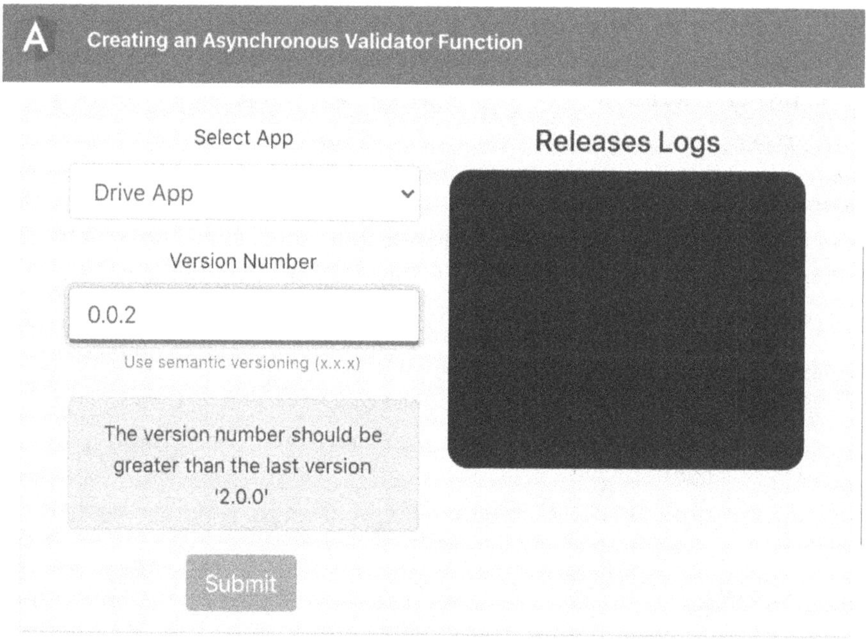

Figure 8.17 – Error being shown when a lower version number is provided

6. One issue right now is that we are able to submit the form while the asynchronous validation is in progress. That's because Angular, by default, marks the error as `null` until the validation is done. To tackle this, we can show a loading message instead of the **submit** button in the template. Modify the `release-form. component.html` file as follows:

```
<form [formGroup]="releaseForm"
(ngSubmit)="formSubmit(releaseForm)">

  <ng-container *ngIf="releaseForm.get('version')
  as versionControl">
    <div class="form-group">

      ...

    </div>
    <div class="form-group">

      ...

    </div>
    <button *ngIf="versionControl.status
    !== 'PENDING'; else loader" type="submit"
    [disabled]="releaseForm.invalid" class="btn
    btn-primary">Submit</button>
  </ng-container>
  <ng-template #loader>
    Please wait...
  </ng-template>
</form>
```

If you refresh the app, select an app, and type a valid version, you should see the **Please wait...** message as follows:

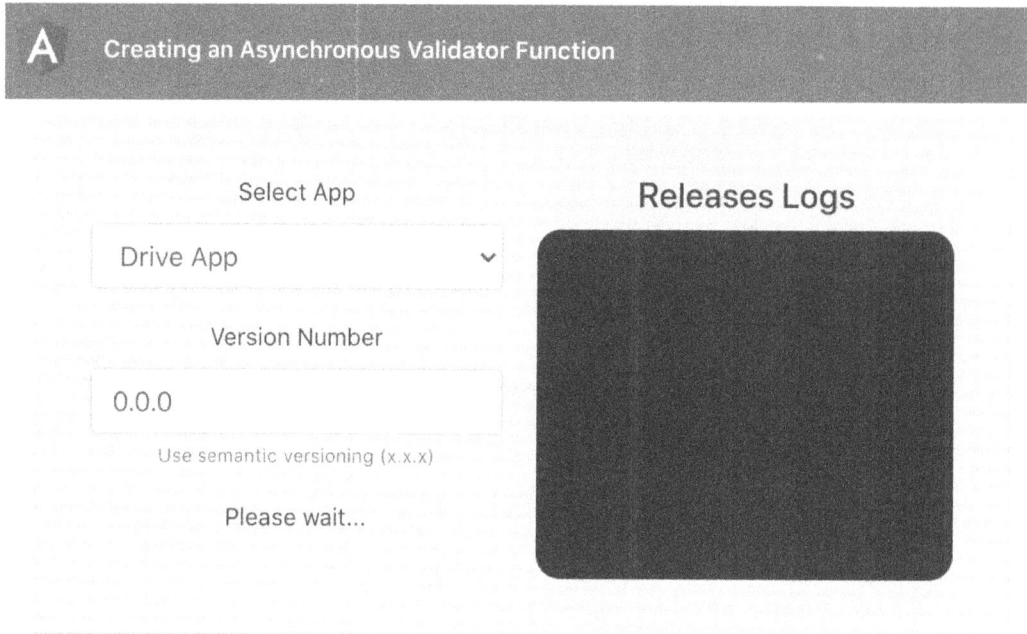

Figure 8.18 – Loader message while async validation is in progress

7. We still have an issue where the user can type and press *Enter* quickly to submit the form. To prevent this from happening, let's add a check in our `formSubmit` method in the `release-form.component.ts` file as follows:

```
formSubmit(form: FormGroup): void {
  if (form.get('version').status === 'PENDING') {
    return;
  }
  const { app, version } = form.value;
  ...
}
```

8. Finally, we have another issue to handle. If we select a valid version number and change the app, we can still submit the form with the entered version number although it is logically wrong. To handle this, we should update the validation of the `'version'` form control whenever the value of the `'app'` form control changes. To do that, modify the `release-form.component.ts` file as follows:

```typescript
import { Component, OnInit, Output, EventEmitter,
OnDestroy } from '@angular/core';
...
import { takeWhile } from 'rxjs/operators';
...
@Component({...})
export class ReleaseFormComponent implements OnInit,
OnDestroy {
  @Output() newReleaseLog = new
  EventEmitter<IReleaseLog>();
  isComponentAlive = false;
  apps = Object.values(Apps);
  ...
  ngOnInit(): void {
    this.isComponentAlive = true;
    this.releaseForm.get
    ('version').setAsyncValidators(...)
    this.releaseForm.get('app').valueChanges
      .pipe(takeWhile(() => this.isComponentAlive))
      .subscribe(() => {
        this.releaseForm.get
        ('version').updateValueAndValidity();
      })
  }
  ngOnDestroy() {
    this.isComponentAlive = false;
  }
  ...
}
```

Cool! So, you now know how to create an asynchronous validator function in Angular for form validation within Reactive forms. Since you've finished the recipe, refer to the next section to see how this works.

How it works...

Angular provides a really easy way to create async validator functions, and they're pretty handy too. In this recipe, we started by creating the validator function named `versionValidator`. Notice that we have an argument named `appNameControl` for the validator function. This is because we want to get the app name for which we are validating the version number. Also notice that we have the return type set to `AsyncValidatorFn`, which is required by Angular. The validator function is supposed to return an `AsyncValidatorFn`, which means it will return a function (let's call it the **inner function**), which receives an `AbstractControl` and returns an `Observable` of `ValidatorErrors`. Inside the inner function, we use the `getVersionLog()` method from `VersionService` to fetch the `data.json` file using the `HttpClient` service. Once we get the version from `data.json` for the specific app selected, we compare the version entered in the form with the value from `data.json` to validate the input. Notice that instead of just returning a `ValidationErrors` object with the `newVersionRequired` property set to `true`, we actually set it to `previousVersion` so that we can use it later to show it to the user.

After creating the validator function, we attached it to the form control for the version name by using the `FormControl.setAsyncValidators()` method in the `ReleaseFormComponent` class. We then used the validation error named `newVersionRequired` in the template to show the error message, along with the version from the `data.json` file.

We also needed to handle the case that while the validation is in progress, the form control is valid until the validation is finished. This allows us to submit the form while the validation for the version name was in progress. We handle it by hiding the submit button during the validation process by checking whether the value of `FormControl.status` is `'PENDING'`. We hide the submit button in that case and show the **Please wait...** message in the meantime. Note that we also add some logic in the `formSubmit` method of the `ReleaseFormComponent` class to check whether `FormControl.status` is `'PENDING'` for the version number, in which case, we just do a `return;`.

One more interesting thing in the recipe is that if we added a valid version number and changed the app, we could still submit the form. We handle that by adding a subscription to `.valueChanges` of the `'app'` form control, so whenever that happens, we trigger another validation on the `'version'` form control using the `.updateValueAndValidity()` method.

See also

- AsyncValidator Angular docs: `https://angular.io/api/forms/AsyncValidator#provide-a-custom-async-validator-directive`

Testing Reactive forms

To make sure we build robust and bug-free forms for end users, it is a really good idea to have tests around your forms. It makes the code more resilient and less prone to errors. In this recipe, you'll learn how to test your template-driven forms using unit tests.

Getting ready

The project for this recipe resides in `chapter08/start_here/testing-reactive-forms`:

1. Open the project in Visual Studio Code.
2. Open the terminal and run `npm install` to install the dependencies of the project.
3. Once done, run `ng serve -o`.

    This should open the app in a new browser tab, and you should see the app as follows:

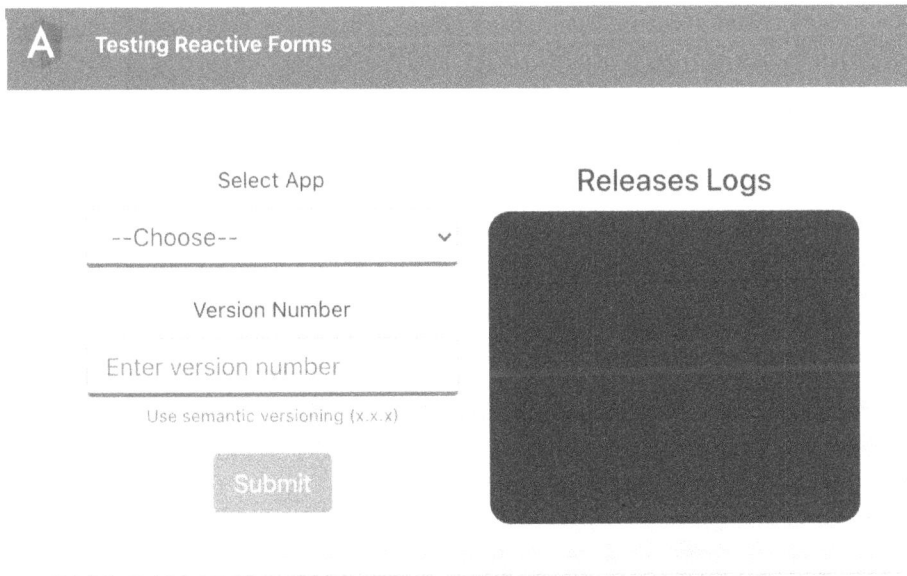

Figure 8.19 – The Testing Reactive Forms app running on http://localhost:4200

Now that we have the app running locally, let's see the steps involved in this recipe in the next section.

How to do it...

We have the Release Logs application that has a Reactive form implemented with some validations. In this recipe, we're going to implement some tests for the forms. Let's get started:

1. First of all, run the following command to run the unit tests in a separate terminal window:

   ```
   yarn test
   ```

   Once the command is run, you should see a new instance of the Chrome window being opened that runs the tests as follows:

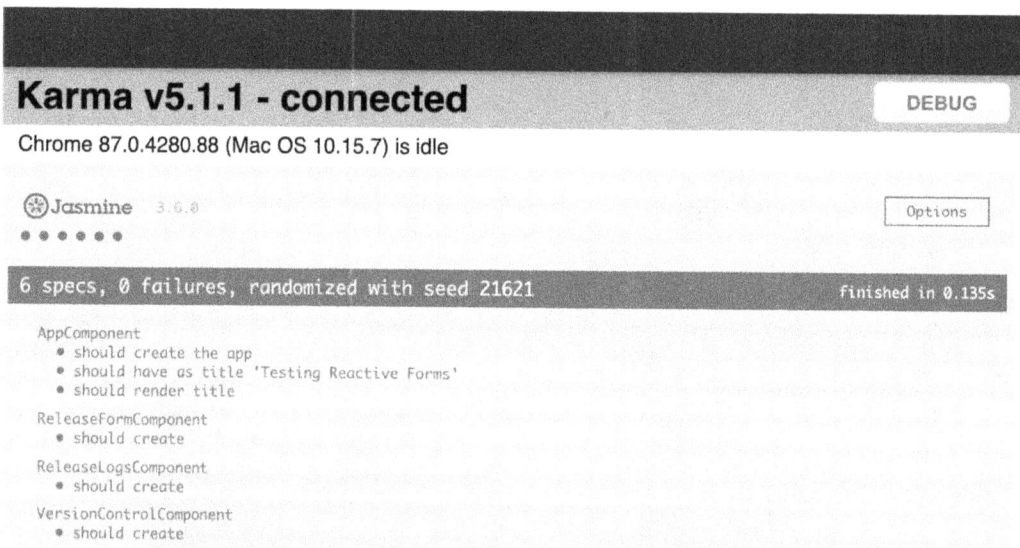

Karma v5.1.1 - connected   DEBUG

Chrome 87.0.4280.88 (Mac OS 10.15.7) is idle

(*) Jasmine   3.6.0   Options
• • • • • •

6 specs, 0 failures, randomized with seed 21621   finished in 0.135s

```
AppComponent
  • should create the app
  • should have as title 'Testing Reactive Forms'
  • should render title
ReleaseFormComponent
  • should create
ReleaseLogsComponent
  • should create
VersionControlComponent
  • should create
```

Figure 8.20 – Unit tests with Karma and Jasmine running in an automated Chrome window

2. Let's add our first test for the case when all the inputs have a valid value. In this case, we should have the form submitted and the form's value emitted through the emitter of the `newReleaseLog` output. Modify the `release-form.component.spec.ts` file as follows:

   ```
   import { ComponentFixture, TestBed } from '@angular/core/
   testing';
   import { ReleaseLog } from 'src/app/classes/release-log';
   ```

```
...
describe('ReleaseFormComponent', () => {
  ...
  it('should submit a new release log with the correct
  input values', (() => {
    const app = component.apps[2];
    const version = '2.2.2';
    const expectedReleaseLog = new ReleaseLog(app,
    version);
    spyOn(component.newReleaseLog, 'emit');
    component.releaseForm.setValue({ app, version });
    component.formSubmit(component.releaseForm);
    expect(component.newReleaseLog.emit)
    .toHaveBeenCalledWith(expectedReleaseLog);
  }));
});
```

If you look at the tests now, you should the new test passing as follows:

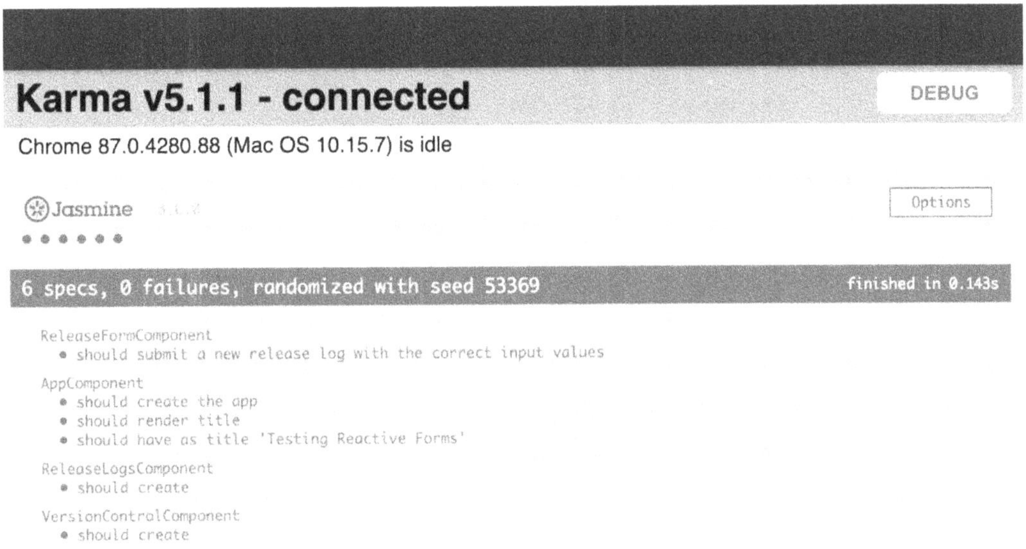

Karma v5.1.1 - connected   DEBUG

Chrome 87.0.4280.88 (Mac OS 10.15.7) is idle

✸ Jasmine   Options

• • • • • •

6 specs, 0 failures, randomized with seed 53369   finished in 0.143s

ReleaseFormComponent
  • should submit a new release log with the correct input values

AppComponent
  • should create the app
  • should render title
  • should have as title 'Testing Reactive Forms'

ReleaseLogsComponent
  • should create

VersionControlComponent
  • should create

Figure 8.21 – Test case passing for the successful input

3. Let's add a test for the case when we have an incorrect version provided in the form. The **submit** button should be disabled and we should have an error thrown by the `formSubmit` method. Add a new test to your `release-form.component.spec.ts` file as follows:

```
. . .
describe('ReleaseFormComponent', () => {
  . . .
  it('should throw an error for a new release log with
  the incorrect version values', (() => {
    const submitButton = fixture.nativeElement.
    querySelector('button[type="submit"]');
    const app = component.apps[2];
    const version = 'x.x.x';
    spyOn(component.newReleaseLog, 'emit');
    const expectedError = 'Invalid version provided.
    Please provide a valid version as (major.minor.
    patch)';
    component.releaseForm.setValue({ app, version });
    expect(() => component.formSubmit(component.
    releaseForm))
      .toThrowError(expectedError);
    expect(submitButton.hasAttribute(
    'disabled')).toBe(true);
    expect(component.newReleaseLog.emit
    ).not.toHaveBeenCalled();
  }));
});
```

4. Let's add our final test, which makes sure that the **submit** button is disabled when we have not selected an app for the release log. Add a new test to the `release-form.component.spec.ts` file as follows:

```
. . .
describe('ReleaseFormComponent', () => {
  . . .
  it('should disable the submit button when we
  don\'t have an app selected', (() => {
    const submitButton = fixture.nativeElement.
    querySelector('button[type="submit"]');
```

```
        spyOn(component.newReleaseLog, 'emit');
        const app = '';
        const version = '2.2.2';
        component.releaseForm.setValue({ app, version });
        submitButton.click();
        fixture.detectChanges();
        expect(submitButton.hasAttribute(
        'disabled')).toBe(true);
        expect(component.newReleaseLog.emit
        ).not.toHaveBeenCalled();
    }));
  });
```

If you look at the Karma tests window, you should see all the new tests passing as follows:

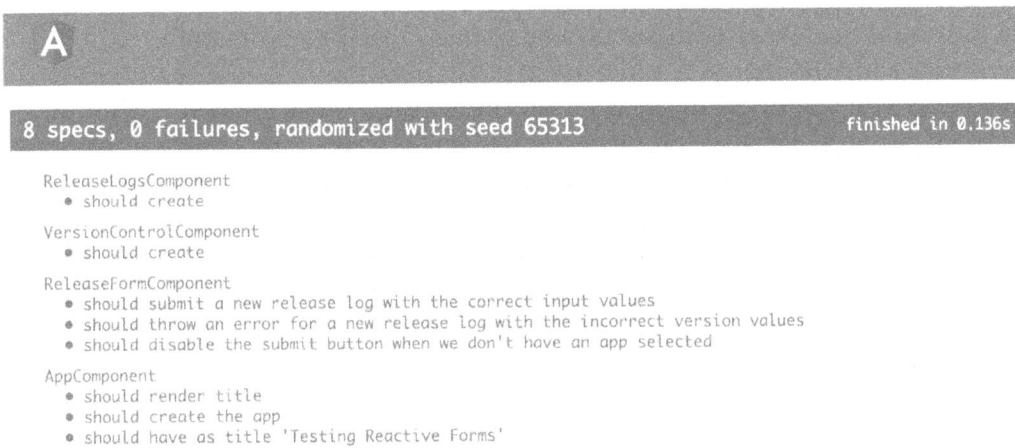

Figure 8.22 – All tests passing for the recipe

Great! You now know how to write some essential tests for Reactive forms. Refer to the next section to understand how it all works.

How it works...

Testing Reactive forms doesn't even require importing `ReactiveFormsModule` into the test module as of Angular 10. For all the tests in our recipe, we spied on the `newReleaseLog` emitter's `emit` event defined in the `ReleaseFormComponent` class. This is because we know that when the inputs are correct, the user should be able to click the **submit** button, and as a result, inside the `formSubmit` method, the `emit` method of the `newReleaseLog` emitter will be called. For the test covering the validity of the `'version'` form control, we rely on `formSubmit` to throw an error. This is because we know that an invalid version will cause an error in the `constructor` of the `ReleaseLog` class when creating a new release log. One interesting thing in this test is that we use the following code:

```
expect(() => component.formSubmit(component.releaseForm))
    .toThrowError(expectedError);
```

The interesting thing here is that we needed to call the `formSubmit` method ourselves with `releaseForm`. We couldn't just do it by writing `expect(component.formSubmit(component.releaseForm)).toThrowError(expectedError);` because that would rather call the function directly there and would result in an error. So we need to pass an anonymous function here that Jasmine will call and would expect this anonymous function to throw an error. And finally, we make sure that our **submit** button is enabled or disabled by first getting the button using `querySelector` on `fixture.nativeElement`. And then we check the `disabled` attribute on the **submit** button using `submitButton.hasAttribute('disabled')`.

See also

- Testing Reactive forms: `https://angular.io/guide/forms-overview#testing-reactive-forms`

Using debounce with Reactive form control

If you're building a medium-to large-scale Angular app with Reactive forms, you'll surely encounter a scenario where you might want to use a debounce on a Reactive form. It could be for performance reasons, or for saving HTTP calls. So, in this recipe, you're going to learn how to use debounce on a Reactive form control.

Getting ready

The project that we are going to work with resides in `chapter08/start_here/using-debounce-with-rfc` inside the cloned repository:

1. Open the project in Visual Studio Code.

2. Open the terminal and run `npm install` to install the dependencies of the project.

3. Once done, run `ng serve -o`.

   This should open the app in a new browser tab and you should see it as follows:

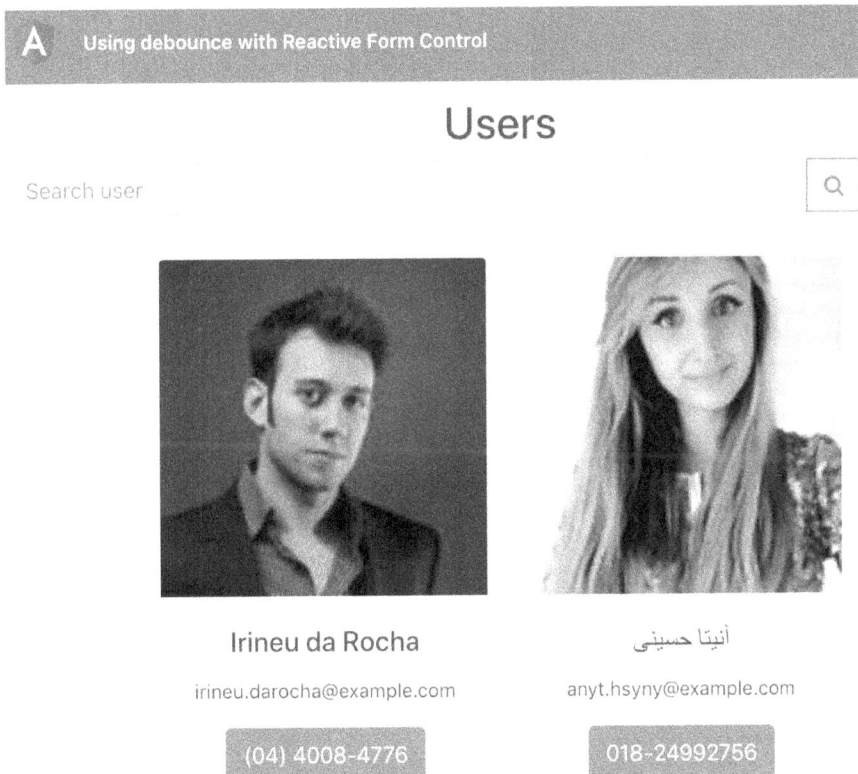

Figure 8.23 – The Using debounce with Reactive Form Control app running on http://localhost:4200

Right now, you'll notice that for each character we type into the input, we send a new HTTP request to the API shown as follows:

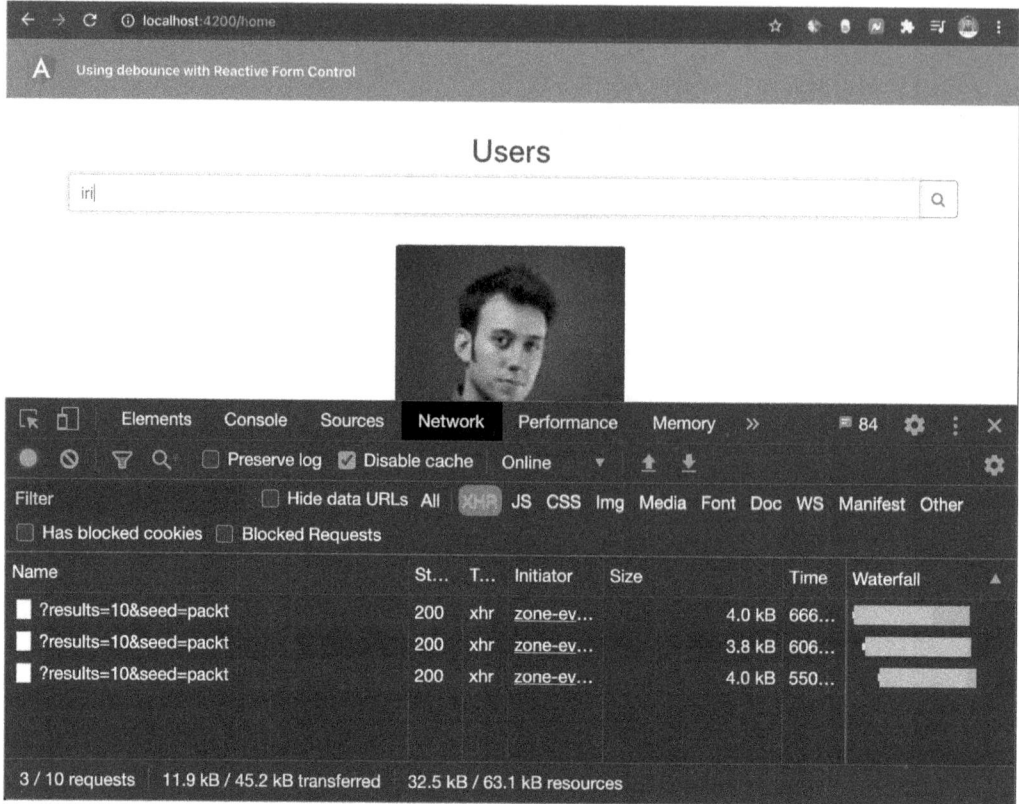

Figure 8.24 – Multiple HTTP calls sent as we type in the input

Now that we have the app running locally, let's see the steps involved in this recipe in the next section.

How to do it...

For this recipe, we're using an application that uses the RandomUser.me API to get users. As you see in *Figure 8.24*, we send a new HTTP call for every change in the input. Let's begin with the recipe to avoid doing that:

1. Adding the debounce to the form is super easy. Let's use the `debounceTime` operator in the `home.component.ts` file as follows:

```
...
import { debounceTime, takeWhile } from 'rxjs/operators';
```

```
@Component({...})
export class HomeComponent implements OnInit, OnDestroy {
  searchDebounceTime = 300;
  ...
  ngOnInit() {
    ...
    this.searchUsers();
    this.searchForm.get('username').valueChanges
      .pipe(
        debounceTime(this.searchDebounceTime),
        takeWhile(() => !!this.componentAlive)
      )
      .subscribe(() => {
        this.searchUsers();
      })
  }
}
```

Well, it's funny that this is it for the recipe as far as the task is concerned. But I do want to give you more out of this book. So we're going to write some interesting tests.

2. We'll add a test now to make sure that our `searchUsers` method isn't called before `searchDebounceTime` has passed. Add the following test to the `home.component.spec.ts` file:

```
import { HttpClientModule } from '@angular/common/http';
import { waitForAsync, ComponentFixture,
discardPeriodicTasks, fakeAsync, TestBed, tick } from '@
angular/core/testing';

import { HomeComponent } from './home.component';

describe('HomeComponent', () => {
  ...
  it('should not send an http request before the
  debounceTime of 300ms', fakeAsync(async () => {
    spyOn(component, 'searchUsers');
```

```
        component.searchForm.get(
        'username').setValue('iri');
        tick(component.searchDebounceTime - 10);
        // less than desired debounce time
        expect(component.searchUsers
        ).not.toHaveBeenCalled();
        discardPeriodicTasks();
    }));
  });
```

3. Now we'll add a test for the case when `searchDebounceTime` has passed and the `searchUsers()` method should have been called. Add a new test to the `home.component.spec.ts` file as follows:

```
  ...
  describe('HomeComponent', () => {
    ...
    it('should send an http request after the debounceTime
    of 300ms', fakeAsync(async () => {
      spyOn(component, 'searchUsers');
      component.searchForm.get(
      'username').setValue('iri');
      tick(component.searchDebounceTime + 10); // more
      than desired debounce time
      expect(component.searchUsers
      ).toHaveBeenCalled();
      discardPeriodicTasks();
    }));
  });
```

If you refresh the Karma test Chrome window, you'll see all the tests passing as follows:

Chrome is being controlled by automated test software.   X

Karma v5.1.1 - connected   DEBUG

Chrome 87.0.4280.88 (Mac OS 10.15.7) is idle

✳Jasmine 3.6.0   Options

● ● ● ● ● ● ● ● ● ● ●

11 specs, 0 failures, randomized with seed 14745   finished in 0.151s

UserDetailComponent
  • should create

HomeComponent
  • should not send an http request before the debounceTime of 300ms
  • should send an http request after the debounceTime of 300ms
  • should create

AppComponent
  • should create the app
  • should have as title 'Using debounce with Reactive Form Control'
  • should render title

UserCardComponent
  • should create

UserService
  • should be created

AppFooterComponent
  • should create

LoaderComponent
  • should create

Figure 8.25 – All tests passing for the recipe

4. Now, run the `npm start` command to spin up the app again. Then, monitor the network calls while you type an input into the search box. You'll see that the `debounceTime` operator causes only 1 call once you stop typing for 300 milliseconds, as shown in the following screenshot:

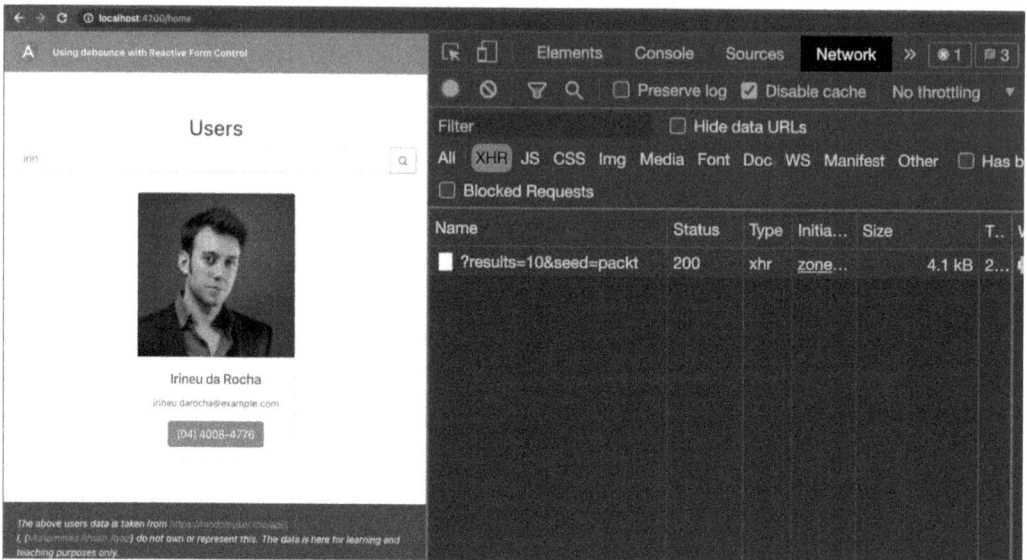

Figure 8.26 – Only one network call sent after a 300ms debounce

Awesome! So, now you know how to use debounce with a Reactive form control and also how to write tests to check whether things work fine with the debounce. And that concludes the recipe. Let's refer to the next section to see how it works.

How it works...

The main task for the recipe was quite easy. We just used the `debounceTime` operator from the `rxjs` package and used it with our Reactive form control's `.valueChanges` Observable. Since we're using it within the `.pipe()` operator before the `.subscribe()` method, every time we change the value of the input, either by entering a value or by pressing the backspace key, it waits for `300ms` according to the `searchDebounceTime` property and then calls the `searchUsers()` method.

We also wrote some tests in this recipe. Notice that we spy on the `searchUsers()` method since that is what it's supposed to be called whenever we change the value of the `'username'` form control. We're wrapping the test functions inside the `fakeAsync` method so we can control the asynchronous behavior of the use cases in our tests. We then set the value of the form control using the `FormControl.setValue()` method, which should trigger the method provided as an argument to the `.subscribe()` method after the time according to `searchDebounceTime` has passed. We then used the `tick()` method with the value of `searchDebounceTime` so it simulates an asynchronous passage of time. Then we write our `expect()` block to check whether the `searchUsers()` method should or shouldn't have been called. Finally, at the end of the tests, we use the `discardPeriodicTasks()` method. We use this method so that we don't face the `Error: 1 periodic timer(s) still in the queue.` error and our tests work.

See also

- RxJS DebounceTime operator: `https://rxjs-dev.firebaseapp.com/api/operators/debounceTime`

Writing your own custom form control using ControlValueAccessor

Angular forms are great. While they support the default HTML tags like input, textarea etc., sometimes, you would want to define your own components that take a value from the user. It would be great if the variables of those inputs were a part of the Angular form you're using already.

In this recipe, you'll learn how to create your own custom Form Control using the ControlValueAccessor API, so you can use the Form Control with both Template Driven forms and Reactive Forms.

Getting ready

The project for this recipe resides in `chapter08/start_here/custom-form-control`:

1. Open the project in Visual Studio Code.
2. Open the terminal and run `npm install` to install the dependencies of the project.
3. Once done, run `ng serve -o`.

This should open the app in a new browser tab and you should see the following view:

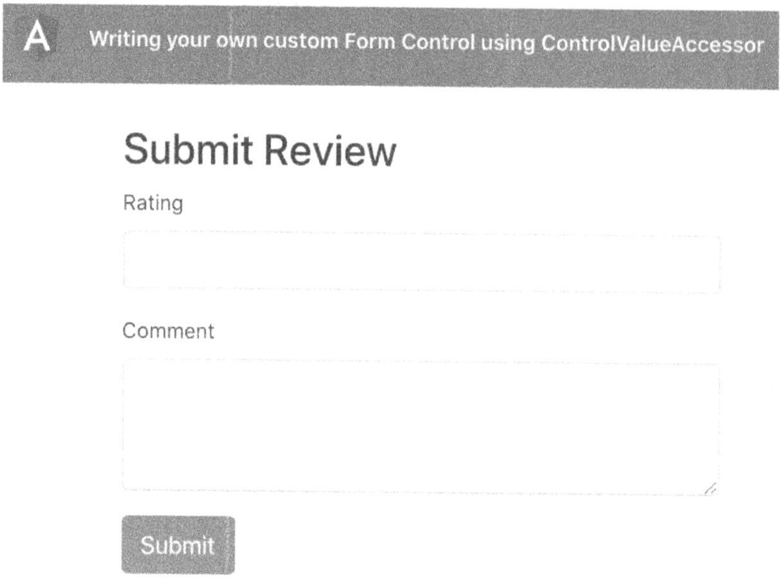

Figure 8.27 – Custom form control app running on http://localhost:4200

Now that we have the app running locally, let's see the steps involved in this recipe in the next section.

How to do it...

We have a simple Angular app. It has two inputs and a **Submit** button. The inputs are for a review and they ask the user to provide a value for the rating of this imaginary item and any comments the user wants to provide. We'll convert the Rating input into a custom Form Control using the ControlValueAccessor API. Let's get started:

1. Let's create a component for our custom form control. Open the terminal in the project root and run the following command:

    ```
    ng g c components/rating
    ```

2. We'll now create the stars UI for the rating component. Modify the `rating.component.html` file as follows:

    ```
    <div class="rating">
      <div
    ```

```
      class="rating__star"
      [ngClass]="{'rating__star--active': (
        (!isMouseOver && value    >= star) ||
        (isMouseOver && hoveredRating    >= star)
      )}"
      (mouseenter)="onRatingMouseEnter(star)"
      (mouseleave)="onRatingMouseLeave()"
      (click)="selectRating(star)"
      *ngFor="let star of [1, 2, 3, 4, 5]; let i = index;">
      <i class="fa fa-star"></i>
    </div>
  </div>
```

3. Add the styles for the rating component to the `rating.component.scss` file as follows:

```
.rating {
  display: flex;
  margin-bottom: 10px;
  &__star {
    cursor: pointer;
    color: grey;
    padding: 0 6px;
    &:first-child {
      padding-left: 0;
    }
    &:last-child {
      padding-right: 0;
    }
    &--active {
      color: orange;
    }
  }
}
```

4. We also need to modify the `RatingComponent` class to introduce the necessary methods and properties. Let's modify the `rating.component.ts` file as follows:

```
...
export class RatingComponent implements OnInit {
  value = 2;
  hoveredRating = 2;
  isMouseOver = false;

  ...

  onRatingMouseEnter(rating: number) {
    this.hoveredRating = rating;
    this.isMouseOver = true;
  }
  onRatingMouseLeave() {
    this.hoveredRating = null;
    this.isMouseOver = false;
  }
  selectRating(rating: number) {
    this.value = rating;
  }
}
```

5. Now we need to use this rating component instead of the input that we already have in the `home.component.html` file. Modify the file as follows:

```
<div class="home">
  <div class="review-container">
    ...
    <form class="input-container" [formGroup]=
    "reviewForm" (ngSubmit)="submitReview(reviewForm)">
      <div class="mb-3">
        <label for="ratingInput" class="form-
        label">Rating</label>
        <app-rating formControlName="rating">
        </app-rating>
      </div>
```

```
        <div class="mb-3">
            ...
        </div>
        <button id="submitBtn" [disabled]="reviewForm.
        invalid" class="btn btn-dark" type="submit">
        Submit</button>
    </form>
    </div>
</div>
```

If you refresh the app now and hover on the stars, you can see the color changing as you hover over the stars. The selected rating is also highlighted as follows:

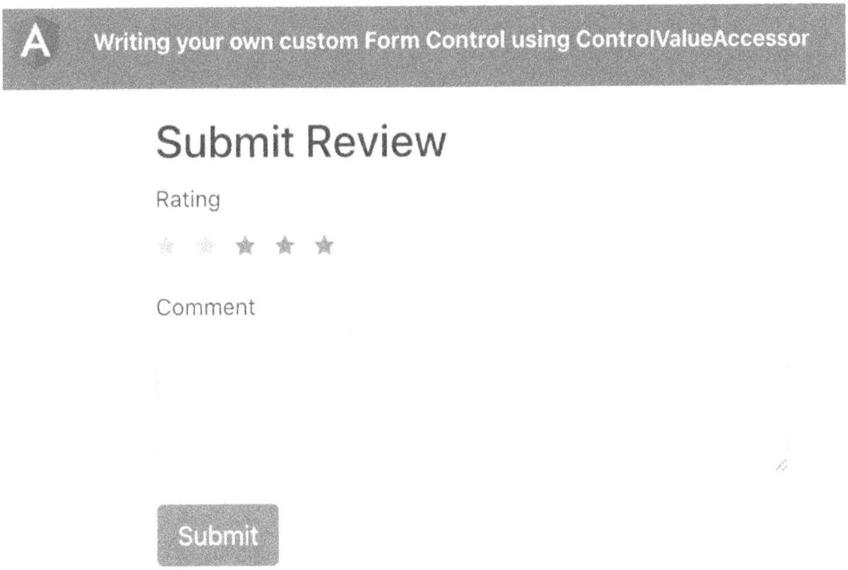

Figure 8.28 – Rating component with hovered stars

6. Let's now implement the `ControlValueAccessor` interface for our rating component. It requires a couple of methods to be implemented and we'll start with the `onChange()` and `onTouched()` methods. Modify the `rating. component.ts` file as follows:

```
import { Component, OnInit } from '@angular/core';
import { ControlValueAccessor } from '@angular/forms';

@Component({...})
```

```
export class RatingComponent implements OnInit,
ControlValueAccessor {

    ...

    constructor() { }


    onChange: any = () => { };
    onTouched: any = () => { };


    ngOnInit(): void {

    }

    ...

    registerOnChange(fn: any){
        this.onChange = fn;
    }


    registerOnTouched(fn: any) {
        this.onTouched = fn;
    }
}
```

7. We'll now add the required methods to disable the input when required and to set the value of the form control, in other words, the setDisabledState() and writeValue() methods. We'll also add the disabled and value properties to our RatingComponent class as follows:

```
import { Component, Input, OnInit } from '@angular/core';
import { ControlValueAccessor } from '@angular/forms';


@Component({...})
export class RatingComponent implements OnInit,
ControlValueAccessor {

    ...

    isMouseOver = false;
    @Input() disabled = false;
    constructor() { }



    ...
```

```
    setDisabledState(isDisabled: boolean): void {
      this.disabled = isDisabled;
    }


    writeValue(value: number) {
      this.value = value;
    }



}
```

8. We need to use the `disabled` property to prevent any UI changes when it is `true`.
    The value of the `value` variable shouldn't be updated either. Modify the `rating.`
    `component.ts` file to do so as follows:

```
  ...
  @Component({...})
  export class RatingComponent implements OnInit,
  ControlValueAccessor {
    ...
    isMouseOver = false;
    @Input() disabled = true;
    ...

    onRatingMouseEnter(rating: number) {
      if (this.disabled) return;
      this.hoveredRating = rating;
      this.isMouseOver = true;
    }
    ...
    selectRating(rating: number) {
      if (this.disabled) return;
      this.value = rating;
    }
    ...
  }
```

9. Let's make sure that we send the value of the `value` variable to `ControlValueAccessor` because that's what we want to access later. Also, let's set the `disabled` property back to `false`. Update the `selectRating` method in the `RatingComponent` class as follows:

```
...
@Component ({...})
export class RatingComponent implements OnInit,
ControlValueAccessor {
  ...
  @Input() disabled = false;
  constructor() { }
  ...
  selectRating(rating: number) {
    if (this.disabled) return;
    this.value = rating;
    this.onChange(rating);
  }
  ...
}
```

10. We need to tell Angular that our `RatingComponent` class has a value accessor, otherwise using the `formControlName` attribute on the `<app-rating>` element will throw errors. Let's add an NG_VALUE_ACCESSOR provider to the `RatingComponent` class's decorator as follows:

```
import { Component, forwardRef, Input, OnInit } from '@
angular/core';
import { ControlValueAccessor, NG_VALUE_ACCESSOR } from
'@angular/forms';

@Component({
  selector: 'app-rating',
  templateUrl: './rating.component.html',
  styleUrls: ['./rating.component.scss'],
  providers: [{
    provide: NG_VALUE_ACCESSOR,
    useExisting: forwardRef(() => RatingComponent),
    multi: true
```

```
    }]
  })
export class RatingComponent implements OnInit,
ControlValueAccessor {
    ...
  }
```

If you refresh the app now, select a rating, and hit the **Submit** button, you should see the values being logged as follows:

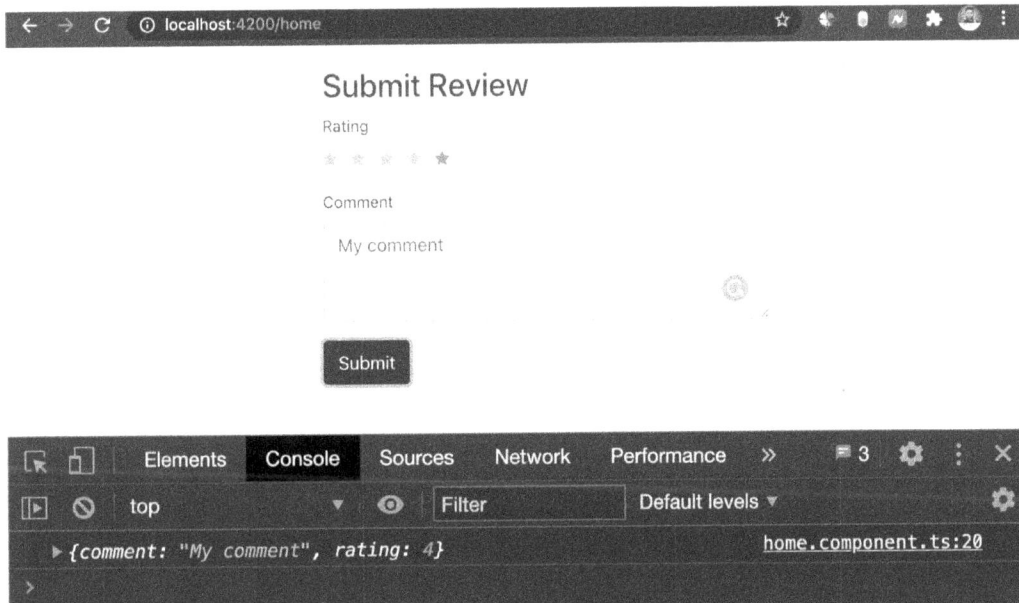

Figure 8.29 – Form value logged using the custom form control

Voilà! You just learned how to create a custom form control using `ControlValueAccessor`. Refer to the next section to understand how it works.

How it works...

We started the recipe by creating a component that we can use to provide a rating for the review we have to submit. We started off by adding the template and the styles for the rating component. Notice that we are using an [ngClass] directive on each of the star elements to add the rating__star--active class conditionally. Let's discuss each condition now:

- (isMouseOver && hoveredRating >= star): This condition relies on the isMouseOver and hoveredRating variables. The isMouseOver variable becomes true as soon as we mouse over any star and is turned back to false when we move away from the star. This means that it is only true when we're hovering over a star. hoveredRating tells us which star we're hovering over at the moment and is assigned the star's value, in other words, a value from 1 to 5. So, this condition is only true when we're doing a mouseover, and the hovered star's rating is greater than the value of the current star. So, if we're hovering over the fourth star, all the stars from value 1 to 4 will be highlighted as they'll have the rating__star--active class conditionally assigned to them.

- (!isMouseOver && value >= star): This condition relies on the isMouseOver variable that we discussed previously and the value variable. The value variable holds the value of the selected rating, which is updated when we click on a star. So, this condition is applied when we're not doing a mouseover and we have the value of the value variable greater than the current star. This is especially beneficial when you have a greater value assigned to the value variable and try to hover over a star with a lesser value, in which case, all the stars with values greater than the hovered star will not be highlighted.

Then we used three events on each star: mouseenter, mouseleave, and click, and then used our onRatingMouseEnter, onRatingMouseLeave, and selectRating methods, respectively, for these events. All of this was designed to ensure that the entire UI is fluent and has a good user experience. We then implemented the ControlValueAccessor interface for our rating component. When we do that, we need to define the onChange and onTouched methods as empty methods, which we did as follows:

```
onChange: any = () => { };
onTouched: any = () => { };
```

Then we used the `registerOnChange` and `registerOnTouched` methods from `ControlValueAccessor` to assign our methods as follows:

```
registerOnChange(fn: any) {
  this.onChange = fn;
}
registerOnTouched(fn: any) {
  this.onTouched = fn;
}
```

We registered these functions because whenever we do a change in our component and want to let `ControlValueAccessor` know that the value has changed, we need to call the onChange method ourselves. We do that in the `selectRating` method as follows, which makes sure that when we select a rating, we set the form control's value to the value of the selected rating:

```
selectRating(rating: number) {
  if (this.disabled) return;
  this.value = rating;
  this.onChange(rating);
}
```

The other way around is when we need to know when the form control's value is changed from outside the component. In this case, we need to assign the updated value to the `value` variable. We do that in the `writeValue` method from the `ControlValueAccessor` interface as follows:

```
writeValue(value: number) {
  this.value = value;
}
```

What if we don't want the user to provide a value for the rating? In other words, we want the rating form control to be disabled. For this, we did two things. First, we used the `disabled` property as an `@Input()`, so we can pass and control it from the parent component when needed. Secondly, we used the `setDisabledState` method from the `ControlValueAccessor` interface, so whenever the form control's `disabled` state is changed, apart from `@Input()`, we set the `disabled` property ourselves.

Finally, we wanted Angular to know that this `RatingComponent` class has a value accessor. This is so that we can use the Reactive forms API, specifically, the `formControlName` attribute with the `<app-rating>` selector, and use it as a form control. To do that, we provide our `RatingComponent` class as a provider to its `@Component` definition decorator using the `NG_VALUE_ACCESSOR` injection token as follows:

```
@Component({
  selector: 'app-rating',
  templateUrl: './rating.component.html',
  styleUrls: ['./rating.component.scss'],
  providers: [{
    provide: NG_VALUE_ACCESSOR,
    useExisting: forwardRef(() => RatingComponent),
    multi: true
  }]
})
export class RatingComponent implements OnInit,
ControlValueAccessor {}
```

Note that we're using the `useExisting` property with the `forwardRef()` method providing our `RatingComponent` class in it. We need to provide `multi: true` because Angular itself registers some value accessors using the `NG_VALUE_ACCESSOR` injection token, and there may also be third-party form controls.

Once we've set everything up, we use `formControlName` on our rating component in the `home.component.html` file as follows:

```
<app-rating formControlName="rating"></app-rating>
```

See also

- Custom form control in Angular by Thoughtram: `https://blog.thoughtram.io/angular/2016/07/27/custom-form-controls-in-angular-2.html`

- ControlValueAccessor docs: `https://angular.io/api/forms/ControlValueAccessor`

9
Angular and the Angular CDK

Angular has an amazing ecosystem of tools and libraries, be it Angular Material, the **Angular command-line interface** (**Angular CLI**), or the beloved **Angular Component Dev Kit** (**Angular CDK**). I call it "beloved" because if you are to implement your own custom interactions and behaviors in Angular apps without having to rely on an entire set of libraries, Angular CDK is going to be your best friend. In this chapter, you'll learn what an amazing combination Angular and the Angular CDK are. You'll learn about some neat components built into the CDK and will also use some CDK **application programming interfaces** (**APIs**) to create amazing and optimized content.

Here are the recipes we're going to cover in this chapter:

- Using Virtual Scroll for huge lists
- Keyboard navigation for lists
- Pointy little popovers with the Overlay API
- Using CDK Clipboard to work with the system clipboard
- Using CDK Drag and Drop to move items from one list to another

- Creating a multi-step game with the CDK Stepper API
- Resizing text inputs with the CDK TextField API

Technical requirements

For the recipes in this chapter, make sure you have **Git** and **Node.js** installed on your machine. You also need to have the `@angular/cli` package installed, which you can do with `npm install -g @angular/cli` from your terminal. The code for this chapter can be found at `https://github.com/PacktPublishing/Angular-Cookbook/tree/master/chapter09`.

Using Virtual Scroll for huge lists

There might be certain scenarios in your application where you might have to show a huge set of items. This could be from either your backend API or the browser's local storage. In either case, rendering a lot of items at once causes performance issues because the **Document Object Model (DOM)** struggles, and also because of the fact that the JS thread gets blocked and the page becomes unresponsive. In this recipe, we'll render a list of 10,000 users and will use the Virtual Scroll functionality from the Angular CDK to improve the rendering performance.

Getting ready

The project that we are going to work with resides in `chapter09/start_here/using-cdk-virtual-scroll`, inside the cloned repository. Proceed as follows:

1. Open the project in **Visual Studio Code (VS Code)**.
2. Open the terminal and run `npm install` to install the dependencies of the project.
3. Once done, run `ng serve -o`.

This should open the app in a new browser tab, and it should look like this:

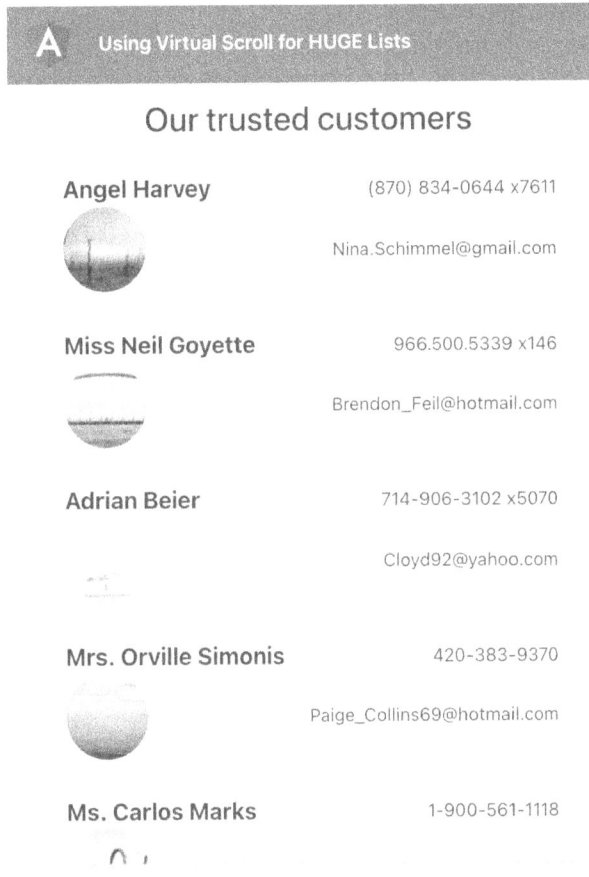

Figure 9.1 – The using-cdk-virtual-scroll app running on http://localhost:4200

Now that we have the app running locally, let's see the steps of the recipe in the next section.

How to do it...

We have a pretty simple Angular app, but with a lot of data. Right now, it shows a loader (button) for about 3 seconds, and then is supposed to show the data. However, you'll notice that right after 3 seconds, the loader keeps showing, the button is unresponsive, and we see a blank screen, as follows:

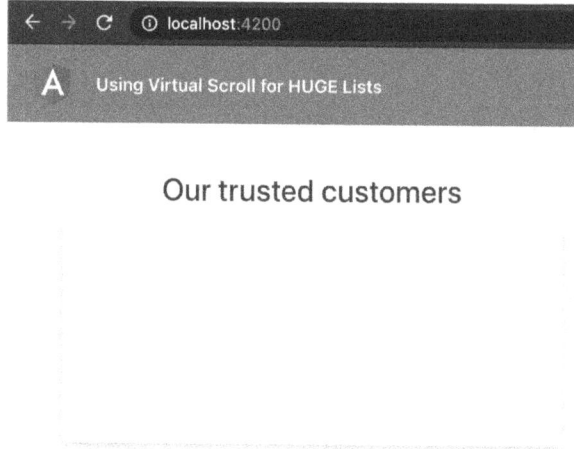

Figure 9.2 – App stuck with a blank screen while rendering list items

In fact, our entire application becomes unresponsive. If you scroll through—or even hover over—the items, you'll see that the hover animation on the list items is not smooth and is a bit laggy. Let's see the steps to use Angular CDK Virtual Scroll to improve the rendering performance, as follows:

1. First, open a new terminal window/tab and make sure you're inside the ch8/ start_here/using-cdk-virtual-scroll folder. Once inside, run the following command to install the Angular CDK:

   ```
   npm install --save @angular/cdk@12.0.0
   ```

2. You'll have to restart your Angular server, so rerun the ng serve command.

3. Add the ScrollingModule class from the @angular/cdk package into your app.module.ts file, as follows:

   ```
   ...
   import { LoaderComponent } from './components/loader/
   loader.component';
   import { ScrollingModule } from '@angular/cdk/scrolling';
   ```

```
@NgModule({
  declarations: [...],
  imports: [
    ...
    HttpClientModule,
    ScrollingModule
  ],
  providers: [],
  bootstrap: [AppComponent]
})
export class AppModule { }
```

4. We now have to implement the virtual scroll, modify the the-amazing-list-item.component.html file to use the *cdkVirtualFor directive instead of the *ngFor directive, and change the container <div> element to a <cdi-virtual-scroll-viewport> element, as follows:

```
<h4 class="heading">Our trusted customers</h4>
<cdk-virtual-scroll-viewport
  class="list list-group"
  [itemSize]="110">
  <div
    class="list__item list-group-item"
    *cdkVirtualFor="let item of listItems">
    <div class="list__item__primary">
      ...
    </div>
    <div class="list__item__secondary">
      ...
    </div>
  </div>
</cdk-virtual-scroll-viewport>
```

Kaboom! Within a few steps, and by using the Angular CDK Virtual Scroll, we were able to fix a big rendering issue within our Angular app. Now that you know how the basic routing is implemented, see the next section to understand how it works.

How it works...

The Angular CDK provides the Scrolling APIs, which include the `*cdkVirtualFor` directive and the `<cdk-virtual-scroll-viewport>` element. It is necessary to have `<cdk-virtual-scroll-viewport>` wrapping the element that has the `*cdkVirtualFor` directive being applied to it. Notice that we have an attribute on the `cdk-virtual-scroll-viewport` element named `[itemSize]`, having its value set to `"110"`. The reason for this is that each list item has a height of approximately 110 pixels, as shown in the following screenshot:

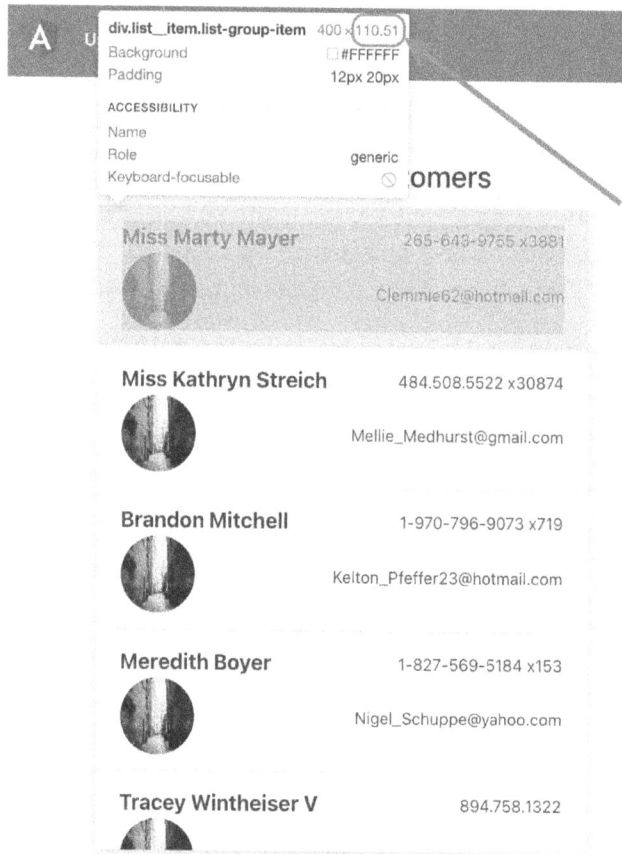

Figure 9.3 – Each list item has a height of approximately 110 pixels

But how does it improve the rendering performance? Glad you asked! In the original code for this recipe, when we loaded the 10,000 users, it would create a separate `<div>` element with the `class="list__item list-group-item"` attribute for each user, thus creating 10,000 DOM elements all being rendered at once. With virtual scroll in place, the CDK only creates a few `<div>` elements, renders them, and just replaces the content of those few `<div>` elements as we scroll through items.

For our example, it creates exactly nine `<div>` elements, as shown in the following screenshot:

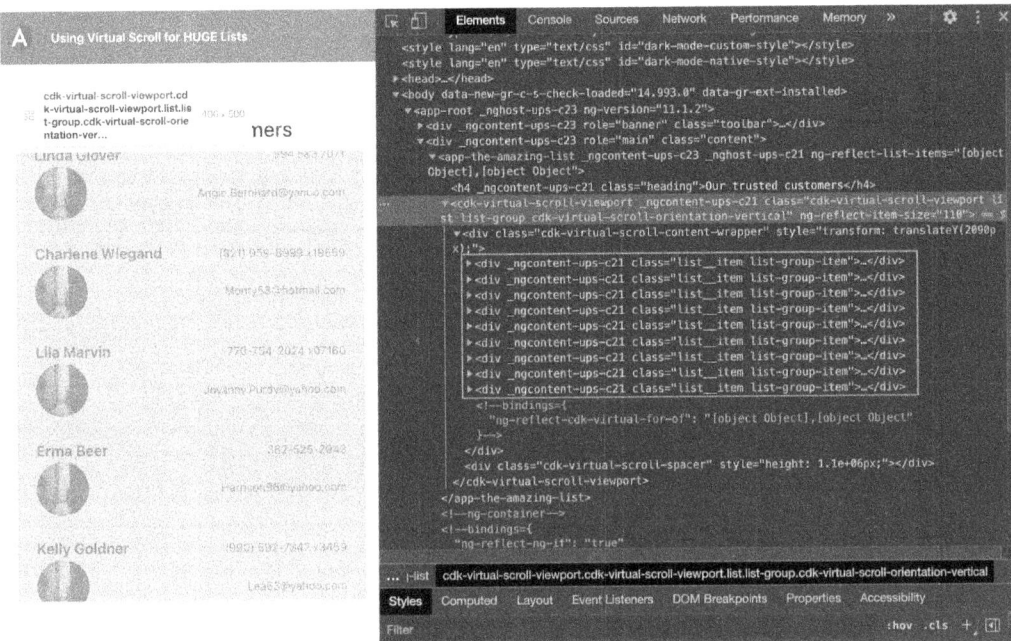

Figure 9.4 – Showing only a few <div> elements rendered on DOM due to virtual scroll

Since we only have a few elements rendered on the DOM, we don't have performance issues anymore, and the hover animation also seems super-smooth now.

> **Tip**
> When implementing virtual scroll in your own applications, make sure that you set a specific height to the `<cdk-virtual-scroll viewport>` element, and also set the `[itemSize]` attribute equal to the expected list-item height in pixels, otherwise the list won't show.

See also

- CDK scrolling examples (`https://material.angular.io/cdk/scrolling/examples`)

Keyboard navigation for lists

Accessibility is one of the most important aspects of building apps with a great user experience. The apps should not only be fast and performant but also accessible. While there are a lot of things to consider when it comes to accessibility, in this recipe, we're going to make lists and list items more accessible by providing keyboard navigation for the items. With Angular CDK, it is super simple. We're going to use the **ListKeyManager** service from Angular to implement keyboard navigation for the users list in our target application.

Getting ready

The project for this recipe resides in `chapter09/start_here/using-list-key-manager`. Proceed as follows:

1. Open the project in VS Code.

2. Open the terminal and run `npm install` to install the dependencies of the project.

3. Once done, run `ng serve -o`.

    This should open the app in a new browser tab, as follows:

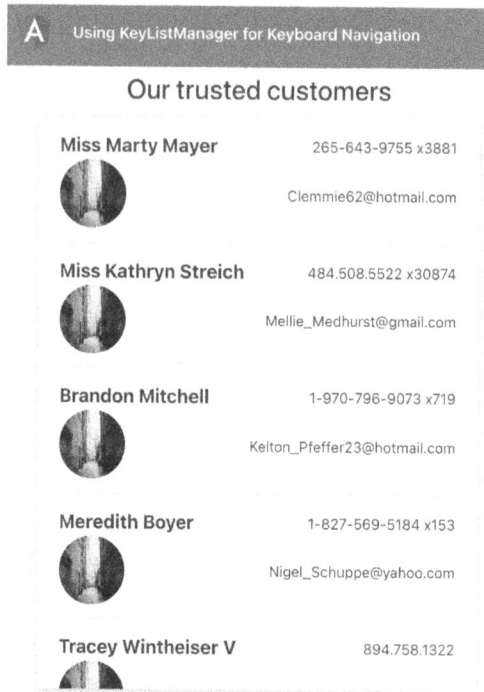

Figure 9.5 – The using-list-key-manager app running on http://localhost:4200

Now that we have the app running locally, let's see the steps of the recipe in the next section.

How to do it...

We have an app that already has some of the goodness of the Angular CDK—that is, it has virtual scroll implemented from the previous recipe. We'll now start making changes to the app to implement keyboard navigation, as follows:

1. First, we need to create a new component for each item in the list as we'll need them to be able to work with the `ListKeyManager` class. Create a component by running the following command in the project:

    ```
    ng g c components/the-amazing-list-item
    ```

2. Now, we'll move the code from the `the-amazing-list-component.html` file to the `the-amazing-list-item.component.html` file for the item's markup. The code in the `the-amazing-list-item.component.html` file should look like this:

    ```html
    <div class="list__item__primary">
      <div class="list__item__primary__info">
        {{ item.name }}
      </div>
      <div class="list__item__primary__info">
        {{ item.phone }}
      </div>
    </div>
    <div class="list__item__secondary">
      <div class="list__item__secondary__info">
        <img src="{{ item.picture }}" />
      </div>
      <div class="list__item__secondary__info">
        {{ item.email }}
      </div>
    </div>
    ```

3. Let's update the respective component as well to include this `item` property used in the template. We'll make it an `@Input()` item for the `TheAmazingListItemComponent` class. Update the `the-amazing-list-item.component.ts` file, as follows:

```
import { Component, Input, OnInit, ViewEncapsulation }
from '@angular/core';
import { AppUserCard } from 'src/interfaces/app-user-
card.interface';

@Component({
  selector: 'app-the-amazing-list-item',
  templateUrl: './the-amazing-list-item.component.html',
  styleUrls: ['./the-amazing-list-item.component.scss'],
  encapsulation: ViewEncapsulation.None
})
export class TheAmazingListItemComponent implements
OnInit {
  @Input() item: Partial<AppUserCard>;
  constructor() { }
  ngOnInit(): void {

  }
}
```

4. Let's add the styles as well. We'll copy the styles from the `the-amazing-list.component.scss` file and paste them into the `the-amazing-list-item.component.scss` file, as follows:

```
.list__item {
  transition: all ease 1s;
  cursor: pointer;
  &:hover, &:focus {
    background-color: #ececec; transform: scale(1.02);
  }
  &__primary,
  &__secondary {
    display: flex;
    justify-content: space-between;
    align-items: center;
```

```scss
    &__info { font-size: small; }
  }
  &__primary {
    &__info {
      &:nth-child(1) { font-weight: bold; font-size:
      larger; }
    }
  }
  img { border-radius: 50%; width: 60px; height: 60px; }
}
```

5. Update the `the-amazing-list.component.scss` file to contain only the styles for the list, as follows:

```scss
.heading {
  text-align: center;
  margin-bottom: 10px;
}
.list {
  box-shadow: rgba(0, 0, 0, 0.24) 0px 3px 8px;
  height: 500px;
  overflow: scroll;
  min-width: 400px;
  max-width: 960px;
  width: 100%;
}
```

6. Now, update the `the-amazing-list.component.html` file to use the `<app-the-amazing-list-item>` component and to pass the `[item]` attribute to it, as follows:

```html
<h4 class="heading">Our trusted customers</h4>
<cdk-virtual-scroll-viewport
  class="list list-group"
  [itemSize]="110">
  <app-the-amazing-list-item
    class="list__item list-group-item"
    *cdkVirtualFor="let item of listItems"
```

```
      [item]="item">
    </app-the-amazing-list-item>
  </cdk-virtual-scroll-viewport>
```

7. The **user interface** (**UI**) is almost done now. We'll now implement the
   FocusableOption interface and some accessibility factors to our
   TheAmazingListItemComponent class, as follows:

```
import { Component, Input, OnInit, ViewEncapsulation }
from '@angular/core';
import { AppUserCard } from 'src/interfaces/app-user-
card.interface';
import { FocusableOption } from '@angular/cdk/a11y';

@Component({
  selector: 'app-the-amazing-list-item',
  templateUrl: './the-amazing-list-item.component.html',
  styleUrls: ['./the-amazing-list-item.component.scss'],
  encapsulation: ViewEncapsulation.None,
  host: {
    tabindex: '-1',
    role: 'list-item',
  },
})
export class TheAmazingListItemComponent implements
OnInit, FocusableOption {
  @Input() item: Partial<AppUserCard>;
  constructor() { }
  focus() { }
  ngOnInit(): void {

  }
}
```

8. We now need to implement what happens in the `focus()` method. We'll use the
    `ElementRef` service to get the `nativeElement` and will set `focus()` on it, as
    follows:

    ```typescript
    import { Component, ElementRef, Input, OnInit,
    ViewEncapsulation } from '@angular/core';
    import { AppUserCard } from 'src/interfaces/app-user-
    card.interface';
    import { FocusableOption } from '@angular/cdk/a11y';


    @Component({...})
    export class TheAmazingListItemComponent implements
    OnInit, FocusableOption {
      @Input() item: Partial<AppUserCard>;
      constructor(private el: ElementRef) { }


      focus() {
        this.el.nativeElement.focus();
      }
      ...
    }
    ```

9. We now need to implement the `FocusKeyManager` class in our
    `TheAmazingListComponent` class. We'll have to query our list items in the
    component to create an instance of the `FocusKeyManager` class. Update the
    `the-amazing-list.component.ts` file, as follows:

    ```typescript
    import { FocusKeyManager } from '@angular/cdk/a11y';
    import { AfterViewInit, Component, Input, OnInit,
    QueryList, ViewChildren } from '@angular/core';
    import { AppUserCard } from 'src/interfaces/app-user-
    card.interface';
    import { TheAmazingListItemComponent } from '../
    the-amazing-list-item/the-amazing-list-item.component';


    @Component({
      ...
      styleUrls: ['./the-amazing-list.component.scss'],
      host: { role: 'list' }
    ```

```
})
export class TheAmazingListComponent implements OnInit,
AfterViewInit {
  @Input() listItems: Partial<AppUserCard>[] = [];

  @ViewChildren(TheAmazingListItemComponent)
  listItemsElements: QueryList
  <TheAmazingListItemComponent>;

  private listKeyManager:
  FocusKeyManager<TheAmazingListItemComponent>;
  constructor() { }

  ...

  ngAfterViewInit() {
    this.listKeyManager = new FocusKeyManager(
      this.listItemsElements
    );
  }
}
```

10. Finally, we need to listen to the keyboard events. For this, you could either use a `keydown` event or a `window:keydown` event. For simplicity of the recipe, we'll go with the `window:keydown` event, as follows:

```
import { FocusKeyManager } from '@angular/cdk/a11y';
import { AfterViewInit, Component, HostListener, Input,
OnInit, QueryList, ViewChildren } from '@angular/core';
...
@Component({...})
export class TheAmazingListComponent implements OnInit,
AfterViewInit {
  ...
  @HostListener('window:keydown', ['$event'])
  onKeydown(event) {
    this.listKeyManager.onKeydown(event);
  }
  constructor() { }
  ...
}
```

Awesomesauce! You've just learned how to implement keyboard navigation using the Angular CDK. See the next section to understand how it works.

How it works...

The Angular CDK provides the `ListKeyManager` class, which allows you to implement keyboard navigation. There are a bunch of techniques we can use with the `ListKeyManager` class, and for this particular recipe, we chose the `FocusKeyManager` class. In order to make it work for a list of items, we need to do the following things:

1. Create a component for each item in the list.

2. Use `ViewChildren()` with `QueryList` in the list component to query all the list-item components.

3. Create a `FocusKeyManager` instance in the list component, providing the type of the list-item component.

4. Add a keyboard listener to the list component and pass the event to the instance of the `FocusKeyManager` class.

When we define the `listKeyManager` property in the `TheAmazingListComponent` class, we define its type as well by specifying it as `FocusKeyManager<TheAmazingListItemComponent>`. This makes it easier to understand that our `FocusKeyManager` class is supposed to work with an array of `TheAmazingListItemComponent` elements. Therefore, in the `ngAfterViewInit()` method, we specify `this.listKeyManager = new FocusKeyManager(this.listItemsElements);`, which provides a queried list of `TheAmazingListItemComponent` elements.

Finally, when we listen to the `window:keydown` event, we take the `keydown` event received in the handler and provide it to the instance of our `FocusKeyManager` class as `this.listKeyManager.onKeydown(event);`. This tells our `FocusKeyManager` instance which key was pressed and what it has to do.

Notice that our `TheAmazingListItemComponent` class implements the `FocusableOption` interface, and it also has the `focus()` method, which the `FocusKeyManager` class uses behind the scenes when we press the keyboard arrow-down or arrow-up keys.

See also

- Angular CDK accessibility documentation (`https://material.angular.io/cdk/a11y/overview`)

Pointy little popovers with the Overlay API

This is one of the advanced recipes in this book, especially for those of you who have already been working with Angular for a while. In this recipe, we'll not only create some popovers using the CDK Overlay API, but we'll also make them pointy, just like tooltips, and that's where the fun lies.

Getting ready

The project for this recipe resides in `chapter09/start_here/pointy-little-popovers`. Proceed as follows:

1. Open the project in VS Code.

2. Open the terminal and run `npm install` to install the dependencies of the project.

3. Once done, run `ng serve -o`.

   This should open the app in a new browser tab, as follows:

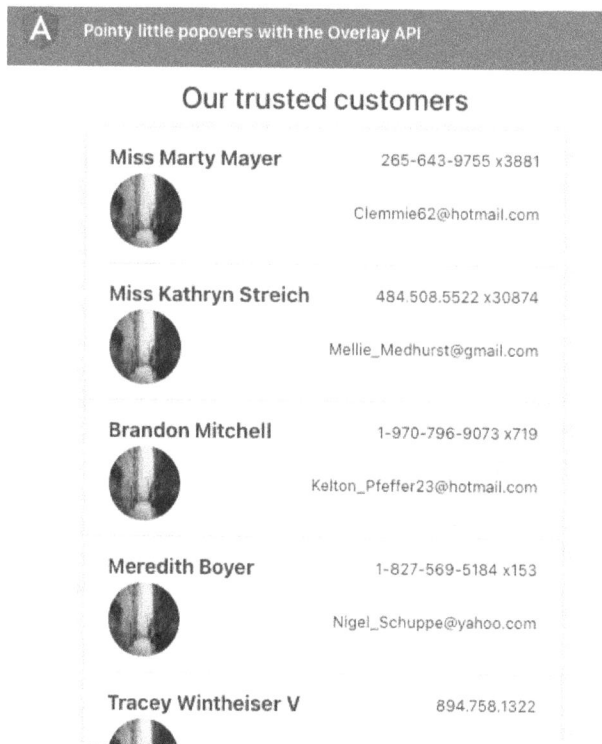

Figure 9.6 – The pointy-little-popovers app running on http://localhost:4200

Now that we have the app running locally, let's see the steps of the recipe in the next section.

How to do it...

Our app has a list of users that we can scroll through on the page. We'll add a popover menu to each item so that a drop-down menu is shown with some actions. We already have the @angular/cdk package installed, so we don't need to worry about that. Let's start with the recipe, as follows:

1. First, we need to install the @angular/cdk as we need to import the OverlayModule class into our AppModule class so that we can use the Overlay API. Update the app.module.ts file, as follows:

    ```
    . . .
    import { TheAmazingListItemComponent } from './
    components/the-amazing-list-item/the-amazing-list-item.
    component';
    import { OverlayModule } from '@angular/cdk/overlay';

    @NgModule({
      declarations: [...],
      imports: [
        . . .
        ScrollingModule,
        OverlayModule
      ],
      providers: [],
      bootstrap: [AppComponent]
    })
    export class AppModule { }
    ```

2. We'll first add the Overlay's default styles so that when the overlay is displayed, it is positioned correctly. Open the src/styles.scss file and update it as per the following gist:

    ```
    https://gist.github.com/AhsanAyaz/
    b039814e898b3ebe471b13880c7b4270
    ```

3. Now, we'll create variables to hold the overlay trigger (for the origin of the positions of the opened overlay) and the actual relative position's settings. Open the `the-amazing-list.component.ts` file and update it, as follows:

```
import { FocusKeyManager } from '@angular/cdk/a11y';
import { CdkOverlayOrigin } from '@angular/cdk/overlay';
...

@Component({...})
export class TheAmazingListComponent implements OnInit,
AfterViewInit {
  @Input() listItems: Partial<AppUserCard>[] = [];
  @ViewChildren(TheAmazingListItemComponent)
  listItemsElements: QueryList
  <TheAmazingListItemComponent>;
  popoverMenuTrigger: CdkOverlayOrigin;
  menuPositions = [
    { offsetY: 4, originX: 'end', originY: 'bottom',
    overlayX: 'end', overlayY: 'top' },
    { offsetY: -4, originX: 'end', originY: 'top',
    overlayX: 'end', overlayY: 'bottom' },
  ];
  private listKeyManager: FocusKeyManager
  <TheAmazingListItemComponent>;
  ...
}
```

4. Now, open the `the-amazing-list.component.html` file and add the `cdkOverlayOrigin` directive to the `<app-the-amazing-list-item>` selector so that we can have each list item as an origin for the pop-up menu, as follows:

```
<h4 class="heading">Our trusted customers</h4>
<cdk-virtual-scroll-viewport
  class="list list-group"
  [itemSize]="110">
  <app-the-amazing-list-item
    cdkOverlayOrigin #itemTrigger="cdkOverlayOrigin"
```

```
      class="list__item list-group-item"
      *cdkVirtualFor="let item of listItems"
      [item]="item">
    </app-the-amazing-list-item>
  </cdk-virtual-scroll-viewport>
```

5. We need to somehow pass the #itemTrigger variable from the template
    to assign its value to the popoverMenuTrigger property in the
    TheAmazingListComponent class. To do so, create a method named
    openMenu() in the the-amazing-list.component.ts file, as follows:

```
    ...

    @Component({...})
    export class TheAmazingListComponent implements OnInit,
    AfterViewInit {
      ...
      ngOnInit(): void {
      }

      openMenu($event, itemTrigger) {
        if ($event) {
          $event.stopImmediatePropagation();
        }
        this.popoverMenuTrigger = itemTrigger;
      }
      ...
    }
```

6. We also need a property to show/hide the popover menu. Let's create it and set it
    to true in the openMenu() method as well. Update the the-amazing-list.
    component.ts file, as follows:

```
    ...
    @Component({...})
    export class TheAmazingListComponent implements OnInit,
    AfterViewInit {
```

```
...
  popoverMenuTrigger: CdkOverlayOrigin;
  menuShown = false;
  ...
  openMenu($event, itemTrigger) {
    if ($event) {
      $event.stopImmediatePropagation();
    }
    this.popoverMenuTrigger = itemTrigger;
    this.menuShown = true;
  }
  ...
}
```

7. We'll now create an actual overlay. To do so, we'll create an `<ng-template>` element with the `cdkConnectedOverlay` directive. Modify your `the-amazing-list.component.html` file, as follows:

```html
<h4 class="heading">Our trusted customers</h4>
<cdk-virtual-scroll-viewport>
  ...
</cdk-virtual-scroll-viewport>
<ng-template cdkConnectedOverlay
[cdkConnectedOverlayOrigin]="popoverMenuTrigger"
  [cdkConnectedOverlayOpen]="menuShown"
  [cdkConnectedOverlayHasBackdrop]="true"
  (backdropClick)="menuShown = false"
  [cdkConnectedOverlayPositions]="menuPositions"
  cdkConnectedOverlayPanelClass="menu-popover"
>
  <div class="menu-popover__list">
    <div class="menu-popover__list__item">
      Duplicate
    </div>
    <div class="menu-popover__list__item">
      Edit
```

```
        </div>
        <div class="menu-popover__list__item">
          Delete
        </div>
      </div>
    </ng-template>
```

8. We need to pass the `#itemTrigger` variable that we have on each list item to the `openMenu()` method on a click of the list item. Update the file, as follows:

```
<h4 class="heading">Our trusted customers</h4>
<cdk-virtual-scroll-viewport
  class="list list-group"
  [itemSize]="110">
  <app-the-amazing-list-item
    class="list__item list-group-item"
    *cdkVirtualFor="let item of listItems"
    (click)="openMenu($event, itemTrigger)"
    cdkOverlayOrigin #itemTrigger="cdkOverlayOrigin"
    [item]="item">
  </app-the-amazing-list-item>
</cdk-virtual-scroll-viewport>

<ng-template>
  ...
</ng-template>
```

9. If you refresh the app now and click on any of the list items, you should see a drop-down menu being shown, as follows:

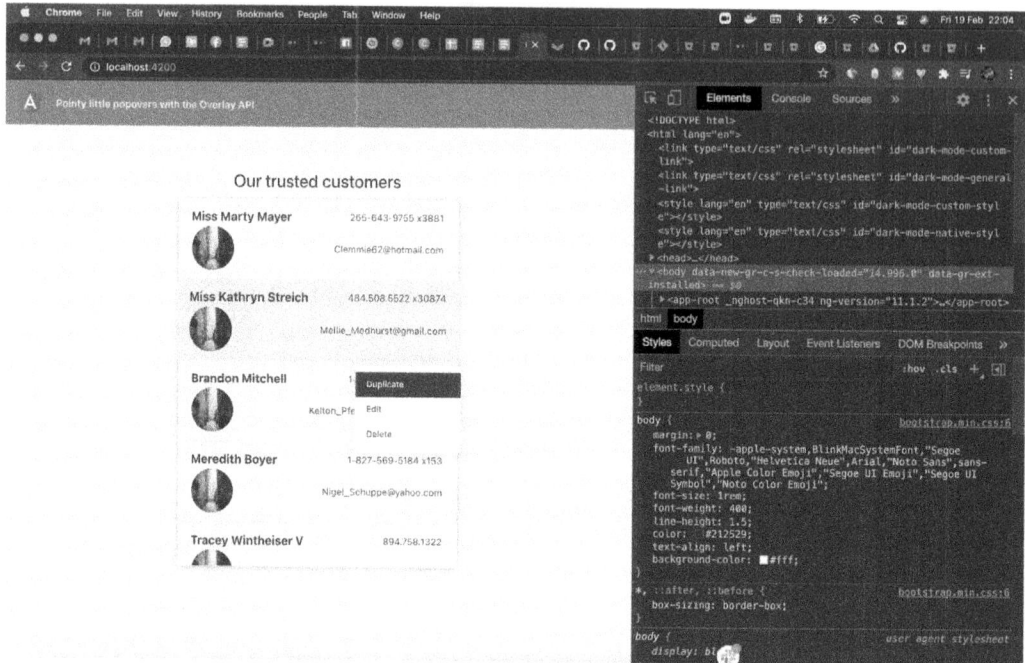

Figure 9.7 – Working drop-down menu for each list item

10. We now have to implement the part where we show a pointy little arrow with the drop-down menu so that we can correlate the drop-down menu with the list item. First, add the following styles to the `.popover-menu` class in the `src/styles.scss` file:

```
...
.menu-popover {
  min-width: 150px;
  height: auto;
  border: 1px solid white;
  border-radius: 8px;

  &::before {
    top: -10px;
    border-width: 0px 10px 10px 10px;
    border-color: transparent transparent white transparent;
  }
}
```

```
    position: absolute;
    content: '';
    right: 5%;
    border-style: solid;
  }

  &__list {...}
}
```

You should now be able to see a pointy arrow on the top right of the drop-down menu, but if you try clicking the last item on the screen, you'll see that the drop-down menu opens upward but still shows the pointer at the top, as follows:

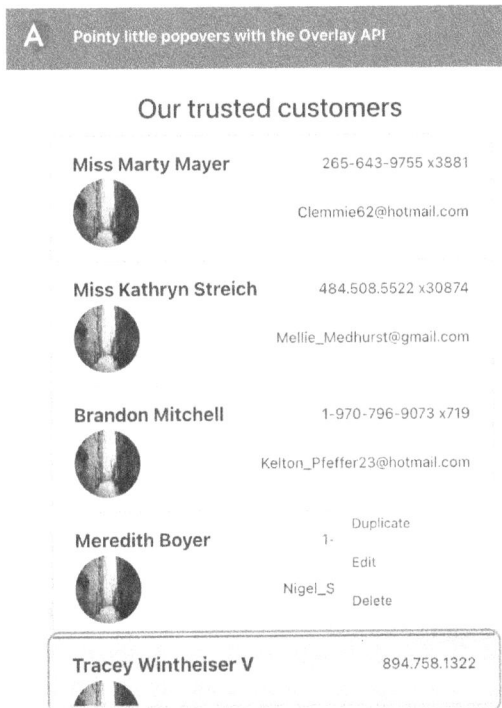

Figure 9.8 – Drop-down arrow pointing at the wrong list item

11. To point to the actual origin of the popover/drop-down menu, we need to implement a custom directive that applies a custom class. Let's start by creating a directive, as follows:

```
ng g directive directives/popover-positional-class
```

12. Update the code in the `popover-positional-class.directive.ts` generated file as per the following gist:

    `https://gist.github.com/AhsanAyaz/f28893e90b71cc03812287016192d294`

13. Now, open the `the-amazing-list.component.html` file to apply our directive to the `cdkConnectedOverlay` directive. Update the `<ng-template>` element in the file, as follows:

```
...
<ng-template cdkConnectedOverlay
[cdkConnectedOverlayOrigin]="popoverMenuTrigger"
  [cdkConnectedOverlayOpen]="menuShown"
  [cdkConnectedOverlayHasBackdrop]="true"
  (backdropClick)="menuShown = false"
  [cdkConnectedOverlayPositions]="menuPositions"
  appPopoverPositionalClass targetSelector=
  ".menu-popover" inverseClass="menu-popover--up"
  [originY]="menuPopoverOrigin.originY"
  (positionChange)="popoverPositionChanged($event,
  menuPopoverOrigin)"
  cdkConnectedOverlayPanelClass="menu-popover"
  >
  <div class="menu-popover__list">
    ...
  </div>
</ng-template>
```

14. We now need to create a `menuPopoverOrigin` property and a `popoverPositionChanged()` method in our `the-amazing-list.component.ts` file. Update this, as follows:

```
...
import { AfterViewInit, ChangeDetectorRef, Component,
HostListener, Input, OnInit, QueryList, ViewChildren }
from '@angular/core';
...
@Component({...})
export class TheAmazingListComponent implements OnInit,
AfterViewInit {
```

```
...
menuPositions = [...];
menuPopoverOrigin = {
  originY: null
}
...
constructor(private cdRef: ChangeDetectorRef) { }

popoverPositionChanged($event, popover) {
  if (popover.originY !== $event.connectionPair.
  originY) {
    popover.originY = $event.connectionPair.originY;
  }
  this.cdRef.detectChanges();
}
...
}
```

15. Finally, let's reverse the popover pointer using this inverse class. Update the `src/
    styles.scss` file to add the following styles:

```
...
.menu-popover {
  ...
  &::before {...}
  &--up {
    transform: translateY(-20px);
    &::before {
      top: unset !important;
      transform: rotate(180deg);
      bottom: -10px;
    }
  }

  &__list {...}
}
```

And voilà! If you now refresh the page and tap each of the list items, you'll see the arrows point in the correct direction. See the following screenshot to view the popover arrow pointing downward for the last item, due to the popover being shown above the item:

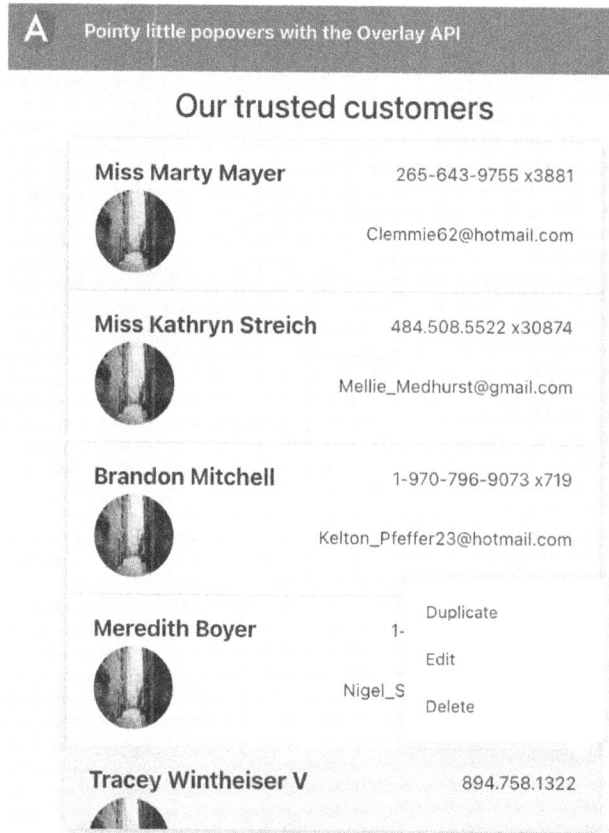

Figure 9.9 – Drop-down arrow pointing at the correct list item (pointing downward)

Great! You now know how to use the Angular CDK to work with overlays to create custom pop-up/drop-down menus. Moreover, you now know how to quickly implement the pointy arrows on the menu using a custom directive. See the next section to understand how it all works.

How it works...

Implementing an overlay using the Angular CDK Overlay API includes a couple of pieces to work with. We first have to import the `OverlayModule` class in our `AppModule` imports. Then, for creating an overlay, we need to have an overlay and an overlay trigger. In this recipe, since we're using the overlay to create a popover menu for each list item, we use the `cdkOverlayOrigin` directive on the `<app-the-amazing-list-item>` elements. Notice that the `<app-the-amazing-list-item>` elements are being rendered through the `*ngFor` directive. Therefore, in order to know which item was clicked or precisely which item we need to show the popover for, we create a `#itemTrigger` template variable on each list-item element, and you'll notice that we also bind the `(click)` event on the list items to call the `openMenu()` method, passing this `itemTrigger` template variable into it.

Now, if you have noticed the `openMenu()` method in the `the-amazing-list.component.ts` file, it looks like this:

```
openMenu($event, itemTrigger) {
    if ($event) {
      $event.stopImmediatePropagation();
    }
    this.popoverMenuTrigger = itemTrigger;
    this.menuShown = true;
}
```

Notice that we assign the `itemTrigger` property to our class's `popoverMenuTrigger` property. This is because this `popoverMenuTrigger` property is being bound with the actual overlay in our template. You can also see that we set the `menuShown` property to `true`, and this is because it will decide whether the overlay should be shown or hidden.

Now, let's see the code for the actual overlay, as follows:

```
<ng-template cdkConnectedOverlay
[cdkConnectedOverlayOrigin]="popoverMenuTrigger"
  [cdkConnectedOverlayOpen]="menuShown"
  [cdkConnectedOverlayHasBackdrop]="true"
  (backdropClick)="menuShown = false"
  [cdkConnectedOverlayPositions]="menuPositions"
  appPopoverPositionalClass targetSelector=".menu-popover"
  inverseClass="menu-popover--up"
  [originY]="menuPopoverOrigin.originY"
  (positionChange)="popoverPositionChanged($event,
```

```
menuPopoverOrigin)"
  cdkConnectedOverlayPanelClass="menu-popover"
  >
  . . .
</ng-template>
```

Let's discuss each of the `cdkConnectedOverlay` directive's attributes, one by one:

- The `cdkConnectedOverlay` attribute: This is the actual overlay directive that makes the `<ng-template>` element an Angular CDK overlay.

- The `[cdkConnectedOverlayOrigin]` attribute: This tells the Overlay API what the origin of this overlay is This is to help the CDK decide where to position the overlay when opened.

- The `[cdkConnectedOverlayOpen]` attribute: This decides whether the overlay should be shown or not.

- The `[cdkConnectedOverlayHasBackdrop]` attribute: This decides whether the overlay should have a backdrop or not—that is, if it has a backdrop, the user shouldn't be able to click anything else apart from the overlay when it is open.

- The `(backdropClick)` attribute: This is the event handler for when we click the backdrop. In this case, we're setting the `menuShown` property to `false`, which hides/closes the overlay.

- The `[cdkConnectedOverlayPositions]` attribute: This provides the positioning configuration to the Overlay API. It is an array of preferred positions that defines whether the overlay should be shown below the origin, on top of the origin, on the left, on the right, how far from the origin, and so on.

- The `[cdkConnectedOverlayPanelClass]` attribute: A **Cascading Style Sheets (CSS)** class to be applied to the generated overlay. This is used for styling.

With all of the attributes set correctly, we are able to see the overlay working when tapping the list items. *"But what about the pointy arrows, Ahsan?"* Well, hold on! We'll discuss them too.

So, the Angular CDK Overlay API already has a lot of things covered, including where to position the overlay based on the available space, and since we want to show the pointy arrows, we'll have to analyze whether the overlay is being shown above the item or below the item. By default, we have the following styles set in the `src/styles.scss` file to show the pointy arrow below the popover:

```scss
.menu-popover {
  ...

  &::before {
    top: -10px;
    border-width: 0px 10px 10px 10px;
    border-color: transparent transparent white  transparent;
    position: absolute;
    content: '';
    right: 5%;
    border-style: solid;
  }

  &--up {...}

  &__list {...}
}
```

And then, we have the `--up` modifier class, as follows, to show the overlay *above* the popover:

```scss
.menu-popover {
  ...
  &::before {...}

  &--up {
    transform: translateY(-20px);
    &::before {
      top: unset !important;
      transform: rotate(180deg);
      bottom: -10px;
    }
  }
```

```
    }


    &__list {...}
}
```

Notice in the preceding code snippet that we rotate the arrow to `180deg` to invert its pointer.

Now, let's talk about how and when this `--up` modifier class is applied. We have created a custom directive named `appPopoverPositionalClass`. This directive is also applied to the `<ng-template>` element we have for the overlay—that is, this directive is applied with the `cdkConnectedOverlay` directive and expects the following input attributes:

- The `appPopoverPositionalClass` attribute: The actual directive selector.

- The `targetSelector` attribute: The query selector for the element that is generated by the Angular CDK Overlay API. Ideally, this should be the same as what we use in `cdkConnectedOverlayPanelClass`.

- The `inverseClass` attribute: The class to be applied when the vertical position (`originY`) of the overlay is changed—that is, from `"top"` to `"bottom"`, and vice versa.

- The `originY` attribute: The `originY` position of the overlay at the moment. The value is either `"top"` or `"bottom"`, based on the overlay position.

We have a `(positionChange)` listener on the CDK Overlay `<ng-template>` element that triggers the `popoverPositionChanged()` method as soon as the overlay position changes. Notice that inside the `popoverPositionChanged()` method, upon getting a new position, we update the `popover.originY` property that is updating `menuPopoverOrigin.originY`, and then we're also passing `menuPopoverOrigin.originY` as the `[originY]` attribute to our `appPopoverPositionalClass` directive. Since we're passing it to the directive, the directive knows if the overlay position is `"top"` or `"bottom"` at any particular time. How? Because we're using the `ngOnChanges` life cycle hook in the directive to listen to the `originY` attribute/input, and as soon as we get a different value for `originY`, we either add the value of `inverseClass` as a CSS class to the Overlay element or remove it based on the value of the `originY` attribute. Also, based on the applied CSS classes, the direction of the popover arrow is decided for the overlay.

See also

- Angular CDK Overlay API (`https://material.angular.io/cdk/overlay/overview`)

- `CdkOverlayOrigin` directive documentation (`https://material.angular.io/cdk/overlay/api#CdkOverlayOrigin`)

Using CDK Clipboard to work with the system clipboard

You may have visited hundreds of websites over time, and you might have seen a feature called **Click to copy** on some of them. This is usually used when you have a long text or a link that you need to copy, and you'll find it way more convenient to just click to copy instead of selecting and then pressing the keyboard shortcuts. In this recipe, we're going to learn how to use the Angular CDK Clipboard API to copy text to the clipboard.

Getting ready

The project for this recipe resides in `chapter09/start_here/using-cdk-clipboard-api`. Proceed as follows:

1. Open the project in VS Code.

2. Open the terminal and run `npm install` to install the dependencies of the project.

3. Once done, run `ng serve -o`.

This should open the app in a new browser tab, as follows:

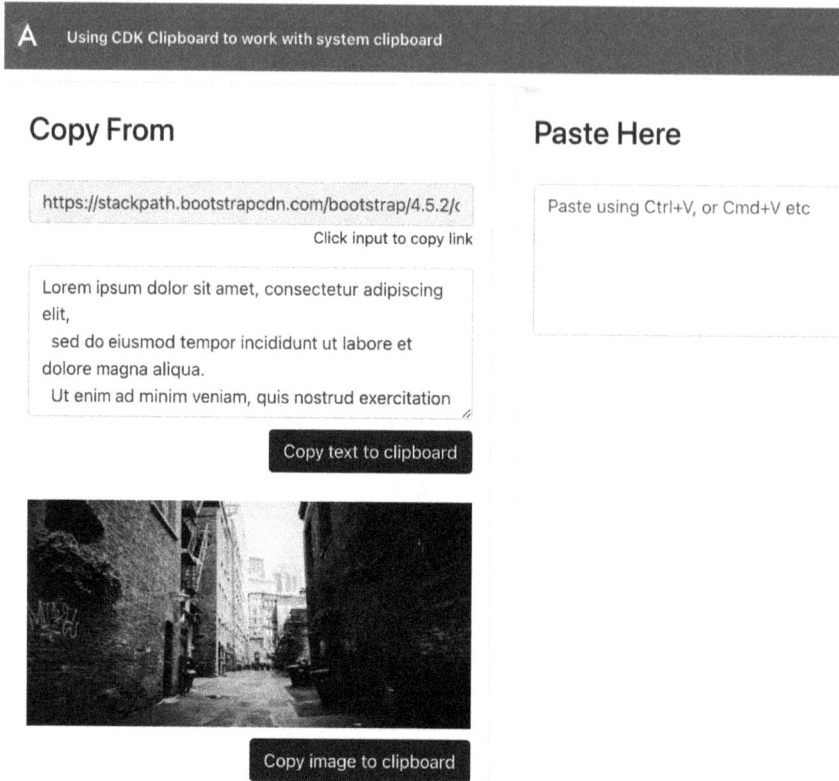

Figure 9.10 – using-cdk-clipboard-api running on http://localhost:4200

Now that we have the app running locally, let's see the steps of the recipe in the next section.

How to do it...

We have an app right now with a couple of options that don't work—that is, we should be able to copy the link, the text from the text area, and the image. In order to do so, we'll use the CDK Clipboard API. Let's get started.

1. First of all, we need to import the `ClipboardModule` class into the `imports` array of our `AppModule` class. Modify the `app.module.ts` file, as follows:

```
...
import { ClipboardModule } from '@angular/cdk/clipboard';
@NgModule({
```

```
    declarations: [...],
    imports: [
      BrowserModule,
      AppRoutingModule,
      ClipboardModule
    ],
    ...
  })
  export class AppModule { }
```

2. Now, we'll apply the click-to-copy functionality to the link. In order to do so, we'll use the cdkCopyToClipboard directive and will apply it on our link input in the app.component.html file, as follows:

```
...
<div class="content" role="main">
  <div class="content__container">
    <div class="content__container__copy-from">
      <h3>Copy From</h3>
      <div class="mb-3 content__container__copy-from__
      input-group">
        <input
          #linkInput
          [cdkCopyToClipboard]="linkInput.value"
          (click)="copyContent($event, contentTypes.
          Link)"
          class="form-control"
          type="text" readonly="true"
          value="...">
        <div class="hint">...</div>
      </div>
      ...
    </div>
  </div>
```

If you click on the link input now and then try to paste it anywhere (within or outside the app), you should see the value of the link.

3. We'll now do something similar for the text input—that is, the `<textarea>`. Update the template again, as follows:

```html
...
<div class="content" role="main">
  <div class="content__container">
    <div class="content__container__copy-from">
      <h3>Copy From</h3>

      ...

      <div class="mb-3 content__container__copy-from__
      input-group">
        <textarea
          #textInput
          class="form-control"
          rows="5">{{loremIpsumText}}</textarea>
        <button
          [cdkCopyToClipboard]="textInput.value"
          (click)="copyContent($event, contentTypes.
          Text)"
          class="btn btn-dark">
          {{ contentCopied === contentTypes.Text ?
          'Text copied' : 'Copy text to clipboard'}}
        </button>
      </div>
      ...
    </div>
  </div>
</div>
```

4. Finally, we'll do something different for the image. Since the CDK Clipboard API only works with strings, we will download the image, convert it into a blob, and copy the blob **Uniform Resource Locator** (**URL**). Let's update the template first with the logic, as follows:

```html
...
<div class="content" role="main">
  <div class="content__container">
    <div class="content__container__copy-from">
      <h3>Copy From</h3>
```

```
      . . .
      <div class="mb-3 content__container__copy-from__
      input-group">
        <img src="assets/landscape.jpg">
        <button
          (click)="copyImageUrl(imageUrl);
          copyContent($event, contentTypes.Image)"
          class="btn btn-dark">
            . . .
        </button>
      </div>
    </div>
    . . .
  </div>
</div>
```

5. Now, let's implement the copyImageUrl() method to fetch the image, convert it into a blob, and copy the URL to the clipboard. Update the app.component.ts file, as follows:

```
import { Clipboard } from '@angular/cdk/clipboard';
import { Component, HostListener, OnInit } from '@
angular/core';
...
@Component({...})
export class AppComponent implements OnInit {
  ...
  constructor(private clipboard: Clipboard) {
    this.resetCopiedHash();
  }
  async copyImageUrl(srcImageUrl) {
    const data = await fetch(srcImageUrl);
    const blob = await data.blob();
    this.clipboard.copy(URL.createObjectURL(blob));
  }
  ...
}
```

Great! With this change, you can try refreshing the app. Now, you should be able to copy the link and the text, as well as the image, by clicking the input link and the buttons respectively. To understand all the magic behind this recipe, see the next section.

How it works...

In the recipe, we've used two main things from the CDK Clipboard API—one is the `cdkCopyToClipboard` directive, and the other is the `Clipboard` service. The `cdkCopyToClipboard` directive binds a click handler to the element this directive is applied to. It works both as the selector of the directive and an `@Input()` item for the directive so that it knows which value is to be copied to the clipboard when the element is clicked. In our recipe, for the link input, notice that we use `[cdkCopyToClipboard]="linkInput.value"`. This binds a click handler to the `<input>` element and also binds the `value` property of the `linkInput` template variable, which points to the value of the input that is the actual link to be copied. When we click the input, it accesses the value of the input using the `linkInput.value` binding, and we do the same for the `<text-area>` input. The only difference is that the `cdkCopyToClipboard` directive is not bound to the `<text-area>` element itself. The reason is that we want to bind the click handler to the button below the text area instead. Therefore, on the button for copying the text, we have the `[cdkCopyToClipboard]="textInput.value"` binding.

For the image, we do something different. We use the `Clipboard` service from the `@angular/cdk/clipboard` package to manually copy the blob URL. We create a method named `copyImageUrl()`, which is called when clicking the button for copying the image. We pass the `imageUrl` property to this method, which in turn downloads the image, reads it as a blob, and generates the blob URL, which is copied to the clipboard using the `copy()` method of the `Clipboard` service.

See also

- CDK Clipboard documentation (`https://material.angular.io/cdk/clipboard/overview`)

Using CDK Drag and Drop to move items from one list to another

Have you ever used the Trello board app, or maybe other apps that also allow you to drag and drop list items from one list to another? Well, you can do this easily using the Angular CDK, and in this recipe, you'll learn about the Angular CDK Drag and Drop API to move items from one list to another. You'll also learn how to reorder the lists.

Getting ready

The project that we are going to work with resides in `chapter09/start_here/ using-cdk-drag-drop`, inside the cloned repository. Proceed as follows:

1. Open the project in VS Code.

2. Open the terminal and run `npm install` to install the dependencies of the project.

3. Once done, run `ng serve -o`.

    This should open the app in a new browser tab, and it should look like this:

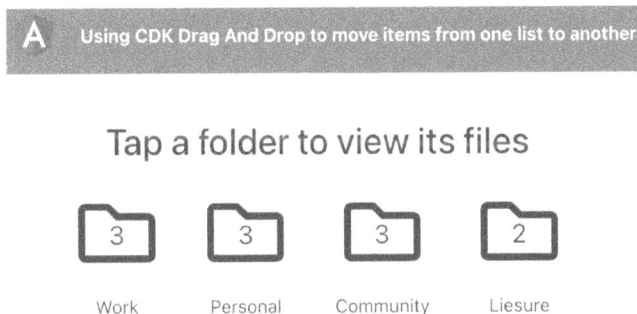

Figure 9.11 – The using-cdk-drag-drop app running on http://localhost:4200

Now that we have the app running locally, let's see the steps of the recipe in the next section.

How to do it...

For this recipe, we have an interesting app that has some folders and files. We're going to implement the drag-and-drop functionality for the files to be dragged to other folders, which should update the folder's file count instantly, and we should be able to see the file in the new folder as well. Let's get started.

1. First of all, we need to import the `DragDropModule` class into the `imports` array of our `AppModule` class. Modify the `app.module.ts` file, as follows:

```
...
import {DragDropModule} from '@angular/cdk/drag-drop';
@NgModule({
  declarations: [...],
  imports: [
    BrowserModule,
    AppRoutingModule,
    FontAwesomeModule,
    DragDropModule
  ],
  ...
})
export class AppModule { }
```

2. Now, we'll apply the `cdkDrag` directive to each of our files, and will apply the `cdkDropList` directive to each of the folders. Update the `folders-list.component.html` file, as follows:

```
<div class="folders">
  ...
  <div class="folders__list">
    <app-folder
      cdkDropList
      ...
      [folder]="folder"
    >
    </app-folder>
  </div>
  <div class="folders__selected-folder-files"
  *ngIf="selectedFolder">
```

```
        <div>
          <app-file
            cdkDrag
            *ngFor="let file of selectedFolder.files"
            [file]="file"
          ></app-file>
        </div>
      </div>
    </div>
```

3. We'll also enable reordering of the files within a folder by adding the cdkDropList directive on the container elements for the files, as follows:

```
<div class="folders">
    . . .
    <div class="folders__selected-folder-files"
    *ngIf="selectedFolder">
      <div cdkDropList>
        <app-file ...></app-file>
      </div>
    </div>
</div>
```

4. We'll now define the origin of the drag-and-drop interaction by specifying the [cdkDragData] attribute on each <app-file> element and the [cdkDropListData] attribute on each <app-folder> element, and on the files container as well. Update the template again, as follows:

```
<div class="folders">
    . . .
    <div class="folders__list">
      <app-folder
        cdkDropList
        [cdkDropListData]="folder.files"
        . . .
      >
      </app-folder>
    </div>
```

```
<div class="folders__selected-folder-files"
*ngIf="selectedFolder">
  <div
    cdkDropList
    [cdkDropListData]="selectedFolder.files"
  >
    <app-file
      cdkDrag
      [cdkDragData]="file"
      ...
    ></app-file>
  </div>
</div>
</div>
```

5. We now need to implement what happens when the file is dropped. To do so, we'll
    use the (cdkDropListDropped) event handler. Update the template, as follows:

```
<div class="folders">
  ...
  <div class="folders__list">
    <app-folder
      cdkDropList
      [cdkDropListData]="folder.files"
      (cdkDropListDropped)="onFileDrop($event)"
      ...
    >
    </app-folder>
  </div>
  <div class="folders__selected-folder-files"
  *ngIf="selectedFolder">
    <div
      cdkDropList
      [cdkDropListData]="selectedFolder.files"
      (cdkDropListDropped)="onFileDrop($event)"
    >
      ...
```

```
            </div>
          </div>
        </div>
```

6. Finally, we need to implement the `onFileDrop` method. Update the `folders-list.component.ts` file, as follows:

```typescript
...
import {
  CdkDragDrop, moveItemInArray, transferArrayItem,
} from '@angular/cdk/drag-drop';

@Component({...})
export class FoldersListComponent implements OnInit {
  ...
  onFileDrop(event: CdkDragDrop<string[]>) {
    if (event.previousContainer === event.container) {
      moveItemInArray(
        event.container.data, event.previousIndex,
        event.currentIndex
      );
    } else {
      transferArrayItem(
        event.previousContainer.data, event.container.
        data,
        event.previousIndex, event.currentIndex
      );
    }
  }
}
```

If you now refresh the app and try to drag a file to a folder, you should see something like this:

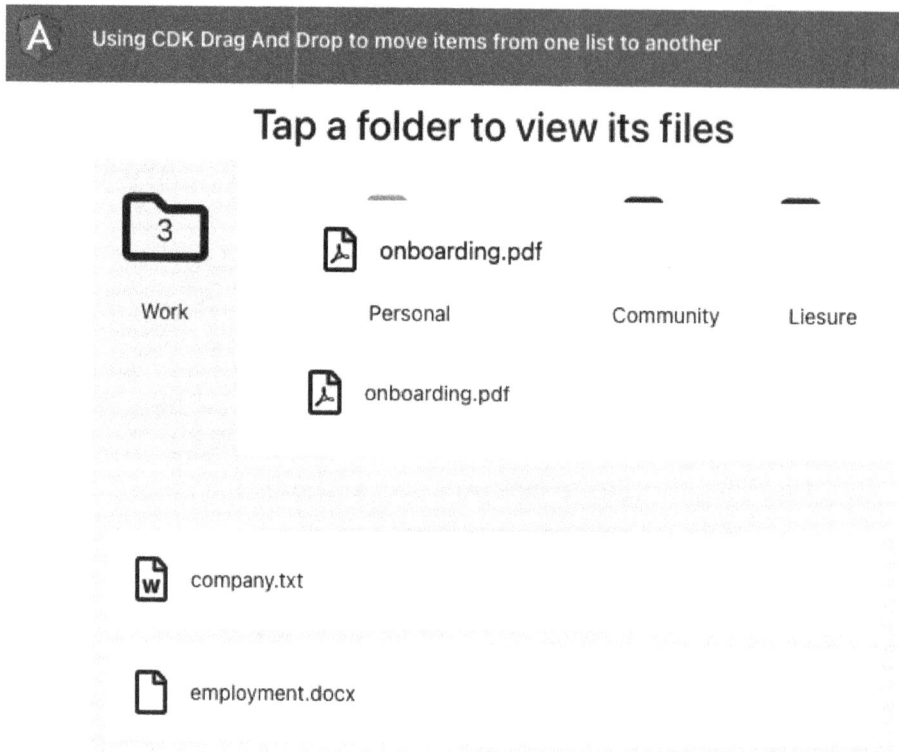

Figure 9.12 – Dragging and dropping a file to another folder

Ugly, isn't it? Well, this is because we have to fix the drag-and-drop previews in the next steps.

7. In order to handle the drag-and-drop previews, we need to enclose them into an element with the `cdkDropListGroup` directive. Update the `folders-list.component.html` file and apply the directive to the element with the `"folders"` class, as follows:

    ```
    <div class="folders" cdkDropListGroup>
    ...
    </div>
    ```

8. To apply a custom drag preview, we use a custom element with the
    `*cdkDragPreview` directive applied to it. Update the `folders-list.`
    `component.html` file, as follows:

```
<div class="folders" cdkDropListGroup>

    ...

    <div class="folders__selected-folder-files"
    *ngIf="selectedFolder">
      <div
        cdkDropList
        ...
      >
        <app-file
          cdkDrag
          ...
        >
          <fa-icon
            class="file-drag-preview"
            *cdkDragPreview
            [icon]="file.icon"
          ></fa-icon>
        </app-file>
      </div>
    </div>
</div>
```

9. We'll also need some styles for the drag-and-drop previews. Update the `folders-`
    `list.component.scss` file, as follows:

```
$folder-bg: #f5f5f5;
$file-preview-transition: transform 250ms cubic-bezier(0,
0, 0.2, 1);

.folders {...}

.file-drag-preview {
  padding: 10px 20px;
```

```
    background: transparent;
    font-size: 32px;
}
.file-drop-placeholder {
    min-height: 60px;
    transition: $file-preview-transition;
    display: flex;
    align-items: center;
    justify-content: center;
    font-size: 32px;
}
```

10. Let's also add some styles to make sure the other list items move smoothly when reordering the items within a folder. Update the src/styles.scss file, as follows:

```
...
* {
    user-select: none;
}


/* Animate items as they're being sorted. */
.cdk-drop-list-dragging .cdk-drag {
    transition: transform 250ms cubic-bezier(0, 0, 0.2, 1);
}


/* Animate an item that has been dropped. */
.cdk-drag-animating {
    transition: transform 300ms cubic-bezier(0, 0, 0.2, 1);
}
```

11. Now, we need to create a drop preview template as well. For this, we use the *cdkDragPlaceholder directive on the preview element. Update the folders-list.component.html file, as follows:

```
<div class="folders" cdkDropListGroup>
    ...
    <div class="folders__selected-folder-files"
```

```html
*ngIf="selectedFolder">
    <div cdkDropList ...>
      <app-file cdkDrag ...>
        <fa-icon class="file-drag-preview"
          *cdkDragPreview ... ></fa-icon>
        <div class="file-drop-placeholder"
        *cdkDragPlaceholder>
          <fa-icon [icon]="upArrow"></fa-icon>
        </div>
      </app-file>
    </div>
  </div>
</div>
```

12. Finally, let's create an upArrow property using the faArrowAltCircleUp icon from the @fortawesome package. Update the folders-list.component.ts file, as follows:

```typescript
import { Component, OnInit } from '@angular/core';
import { APP_DATA } from '../constants/data';
import { IFolder } from '../interfaces';
import { faArrowAltCircleUp } from '@fortawesome/free-regular-svg-icons';
import {
  CdkDragDrop,
  moveItemInArray,
  transferArrayItem,
} from '@angular/cdk/drag-drop';
import { FileIconService } from '../core/services/file-icon.service';

@Component({...})
export class FoldersListComponent implements OnInit {
  folders = APP_DATA;
  selectedFolder: IFolder = null;
  upArrow = faArrowAltCircleUp;
```

```
constructor(private fileIconService: FileIconService)
{...}

    ...

}
```

And boom! We now have a seamless **user experience (UX)** for the entire drag-and-drop flow. Like it? Make sure that you share a snapshot on your Twitter and tag me at @muhd_ahsanayaz.

Now that we've finished the recipe, let's see in the next section how it all works.

How it works...

There were a couple of interesting directives in this recipe, and we'll go through them all one by one. First of all, as good Angular developers, we import the DragDropModule class into the imports array of our AppModule, just to make sure we don't end up with errors. Then, we start making the files draggable. We do this by adding the cdkDrag directive to each file element by applying the *ngFor directive to it. This tells the Angular CDK that this element will be dragged and, therefore, the Angular CDK binds different handlers to each element to be dragged.

> **Important note**
>
> Angular components by default are not block elements. Therefore, when applying the cdkDrag directive to an Angular component such as the <app-file> component, it might restrict the animations from the CDK being applied when we're dragging the elements. In order to fix this, we need to set a display: block; for our component elements. Notice that we're applying the required styles in the folders-list.component.scss file (*line 25*) for the .folders__selected-folder-files__file class.

After configuring the drag elements, we use the cdkDropList directive to each container DOM element where we're supposed to drop the file. In our recipe, that is each folder that we see on the screen, and we can also reorder the files within a folder. Therefore we apply the cdkDropList directive to the wrapper element of the currently displayed files, as well as to each <app-folder> item with the *ngFor looping over the folders array.

Then, we specify the `data` that we're dragging by specifying `[cdkDragData]="file"` for each draggable file. This helps us identify it in the later process, when we drop it either within the current folder or within other folders. We also specify in which array this dragged item will be added when dropped upon the particular list, and we do this by specifying `[cdkDropListData]="ARRAY"` statements on the elements that we've applied the `cdkDropList` directive to. When the Angular CDK combines the information from the `cdkDragData` and the `cdkDropListData` attributes, it can easily identify if the item was dragged and then dropped within the same list or in another list.

To handle what happens when we drop the dragged file, we use the `(cdkDropListDropped)` method from the Angular CDK on the element with the `cdkDropList` directive. We take the `$event` emitted from the CDK and pass it to our `onFileDrop()` method. What's great is that within the `onFileDrop()` method, we use the `moveItemInArray()` and `transferArrayItem()` helper methods from the Angular CDK, with a really simple logic to compare the containers. That is, the Angular CDK provides us enough information that we can get away with the whole functionality really easily.

Toward the end of the recipe, we customize how our drag preview should look when we are dragging a file using a custom template, by using the `*cdkDragPreview` directive on it. This tells the Angular CDK to not render it right away but to show it with the mouse when we start dragging a file. For our recipe, we only show the icon of the file as the drag preview. And finally, we also customize the drop preview (or drag placeholder) using the `*cdkDragPlaceholder` directive, which shows a transparent rectangle with an upward-arrow icon to reflect where the item is going to be added when dropped. Of course, we had to add some custom styles for both the drag preview and the drop preview.

See also

- Angular CDK Drag and Drop documentation (`https://material.angular.io/cdk/drag-drop/overview`)

Creating a multi-step game with the CDK Stepper API

If you try finding examples of the CDK Stepper over the internet, you'll find a bunch of articles revolving around creating multi-step forms using the CDK Stepper API, but since it is a stepper at its base, it can be used for various use cases. In this recipe, we're going to build a guessing game using the Angular CDK Stepper API, in which the user will guess what the output of a rolled dice will be.

Getting ready

The project that we are going to work with resides in `chapter09/start_here/using-cdk-stepper`, inside the cloned repository. Proceed as follows:

1. Open the project in VS Code.

2. Open the terminal and run `npm install` to install the dependencies of the project.

3. Once done, run `ng serve -o`.

    This should open the app in a new browser tab, and you should see something like this:

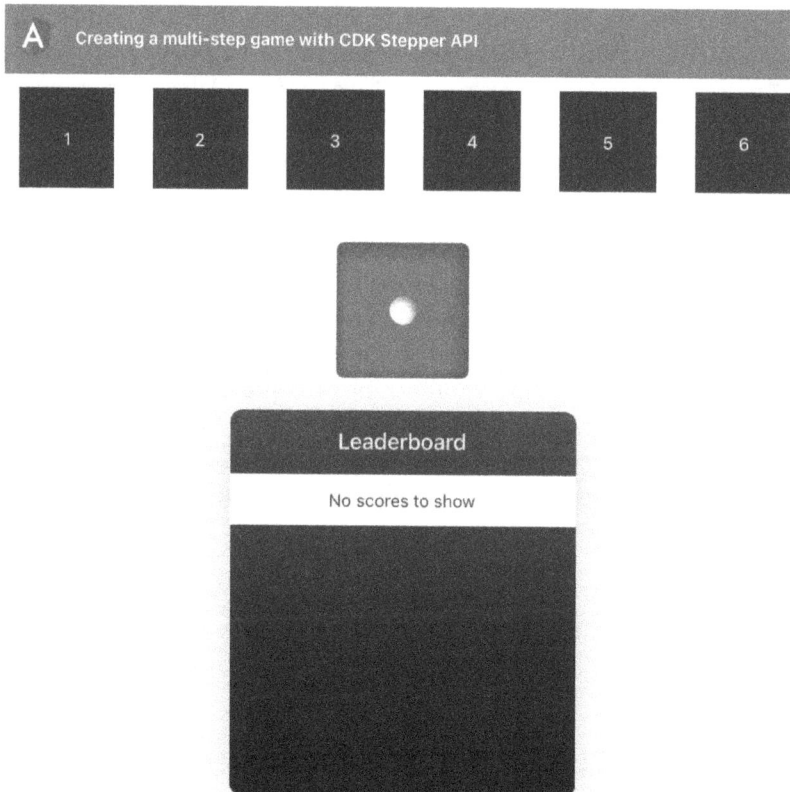

Figure 9.13 – The using-cdk-stepper app running on http://localhost:4200

Now, let's look at how to create a multi-step game with the CDK Stepper API in the next section.

How to do it...

We have a really simple yet interesting application at hand that has a couple of components built already, including the dice component, the value-guess component, and the leaderboard component. We'll create this game as a multi-step game using the Stepper API. Proceed as follows:

1. First, open a new terminal window/tab and make sure you're inside the `ch8/start_here/using-cdk-stepper` folder. Once inside, run the following command to install the Angular CDK:

   ```
   npm install --save @angular/cdk@12.0.0
   ```

2. You'll have to restart your Angular server, so rerun the `ng serve` command.

3. Now, import the `CdkStepperModule` class from the `@angular/cdk` package in your `app.module.ts` file, as follows:

   ```
   ...
   import { LeaderBoardComponent } from './components/
   leader-board/leader-board.component';
   import { CdkStepperModule } from '@angular/cdk/stepper';
   ...
   @NgModule({
     declarations: [...],
     imports: [BrowserModule, AppRoutingModule,
     ReactiveFormsModule, CdkStepperModule],
     providers: [],
     bootstrap: [AppComponent],
   })
   export class AppModule {}
   ```

4. Let's create our stepper component now. Run the following command in the project folder:

   ```
   ng g c components/game-stepper
   ```

5. To make our component a `CdkStepper`, we need to provide it using the
    `CdkStepper` token and have to extend our component class from `CdkStepper`
    as well. We can remove the `constructor`, the `OnInit` implementation, and the
    `ngOnInit` method. Modify the `game-stepper.component.ts` file, as follows:

    ```
    import { Component } from '@angular/core';
    import { CdkStepper } from '@angular/cdk/stepper';

    @Component({
      selector: 'app-game-stepper',
      templateUrl: './game-stepper.component.html',
      styleUrls: ['./game-stepper.component.scss'],
      providers: [{ provide: CdkStepper, useExisting:
      GameStepperComponent }],
    })
    export class GameStepperComponent extends CdkStepper {
    }
    ```

    Notice that we have removed the usage of `ngOnInit` and the `OnInit` life cycle
    since we don't want these for this component.

6. Let's add the template for our `<game-stepper>` component. We'll start by
    adding the header that will show the step label. Update your `game-stepper.`
    `component.html` file, as follows:

    ```
    <section class="game-stepper">
      <header>
        <h3>
          <ng-container
            *ngIf="selected.stepLabel; else showLabelText"
            [ngTemplateOutlet]="
            selected.stepLabel.template"
          >
          </ng-container>
          <ng-template #showLabelText>
            {{ selected.label }}
          </ng-template>
        </h3>
    ```

```
    </header>
  </section>
```

7. Now, we'll add the template to show our main content for the selected step—this is pretty simple to do. We need to add a div with the `[ngTemplateOutlet]` attribute, where we'll show the content. Update the `game-stepper.component.html` file, as follows:

```html
<section class="game-stepper">
  <header>

    . . .

  </header>
  <section class="game-stepper__content">
    <div [ngTemplateOutlet]="selected ? selected.content
    : null"></div>
  </section>

  . . .

</section>
```

8. Finally, we'll add a footer element that'll contain the navigation buttons for our stepper— that is, we should be able to jump to the next and the previous step using those navigation buttons. Update the `game-stepper.component.html` file further, as follows:

```html
<section class="game-stepper">

  . . .

  <section class="game-stepper__content">
    <div [ngTemplateOutlet]="selected ? selected.content
    : null"></div>
  </section>
  <footer class="game-stepper__navigation">
    <button
      class="game-stepper__navigation__button btn
      btn-primary"
      cdkStepperPrevious
      [style.visibility]="steps.get(selectedIndex - 1) ?
      'visible' : 'hidden'"
    >
```

```
        &larr;
      </button>

      <button
        class="game-stepper__navigation__button btn
        btn-primary"
        cdkStepperNext
        [style.visibility]="steps.get(selectedIndex + 1) ?
        'visible' : 'hidden'"
      >
        &rarr;
      </button>
    </footer>
  </section>
```

9. Let's add some styles to our game-stepper component. Modify the game-stepper.component.scss file, as follows:

```scss
.game-stepper {
  display: flex;
  flex-direction: column;
  align-items: center;

  &__navigation {
    width: 100%;
    display: flex;
    align-items: center;
    justify-content: space-between;
    > button {
      margin: 0 8px;
    }
  }

  &__content {
    min-height: 350px;
    display: flex;
```

```
    justify-content: center;
    align-items: center;
    flex-direction: column;
  }

  header,
  footer {
    margin: 10px auto;
  }
}
```

10. We'll now wrap our entire template in the `game.component.html` file with the `<app-game-stepper>` component. Update the file, as follows:

```
<app-game-stepper>
  <form (ngSubmit)="submitName()" [formGroup]="nameForm">
    . . .
  </form>
  <app-value-guesser></app-value-guesser>
  <app-dice></app-dice>
  <app-leader-board></app-leader-board>
</app-game-stepper>
```

11. We'll now modify our `game.component.html` file to break down the inner template into steps. For that, we'll use the `<cdk-step>` element to wrap around the content for each step. Update the file, as follows:

```
<app-game-stepper>
  <cdk-step>
    <form (ngSubmit)="submitName()"
    [formGroup]="nameForm">
      . . .
    </form>
  </cdk-step>
  <cdk-step>
    <app-value-guesser></app-value-guesser>
    <app-dice></app-dice>
```

```
    </cdk-step>
    <cdk-step>
      <app-leader-board></app-leader-board>
    </cdk-step>
  </app-game-stepper>
```

12. Now, we'll add a label for each step to show our main content for the selected step—
    this is pretty simple to do. We need to add an `<ng-template>` element within
    each `<cdk-step>` element. Update the `game.component.html` file, as follows:

```
<app-game-stepper>
  <cdk-step>
    <ng-template cdkStepLabel>Enter your
    name</ng-template>
    <form (ngSubmit)="submitName()"
    [formGroup]="nameForm">

      ...

    </form>
  </cdk-step>
  <cdk-step>
    <ng-template cdkStepLabel>Guess what the value
    will be when the die is rolled</ng-template>
    <app-value-guesser></app-value-guesser>
    <app-dice></app-dice>
  </cdk-step>
  <cdk-step>
    <ng-template cdkStepLabel> Results</ng-template>
    <app-leader-board></app-leader-board>
  </cdk-step>
</app-game-stepper>
```

If you refresh the app, you should see the first step as the visible step, as well as the bottom navigation button, as follows:

Figure 9.14 – The first step and the navigation button using CDKStepper

13. Now, we need to make sure that we can only move forward to the second step once we have entered a name in the first step. Make the following changes to the game.component.html file:

```
<app-game-stepper [linear]="true">
  <cdk-step [completed]="!!nameForm.get('name').value">
    <ng-template cdkStepLabel> Enter your
    name</ng-template>
    <form (ngSubmit)="submitName()"
    [formGroup]="nameForm">
      <div class="mb-3" *ngIf="nameForm.get('name')
      as nameControl">
        . . .
      </div>
      <button ← REMOVE THIS
        type="submit"
        [disabled]="!nameForm.valid"
        class="btn btn-primary"
      >
        Submit
      </button>
    </form>
  </cdk-step>
```

```
    . . .
  </app-game-stepper>
```

14. We also need to disable the next button on the first step until we have entered a value for the player name. To do so, update the game-stepper.component. html file—specifically, the element with the cdkStepperNext attribute—as follows:

```
<section class="game-stepper">
  . . .
  <footer class="game-stepper__navigation">
    . . .
    <button
      class="game-stepper__navigation__button btn
      btn-primary"
      cdkStepperNext
      [disabled]="!selected.completed"
      [style.visibility]="steps.get(selectedIndex + 1) ?
      'visible' : 'hidden'"
    >
      &rarr;
    </button>
  </footer>
</section>
```

15. To handle the case when the user provides the name and hits the *Enter* key, resulting in the form submission, we can handle moving to the next step using a @ ViewChild() in the GameComponent class. Modify the game.component.ts file as follows, and try entering the name and then pressing the *Enter* key:

```
import { CdkStepper } from '@angular/cdk/stepper';
import { Component, OnInit, ViewChild } from '@angular/
core';
import { FormControl, FormGroup, Validators } from '@
angular/forms';

@Component({...})
export class GameComponent implements OnInit {
  @ViewChild(CdkStepper) stepper: CdkStepper;
```

```
    nameForm = new FormGroup({
      name: new FormControl('', Validators.required),
    });
  ...
    submitName() {
      this.stepper.next();
    }
}
```

16. Now, let's write the flow for guessing the number. Update the game.component. ts file, as follows:

```
  ...
  import { DiceComponent } from '../components/dice/dice.
  component';
  import { ValueGuesserComponent } from '../components/
  value-guesser/value-guesser.component';
  import { IDiceSide } from '../interfaces/dice.interface';

  @Component({...})
  export class GameComponent implements OnInit {
    @ViewChild(CdkStepper) stepper: CdkStepper;
    @ViewChild(DiceComponent) diceComponent: DiceComponent;
    @ViewChild(ValueGuesserComponent)
    valueGuesserComponent: ValueGuesserComponent;
    guessedValue = null;
    isCorrectGuess = null;
    ...
    submitName() {...}

    rollTheDice(guessedValue) {
      this.isCorrectGuess = null;
      this.guessedValue = guessedValue;
      this.diceComponent.rollDice();
    }

    showResult(diceSide: IDiceSide) {
      this.isCorrectGuess = this.guessedValue === diceSide.
```

```
value;
    }
}
```

17. Now that we have the functions in place, let's update the template to listen to the event listeners from the `<app-value-guesser>` and `<app-dice>` components and to act accordingly. We'll also add `.alert` elements to show messages in case of a successful or wrong guess. Update the `game.component.html` file, as follows:

```
<app-game-stepper [linear]="true">
  <cdk-step [completed]="!!nameForm.get('name').value">
    ...
  </cdk-step>
  <cdk-step [completed]="isCorrectGuess !== null">
    <ng-template cdkStepLabel
      >Guess what the value will be when the die is
      rolled</ng-template
    >
    <app-value-guesser (valueGuessed)="rollTheDice
    ($event)"></app-value-guesser>
    <app-dice (diceRolled)="showResult($event)">
    </app-dice>
    <ng-container [ngSwitch]="isCorrectGuess">
      <div class="alert alert-success"
      *ngSwitchCase="true">
        You rock {{ nameForm.get('name').value }}!
        You got 50 points
      </div>
      <div class="alert alert-danger"
      *ngSwitchCase="false">
        Oops! Try again!
      </div>
    </ng-container>
  </cdk-step>
  <cdk-step>...</cdk-step>
</app-game-stepper>
```

18. Finally, we need to populate the leaderboards. Update the `game.component.ts` file to use the `LeaderboardService` class, as follows:

```
...
import { LeaderboardService } from '../core/services/
leaderboard.service';
import { IDiceSide } from '../interfaces/dice.interface';
import { IScore } from '../interfaces/score.interface';
@Component({...})
export class GameComponent implements OnInit {
  ...
  scores: IScore[] = [];
  constructor(private leaderboardService:
  LeaderboardService) {}
  ngOnInit(): void {
    this.scores = this.leaderboardService.getScores();
  }
  ...
  showResult(diceSide: IDiceSide) {
    this.isCorrectGuess = this.guessedValue ===
    diceSide.value;
    if (!this.isCorrectGuess) {
      return;
    }
    this.scores = this.leaderboardService.setScores({
      name: this.nameForm.get('name').value,
      score: 50,
    });
  }
}
```

19. Now, update the `game.component.html` file to pass the scores as an attribute to the `<app-leader-board>` component, as follows:

```
<app-game-stepper [linear]="true">
  <cdk-step [completed]="!!nameForm.get('name').value">
    ...
  </cdk-step>
```

```
<cdk-step [completed]="isCorrectGuess !== null">
  ...
</cdk-step>
<cdk-step>
  <ng-template cdkStepLabel>Results</ng-template>
  <app-leader-board [scores]="scores"></app-leader-
  board>
</cdk-step>
</app-game-stepper>
```

If you refresh the app now and play the game, you should be able to see the leaderboard, as follows:

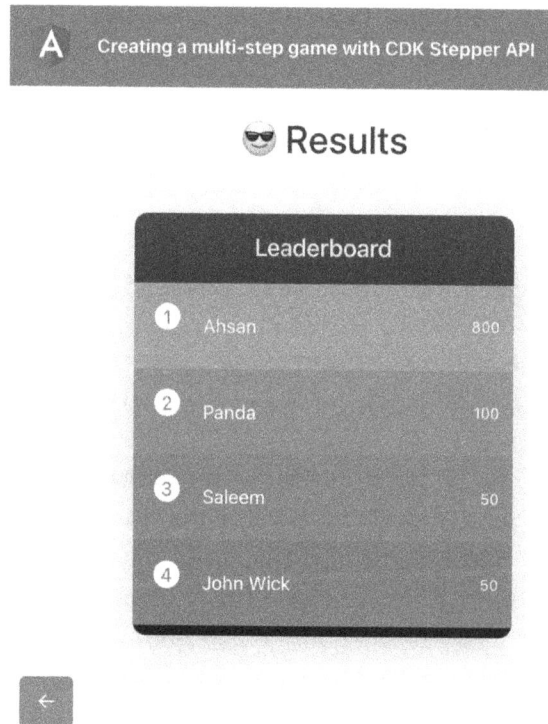

Figure 9.15 – Displaying results in the leaderboard at Step 3

Phew! That was a LONG recipe! Well, perfection requires time and dedication. Feel free to use this game yourself or even with your friends, and if you improve it, do let me know on my socials.

Now that you've finished the recipe, see the next section on how this works.

How it works...

There are a lot of moving parts in this recipe, but they're super-easy. First, we import the `CdkStepperModule` class into the `imports` array of our `AppModule` class. Then, we create a component that extends the `CdkStepper` class. The reason for extending the `CdkStepper` class is to be able to create this `GameStepperComponent` component so that we can create a reusable template with some styles, and even some custom functionality.

To start using the `GameStepperComponent` component, we wrap the entire template within the `<app-game-stepper>` element in the `game.component.html` file. Since the component extends the `CdkStepper` API, we can use all the functionality of the `CdkStepper` component here. For each step, we use the `<cdk-step>` element from the CDK and wrap the template of the step inside it. Notice that in the `game-stepper.component.html` file, we use the `[ngTemplateOutlet]` attribute for both the step's label and the step's actual content. This is a reflection of how amazing the `CdkStepper` API is. It automatically generates the `step.label` property and the `content` property on each step based on the values/template we provide for each step. Since we provide an `<ng-template cdkStepLabel>` inside each `<cdk-step>` element, the CDK generates a `step.stepLabel.template` automatically, which we then use inside the `game-stepper.component.html` file, as mentioned. If we didn't provide it, it would then use the `step.label` property as per our code.

For the bottom navigation buttons, you notice that we use `<button>` elements with the `cdkStepperPrevious` and `cdkStepperNext` directives for going to the previous step and the next step respectively. We also show/hide the next and previous button based on the conditions to check if there is a step to go to. We hide the navigation button using the `[style.visibility]` binding, as you see in the code.

One interesting thing about the `CdkStepper` API is that we can tell whether the user should be able to go to the next steps and backward, regardless of the state of the current step, or whether the user should first do something in the current step to go to the next one. The way we do it is by using the `[linear]` attribute on our `<app-game-stepper>` element, by setting its value to `true`. This tells the `CdkStepper` API to not move to the next step using the `cdkStepperNext` button, until the current step's `completed` property is `true`. While just providing `[linear]="true"` is enough to handle the functionality, we improve the UX by disabling the **Next** button—in this case, by using `[disabled]="!selected.completed"` on the `cdkStepperNext` button, as it makes more sense to just disable the button if it isn't going to do anything on click.

Also, we needed to decide when a step is considered complete. For the first step, it is obvious that we should have a name entered in the input to consider the step completed—or, in other words, the `FormControl` for the `'name'` property in the `nameForm` `FormGroup` should have a value. For the second step, it makes sense that after the user has guessed a number, regardless of whether the guess is correct or not, we mark the step as completed and let the user go to the next step (the leaderboard) if the user wants to. And that's pretty much about it.

See also

- Angular CDK Stepper examples (`https://material.angular.io/cdk/stepper/examples`)

Resizing text inputs with the CDK TextField API

Text inputs are an essential part of our everyday computer usage. Be it filling a form, searching some content on Google, or finding your favorite YouTube video, we all interact with text inputs, and when we have to write a bunch of content into a single text input, it really is necessary to have a good UX. In this recipe, you'll learn how to automatically resize the `<textarea>` inputs based on the input value, using the CDK TextField API.

Getting ready

The project for this recipe resides in `chapter09/start_here/resizable-text-inputs-using-cdk`. Proceed as follows:

1. Open the project in VS Code.

2. Open the terminal and run `npm install` to install the dependencies of the project.

3. Once done, run `ng serve -o`.

This should open the app in a new browser tab, and you should be able to see the app. Try typing a long text, and you'll see the text area displaying as follows:

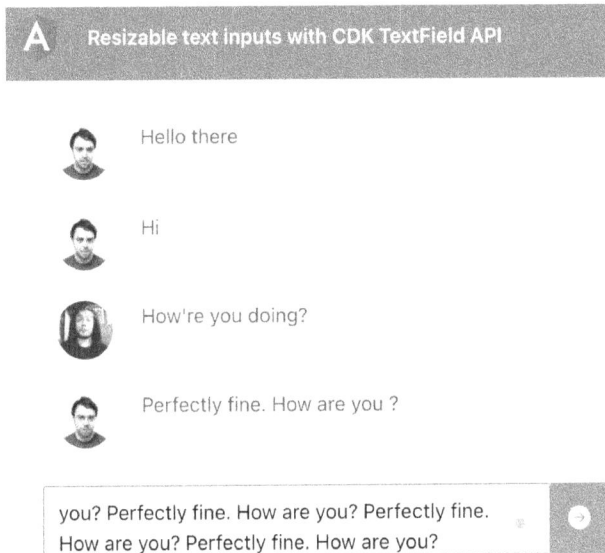

Figure 9.16 – The resizable-text-inputs-using-cdk app running on http://localhost:4200

Now that we have the app running locally, let's see the steps of the recipe in the next section.

How to do it...

In *Figure 9.16*, you will notice that we can't see the entire content of the input—this is somewhat annoying at the best of times because you can't really review it before pressing the **Action** button. Let's use the CDK TextField API by following these steps:

1. First, open a new terminal window/tab and make sure you're inside the `chapter09/start_here/resizable-text-inputs-using-cdk` folder. Once inside, run the following command to install the Angular CDK:

   ```
   npm install --save @angular/cdk@12.0.0
   ```

2. You'll have to restart your Angular server, so rerun the `ng serve` command.

3. Now, we need to import the `TextFieldModule` class into the `imports` array of our `AppModule` class. Modify the `app.module.ts` file, as follows:

```
...
import { TextFieldModule } from '@angular/cdk/text-
field';
@NgModule({
  declarations: [...],
  imports: [
    BrowserModule,
    AppRoutingModule,
    TextFieldModule
  ],
  ...
})
export class AppModule { }
```

4. Now, we'll apply the `cdkTextareaAutosize` directive to our `<text-area>` element so that it can be resized automatically based on the content. Update the `write-message.component.html` file, as follows:

```
<div class="write-box-container">
  <div class="write-box">
    <textarea
      cdkTextareaAutosize
      placeholder="Enter your message here"
      class="chat-input"
      [(ngModel)]="chatInput"
      rows="1"
      (keyup.enter)="sendMessage()"
    ></textarea>
  </div>
  <div class="send-button">
    ...
  </div>
</div>
```

If you now enter some long phrases in the text input, you should see it being resized properly, as follows:

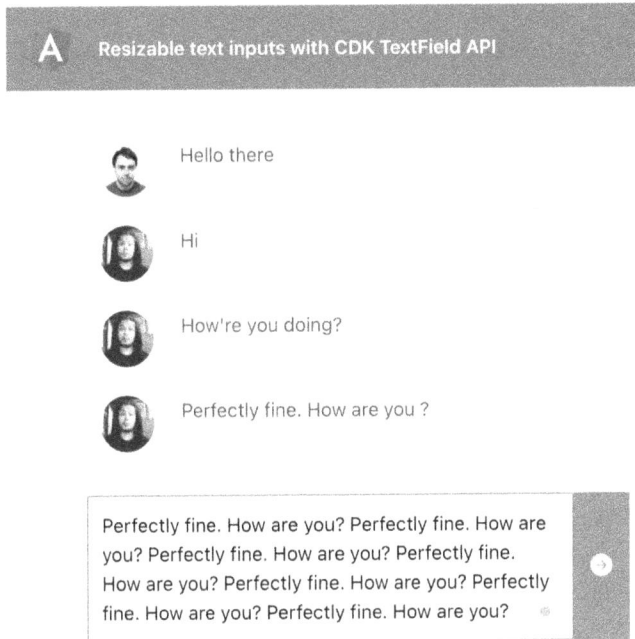

Figure 9.17 – text-area being resized based on the content

While this is awesome, you might notice that once the message is sent (that is, added to the messages list), the size of the `<text-area>` element isn't reset to its initial state.

5. In order to reset the size of the `<text-area>` element to its initial size, we'll use the `reset()` method of the `CdkTextareaAutosize` directive. To do so, we'll get the directive as a `ViewChild` and will then trigger the `reset()` method. Modify the `write-message.component.ts` file, as follows:

```
import { CdkTextareaAutosize } from '@angular/cdk/text-
field';
import {
  ...
  EventEmitter,
  ViewChild,
} from '@angular/core';
...
@Component({...})
```

```
export class WriteMessageComponent implements OnInit {
  @Output() public onMessageSent = new
  EventEmitter<any>();
  @ViewChild(CdkTextareaAutosize) newMessageInput:
  CdkTextareaAutosize;
  public chatInput = '';
  ...
  /**
   * @author Ahsan Ayaz
   * Creates a new message and emits to parent component
   */
  sendMessage() {
    if (this.chatInput.trim().length) {
      ...
      this.chatInput = '';
      this.newMessageInput.reset();
    }
  }
}
```

Great! With this change, when you refresh the page, enter a really long sentence in the input and hit the *Enter* key. You'll see the size of the `<text-area>` element being reset, as follows:

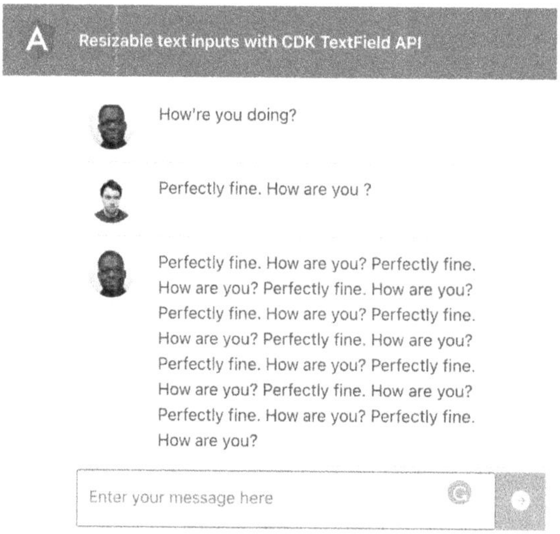

Figure 9.18 – <text-area> size being reset on new message creation

Now that you have finished the recipe, see the next section to understand how it works.

How it works...

In the recipe, we've used two main things from the CDK Clipboard API—one is the `cdkCopyToClipboard` directive, and the other is the `Clipboard` service. The `cdkCopyToClipboard` directive binds a click handler to the element this directive is applied to. It works both as the `selector` of the directive as well as an `@Input()` item for the directive so that it knows which value is to be copied to the clipboard when the element is clicked. In our recipe, for the link input, notice that we use `[cdkCopyToClipboard]="linkInput.value"`. This binds a click handler to the `<input>` element and also binds the `value` property of the `linkInput` template variable that points to the value of the input—that is, the actual link to be copied. When we click the input, it accesses the value of the input using the `linkInput.value` binding, and we do the same for the `<text-area>` input. The only difference is that the `cdkCopyToClipboard` directive is not bound to the `<text-area>` element itself. The reason is that we want to bind the click handler to the button below the text area instead. Therefore, on the button for copying the text, we have the `[cdkCopyToClipboard]="textInput.value"` binding.

For the image, we do something different. We use the `Clipboard` service from the `@angular/cdk/clipboard` package to manually copy the blob URL. We create a method named `copyImageUrl()` that is called when clicking the button for copying the image. We pass the `imageUrl` property to this method, which in turn downloads the image, reads it as a blob, and generates the blob URL, which is copied to the clipboard using the `copy()` method of the `Clipboard` service.

See also

- CDK `text-field` documentation (`https://material.angular.io/cdk/text-field/overview`)

10
Writing Unit Tests in Angular with Jest

"It works on my machine…" is a phrase that won't lose its beauty with time. It is a shield for many engineers and a nightmare for the QAs. But honestly, what's a better way than writing tests for your application's robustness, right? And when it comes to writing unit tests, my personal favorite is Jest. That is because it is super fast, lightweight, and has an easy API to write tests. More importantly, it is faster than the Karma and Jasmine setup that comes out of the box with Angular. In this chapter, you'll learn how to configure Jest with Angular to run these tests in parallel. You'll learn how to test components, services, and pipes with Jest. You'll also learn how to mock dependencies for these tests.

In this chapter, we're going to cover the following recipes:

- Setting up unit tests in Angular with Jest
- Providing global mocks for Jest
- Mocking services using stubs
- Using spies on an injected service in a unit test
- Mocking child components and directives using the `ng-mocks` package

- Creating even easier component tests with Angular CDK component harnesses
- Unit testing components with Observables
- Unit testing Angular pipes

Technical requirements

For the recipes in this chapter, ensure you have both **Git** and **NodeJS** installed on your machine. You also need to have the `@angular/cli` package installed, which you can do using `npm install -g @angular/cli` from your Terminal. The code for this chapter can be found at `https://github.com/PacktPublishing/Angular-Cookbook/tree/master/chapter10`.

Setting up unit tests in Angular with Jest

By default, a new Angular project comes bundled with a lot of goodness, including the configuration and tooling in which to run unit tests with Karma and Jasmine. While working with Karma is relatively convenient, many developers find that in large-scale projects, the whole testing process becomes much slower if there are a lot of tests involved. This is mainly because you can't run tests in parallel. In this recipe, we'll set up Jest for unit testing in an Angular app. Additionally, we'll migrate existing tests from the Karma syntax to the Jest syntax.

Getting ready

The project that we are going to work with resides in `chapter10/start_here/setting-up-jest`, which is inside the cloned repository. To begin, perform the following steps:

1. Open the project in Visual Studio Code.
2. Open the Terminal and run `npm install` to install the dependencies of the project.
3. Once done, run `ng serve -o`.

This should open the app in a new browser tab. You should see something similar to the following screenshot:

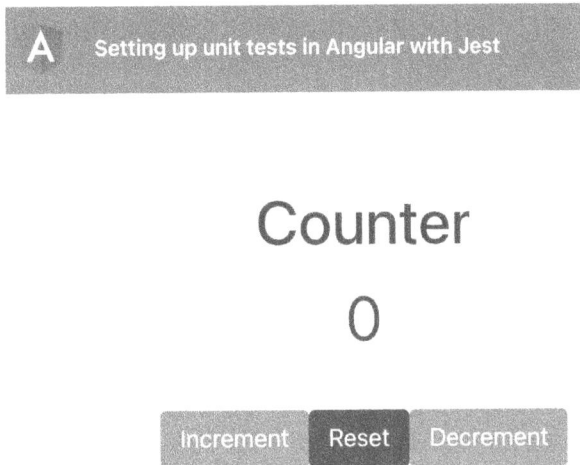

Figure 10.1 – The setting-up-jest app running on http://localhost:4200

Next, try to run the test and monitor how much time it takes for the entire process to run. Run the ng test command from your Terminal; within a few seconds, a new Chrome window should open, as follows:

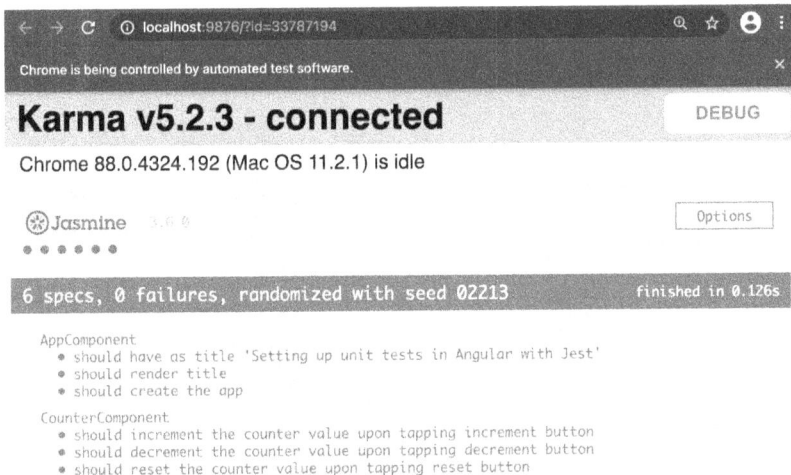

Figure 10.2 – The tests' results with Karma and Jasmine

Looking at the preceding screenshot, you might say *"Pfffttt Ahsan, it says 'finished in 0.126s!' What else do you want?"* Well, that time only covers how long it took Karma to run the tests *in the browser after the Chrome window was created*. It doesn't count the time it took to actually start the process, start the Chrome window, and then load the tests. On my machine, it takes about *15 seconds* to run the entire process. That's why we're going to replace it with Jest. Now that you understand the issue, in the next section, let's take a look at the steps of the recipe.

How to do it...

Here, we have an Angular app with a really simple **Counter** component. It shows the value of the counter and has three action buttons: one to increment the value of the counter, one to decrement the value, and one to reset the value. Additionally, there are some tests written with Karma and Jasmine, and all of the tests pass if you run the ng test command. We'll start by setting up Jest instead. Perform the following steps:

1. First, open a new Terminal window/tab and make sure you're inside the
    chapter10/start_here/setting-up-jest folder. Once inside, run the
    following command to install the packages that are required to test with Jest:

    ```
    npm install --save-dev jest jest-preset-angular @types/
    jest
    ```

2. We can now uninstall Karma and the unwanted dependencies. Now run the
    following command in your Terminal:

    ```
    npm uninstall karma karma-chrome-launcher karma-jasmine-
    html-reporter @types/jasmine @types/jasminewd2 jasmine-
    core jasmine-spec-reporter karma-coverage-istanbul-
    reporter karma-jasmine
    ```

3. We also need to get rid of some extra files that we don't require. Delete the karma.
    conf.js file and the src/test.ts file from the project.

4. Now update the test configuration in the angular.json file, as follows:

    ```
    {
      . . .
      "projects": {
        "setting-up-jest": {
          ". . .
          "prefix": "app",
          "architect": {
    ```

```
            "build": {...},
            "serve": {...},
            "extract-i18n": {...},
            "test": {
              "builder": "@angular-builders/jest:run",
              "options": {
                "tsConfig": "<rootDir>/src/tsconfig.test.
                json",
                "collectCoverage": false,
                "forceExit": true
              }
            },
            "lint": {...},
            "e2e": {...}
          }
        }
      },
      "defaultProject": "setting-up-jest"
  }
```

5. We will now create a file to configure Jest for our project. Create a file named
    jestSetup.ts inside the project's root folder and paste the following content
    inside:

    ```
    import 'jest-preset-angular /setup-jest';
    ```

6. Now, let's modify tsconfig.spec.json to use Jest instead of Jasmine. After the
    modification, your entire file should appear as follows:

    ```
    {
      "extends": "./tsconfig.json",
      "compilerOptions": {
        "outDir": "./out-tsc/spec",
        "types": ["jest", "node"],
        "esModuleInterop": true,
        "emitDecoratorMetadata": true
      },
      "files": ["src/polyfills.ts"],
    ```

```
        "include": ["src/**/*.spec.ts", "src/**/*.d.ts"]
    }
```

7. We'll now modify `package.json` to add the npm scripts that'll run the Jest tests:

```
    {
        "name": "setting-up-jest",
        "version": "0.0.0",
        "scripts": {
            ...
            "build": "ng build",
            "test": "jest",
            "test:coverage": "jest --coverage",
            ...
        },
        "private": true,
        "dependencies": {...},
        "devDependencies": {...},
    }
```

8. Finally, let's wrap up the entire configuration for our Jest tests by adding the Jest configuration in the `package.json` file, as follows:

```
    {
        ...
        "dependencies": {...},
        "devDependencies": {...},
        "jest": {
            "preset": "jest-preset-angular",
            "setupFilesAfterEnv": [
                "<rootDir>/jestSetup.ts"
            ],
            "testPathIgnorePatterns": [
                "<rootDir>/node_modules/",
                "<rootDir>/dist/"
            ],
            "globals": {
```

```
      "ts-jest": {
        "tsconfig": "<rootDir>/tsconfig.spec.json",
        "stringifyContentPathRegex": "\\.html$"
      }
    }
  }
}
```

9. Now that we have set everything up, simply run the `test` command, as follows:

    npm run test

Once the tests are finished, you should be able to see the following output:

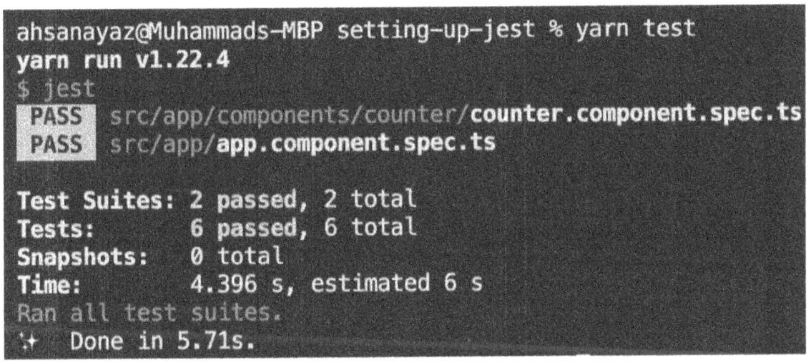

```
ahsanayaz@Muhammads-MBP setting-up-jest % yarn test
yarn run v1.22.4
$ jest
 PASS  src/app/components/counter/counter.component.spec.ts
 PASS  src/app/app.component.spec.ts

Test Suites: 2 passed, 2 total
Tests:       6 passed, 6 total
Snapshots:   0 total
Time:        4.396 s, estimated 6 s
Ran all test suites.
✨  Done in 5.71s.
```

Figure 10.3 – The results of the tests with Jest

Kaboom! You will notice that the entire process of running the tests with Jest takes about 6 seconds. It might take more time when you run it for the first time, but the subsequent runs should be faster. Now that you know how to configure an Angular app to use Jest for unit tests, please refer to the next section for resources in which to learn more.

See also

- *Testing Web Frameworks with Jest* (https://jestjs.io/docs/en/testing-frameworks)

- *Getting Started with Jest* (https://jestjs.io/docs/en/getting-started)

Providing global mocks for Jest

In the previous recipe, we learned how to set up Jest for Angular unit tests. There might be some scenarios in which you'd want to use a browser API that might not be part of your actual Angular code; for instance, using `localStorage` or `alert()`. In such cases, we need to provide some global mocks for the functions we want to return mock values from. This is so that we can perform tests involving them as well. In this recipe, you'll learn how to provide global mocks to Jest.

Getting ready

The project for this recipe resides in `chapter10/start_here/providing-global-mocks-for-jest`. Perform the following steps:

1. Open the project in Visual Studio Code.

2. Open the Terminal and run `npm install` to install the dependencies of the project.

3. Once done, run `ng serve -o`.

This should open the app in a new browser tab. The app should appear as follows:

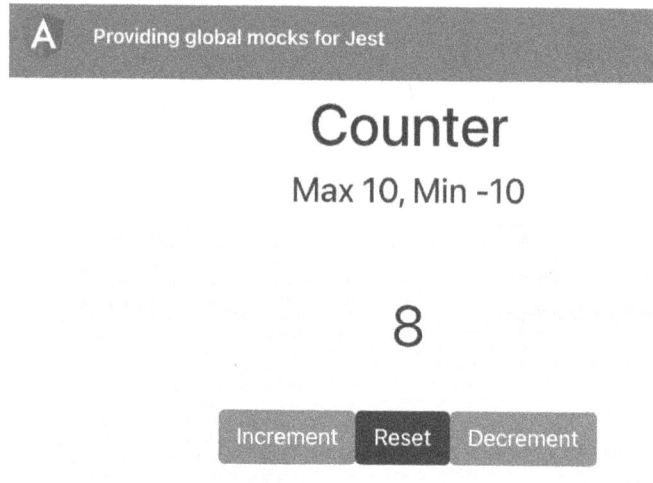

Figure 10.4 – The providing-global-mocks-for-jest app running on http://localhost:4200

Now that we have the app running locally, in the next section, let's go through the steps of the recipe.

How to do it...

The app we're using for this recipe uses two global APIs: window.localStorage and window.alert(). Note that when the app starts, we fetch the counter value from localStorage, and then upon increment, decrement, and reset, we store it in localStorage. When the counter value becomes greater than the MAX_VALUE or lower than the MIN_VALUE, we show the alert using the alert() method. Let's begin the recipe by writing some cool unit tests:

1. First, we'll write our test cases to show the alert when the counter value goes beyond MAX_VALUE and MIN_VALUE. Modify the counter.component.spec.ts file as follows:

```
. . .
describe('CounterComponent', () => {
  . . .
  it('should show an alert when the counter value goes
  above the MAX_VALUE', () => {
    spyOn(window, 'alert');
    component.counter = component.MAX_VALUE;
    component.increment();
    expect(window.alert).toHaveBeenCalledWith('Value too
    high');
    expect(component.counter).toBe(component.MAX_VALUE);
  });
  it('should show an alert when the counter value goes
  above the MAX_VALUE', () => {
    spyOn(window, 'alert');
    component.counter = component.MIN_VALUE;
    component.decrement();
    expect(window.alert).toHaveBeenCalledWith('Value too
    low');
    expect(component.counter).toBe(component.MIN_VALUE);
  });
});
```

Here, you can see that the tests pass. But what if we wanted to check whether the value from localStorage is being saved and retrieved properly?

2. We'll create a new test to make sure the `localStorage.getItem()` method is called to retrieve the last saved value from the `localStorage` API. Add the test to the `counter.component.spec.ts` file, as follows:

```
...
describe('CounterComponent', () => {
    ...
    it.only('should call the localStorage.getItem method on
    component init', () => {
        spyOn(localStorage, 'getItem');
        component.ngOnInit();
        expect(localStorage.getItem).toBeCalled();
    });
});
```

Notice that we're using `it.only` for this test case. This is to ensure that we're only running this test (for now). If you run the tests, you should be able to see something similar to the following screenshot:

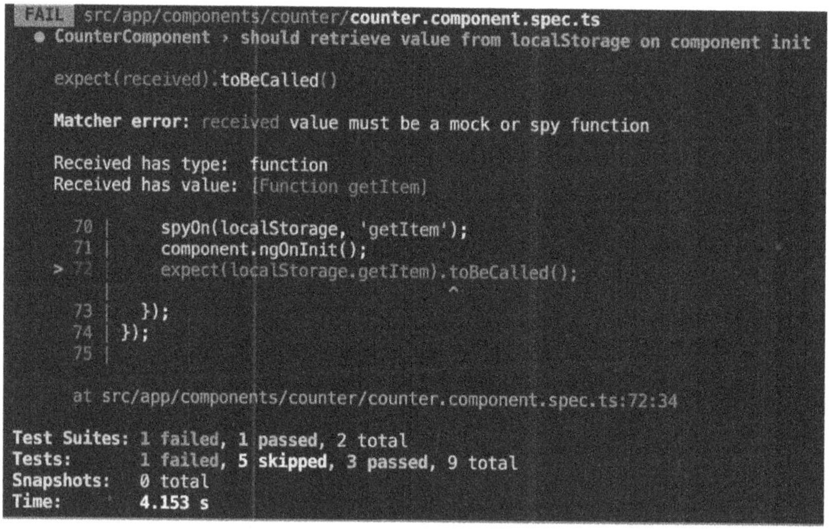

Figure 10.5 – The test that is covering the localStorage API has failed

Notice the `Matcher error: received value must be a mock or a spy function` message. This is what we're going to do next, that is, provide a mock.

3. Create a file in the project's root, called `jest-global-mocks.ts`. Then, add the following code to mock the `localStorage` API:

```
const createLocalStorageMock = () => {
  let storage = {};
  return {
    getItem: (key) => {
      return storage[key] ? storage[key] : null;
    },
    setItem: (key, value) => {
      storage[key] = value;
    },
  };
};

Object.defineProperty(window, 'localStorage', {
  value: createLocalStorageMock(),
});
```

4. Now import this file into the `jestSetup.ts` file, as follows:

```
import 'jest-preset-angular';
import './jest-global-mocks';
```

Now if you rerun the tests, they should pass.

5. Let's add another test to ensure we retrieve the last saved value from `localStorage` in the component initiation. Modify the `counter.component.spec.ts` file, as follows:

```
...
describe('CounterComponent', () => {
  ...
  it('should call the localStorage.getItem method on
  component init', () => {
    spyOn(localStorage, 'getItem');
    component.ngOnInit();
    expect(localStorage.getItem).toBeCalled();
  });
```

```
  it('should retrieve the last saved value from
  localStorage on component init', () => {
    localStorage.setItem('counterValue', '12');
    component.ngOnInit();
    expect(component.counter).toBe(12);
  });
});
```

6. Finally, let's make sure that we save the counter value to localStorage whenever we trigger the increment(), decrement(), or reset() methods. Update the counter.component.spec.ts as follows:

```
...
describe('CounterComponent', () => {

  ...

  it('should save the new counterValue to localStorage
  on increment, decrement and reset', () => {
    spyOn(localStorage, 'setItem');
    component.counter = 0;
    component.increment();
    expect(localStorage.setItem).
    toHaveBeenCalledWith('counterValue', '1');
    component.counter = 20;
    component.decrement();
    expect(localStorage.setItem).
    toHaveBeenCalledWith('counterValue', '19');
    component.reset();
    expect(localStorage.setItem).
    toHaveBeenCalledWith('counterValue', '0');
  });
});
```

Awesome sauce! You've just learned how to provide global mocks to Jest for testing. Please refer to the next section to understand how this works.

How it works...

Jest provides a way in which to define a list of paths to the files that we want to load for each test. If you open the `package.json` file and see the `jest` property, you can view the `setupFilesAfterEnv` property, which takes an array of paths to the files. We already have the path defined there for the `jestSetup.ts` file. And one way to define global mocks is to create a new file and then import it into `jestSetup.ts`. This is because it is going to be called in the test environment anyway. And that's what we do in this recipe.

Notice that we use the `Object.defineProperty` method in the `window` object to provide a mock implementation for the `localStorage` object. This is actually the same for any API that is not implemented in the JSDOM. Similarly, you can provide a global mock for each API that you use in your tests. Notice that in the `value` property, we use the `createLocalStorageMock()` method. Essentially, this is one way to define mocks. We create the `createLocalStorageMock()` method, and in there we have a private/encapsulated object named `storage` that mimics the `localStorage` object. We have also defined the `getItem()` and `setItem()` methods in there so that we can set values to this storage and get values from it. Notice that we do not have the implementations of the `removeItem()` and `clear()` methods that we have in the original `localStorage` API. We don't have to do it because we're not using these methods in our tests.

In the `'should call the localStorage.getItem method on component init'` test, we simply spy on the `localStorage` object's `getItem()` method, call the `ngOnInit()` method ourselves, and then expect it to have been called. Easy peasy.

In the `'should retrieve the last saved value from localStorage on component init'` test, we save a value in the `localStorage` object for the counter value as `'12'` using the `setItem()` method. Essentially, calling the `setItem()` method calls our mock implementation method and not the actual `localStorage` API's `setItem()` method. Notice that, here, we *do not* spy on the `getItem()` method; this is because later on, we want the value of the component's `counter` property to be 12.

> **Important note**
>
> Whenever we spy on a method, remember that any statements in the actual function will no longer be executed. This is why we do not spy on the `getItem()` method in the preceding test. If we do so, the `getItem()` method from the mock implementation *will not return anything*. Therefore, our expected value for the counter property will not be `12`.
>
> Put simply, if you have to rely on the outcome of a function's implementation, or the statements executed within a function, do not spy on that function and write your test accordingly.
>
> PS: I always end up learning this the hard way after debugging and bashing my head for a while. Just kidding!

The final test is an easy one. In the `'should save the new counterValue to localStorage on increment, decrement and reset'` test, we simply spy on the `setItem()` method as we're not concerned about its implementation. Then, we manually set the value of the counter property multiple times before we run the `increment()`, `decrement()`, and `reset()` methods, respectively. Additionally, we expect the `setItem()` method to have been called with the right arguments to save the value to the store. Note that we do not check the store's value after saving it. As I mentioned earlier, since we have spied on the `setItem()` method, its internal statement won't trigger and the value won't be saved; therefore, we can't retrieve the saved value afterward.

See also

- The Jest documentation for `setupFiles` (`https://jestjs.io/docs/en/configuration#setupfiles-array`)
- *Manual Mocks with Jest* (`https://jestjs.io/docs/en/manual-mocks`)

Mocking services using stubs

There's rarely an Angular app that doesn't have a `Service` created inside it. And where the overall business logic is concerned, services hold a great deal of the business logic, particularly when it comes to interacting with APIs. In this recipe, you'll learn how to mock services using stubs.

Getting ready

The project for this recipe resides in `chapter10/start_here/mocking-services-using-stubs`. Perform the following steps:

1. Open the project in Visual Studio Code.

2. Open the Terminal and run `npm install` to install the dependencies of the project.

3. Once done, run `ng serve -o`.

This should open the app in a new browser tab. You should see something like the following screenshot:

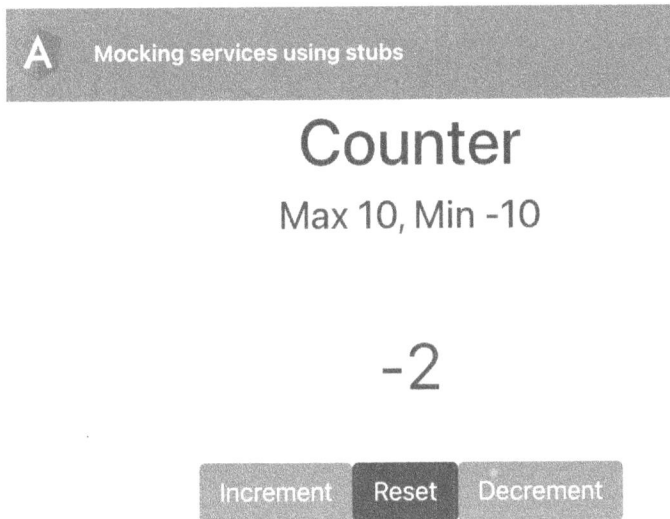

Figure 10.6 – The mocking-services-using-stubs app running on http://localhost:4200

Now that we have the app running locally, in the next section, let's take a look at the steps of the recipe.

How to do it...

We have the same application as the previous recipe; however, we've moved the logic of saving and retrieving data from `localStorage` to the `CounterService` we've created. Now all the tests pass. However, what if we wanted to hide/encapsulate the logic of where the counter value is stored? Perhaps we want to send a backend API call for it. To do this, it makes more sense to spy on the service's methods. Let's follow the recipe to provide a mock stub for our service:

1. First of all, let's create a folder inside the `src` folder, named `__mocks__`. Inside it, create another folder named `services`. Then, again inside this folder, create the `counter.service.mock.ts` file with the following content:

    ```
    const CounterServiceMock = {
      storageKey: 'counterValue',
      getFromStorage: jest.fn(),
      saveToStorage: jest.fn(),
    };


    export default CounterServiceMock;
    ```

2. Now provide the mock service instead of the actual service in the `counter.component.spec.ts`, as follows:

    ```
    import { ComponentFixture, TestBed } from '@angular/core/
    testing';
    import { CounterService } from 'src/app/core/services/
    counter.service';
    import CounterServiceMock from 'src/__mocks__/services/
    counter.service.mock';
    . . .
    describe('CounterComponent', () => {
      . . .
      beforeEach(async () => {
        await TestBed.configureTestingModule({
          declarations: [CounterComponent],
          providers: [
            {
              provide: CounterService,
              useValue: CounterServiceMock,
    ```

```
        },
      ],
    }).compileComponents();
  });
  ...
});
```

With the preceding change, you should see the following error that says the `localStorage.setItem` hasn't been called. This is because we're now spying on the methods on our mock stub for the service:

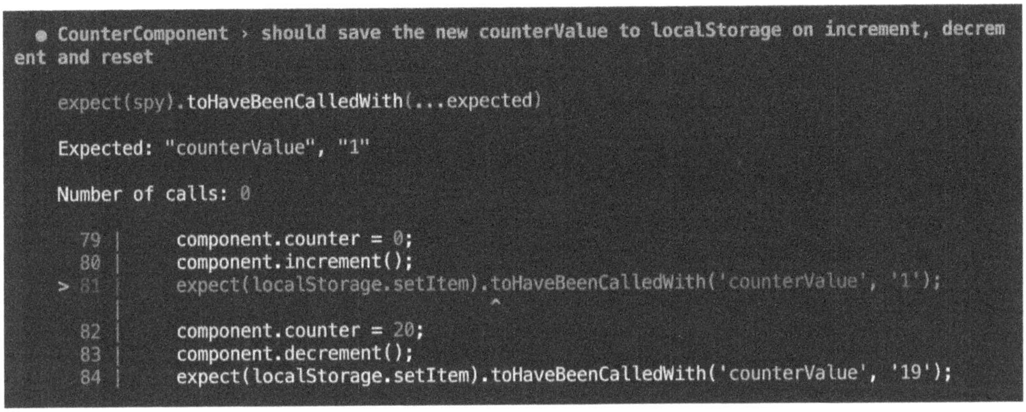

```
● CounterComponent › should save the new counterValue to localStorage on increment, decrem
ent and reset

  expect(spy).toHaveBeenCalledWith(...expected)

  Expected: "counterValue", "1"

  Number of calls: 0

    79 |     component.counter = 0;
    80 |     component.increment();
  > 81 |     expect(localStorage.setItem).toHaveBeenCalledWith('counterValue', '1');
       |                                  ^
    82 |     component.counter = 20;
    83 |     component.decrement();
    84 |     expect(localStorage.setItem).toHaveBeenCalledWith('counterValue', '19');
```

Figure 10.7 – localStorage.setItem is not called because of the methods being spied on

3. Now, instead of expecting the `localStorage` object's methods to be called, let's expect our service's methods to be called in our tests. Update the `counter.component.spec.ts` file as follows:

```
...
describe('CounterComponent', () => {

  ...

  it('should call the CounterService.getFromStorage
  method on component init', () => {
    component.ngOnInit();
    expect(CounterServiceMock.getFromStorage).
    toBeCalled();
  });


  it('should retrieve the last saved value from
  CounterService on component init', () => {
```

```
        CounterServiceMock.getFromStorage.
        mockReturnValue(12);

        component.ngOnInit();
        expect(component.counter).toBe(12);
    });

    it('should save the new counterValue via CounterService
    on increment, decrement and reset', () => {
        component.counter = 0;
        component.increment();
        expect(CounterServiceMock.saveToStorage).
        toHaveBeenCalledWith(1);
        component.counter = 20;
        component.decrement();
        expect(CounterServiceMock.saveToStorage).
        toHaveBeenCalledWith(19);
        component.reset();
        expect(CounterServiceMock.saveToStorage).
        toHaveBeenCalledWith(0);
    });
});
```

Great! You now know how to mock services to test components with service dependencies. Please refer to the next section to understand how it all works.

How it works...

Providing stubs for Angular services is already a breeze. This is thanks to Angular's out-of-the-box methods and tooling from the @angular/core package, especially @angular/core/testing. First, we create the stub for our CounterService and use jest.fn() for every method within CounterService.

Using jest.fn() returns a new, unused mock function that Jest automatically spies upon as well. Optionally, we can also pass a mock implementation method as a parameter to jest.fn. View the following example from the official documentation for jest.fn():

```
const mockFn = jest.fn();
mockFn();
expect(mockFn).toHaveBeenCalled(); // test passes

// With a mock implementation:
const returnsTrue = jest.fn(() => true);
console.log(returnsTrue()); // true;
expect(returnsTrue()).toBe(true); // test passes
```

Once we create the stub, we pass it to the TestBed configuration in the provider's array against the CounterService – but with the useValue property set to the CounterServiceMock. This tells Angular to use our stub as it is for CounterService.

Then, in the test where we expect CounterService.getFromStorage() to be called when the component initiates, we use the following statement:

```
expect(CounterServiceMock.getFromStorage).toBeCalled();
```

Notice that in the preceding code, we are able to directly use expect() on CounterServiceMock.getFromStorage. While this isn't possible in Karma and Jasmine, it is possible with Jest, since we're using jest.fn() for each underlying method.

Then, for a test in which we want to check whether the getFromStorage() method is called and returns a saved value, we first use the CounterServiceMock.getFromStorage.mockReturnValue(12); statement. This ensures that when the getFromStorage() method is called, it returns the value of 12. Then, we just run the ngOnInit() method in the test and expect that our component's counter property has now been set to 12. This actually means that the following things happen:

1. ngOnInit() calls the getFromStorage() method.

2. getFromStorage() returns the previously saved value (in our case, that's 12, but in reality, that'll be fetched from localStorage).

3. The component's counter property is set to the retrieved value, which, in our case, is 12.

Now, for the final test, we just expect that the `saveToStorage` method of our `CounterService` is called in each necessary case. For this, we use the following types of `expect()` statements:

```
expect(CounterServiceMock.saveToStorage).
toHaveBeenCalledWith(1);
```

That's pretty much about it. Unit tests are fun, aren't they? Now that you've understood how it all works, please refer to the next section for some helpful resources that you can use for further reading.

See also

- The official documentation for `jest.fn()` (`https://jestjs.io/docs/en/jest-object.html#jestfnimplementation`)

- Angular's *Component testing scenarios* (`https://angular.io/guide/testing-components-scenarios`)

Using spies on an injected service in a unit test

While you can provide stubs for your services in the unit tests with Jest, sometimes, it might feel like an overhead creating a mock for every new service. Let's suppose that if the service's usage is limited to one test file, it might make more sense to just use spies on the actual injected service. In this recipe, that's exactly what we're going to do.

Getting ready

The project for this recipe resides in `chapter10/start_here/using-spies-on-injected-service`.

1. Open the project in Visual Studio Code.

2. Open the Terminal and run `npm install` to install the dependencies of the project.

3. Once done, run `npm run test`.

This should run the unit tests on the console using Jest. You should see something similar to the following output:

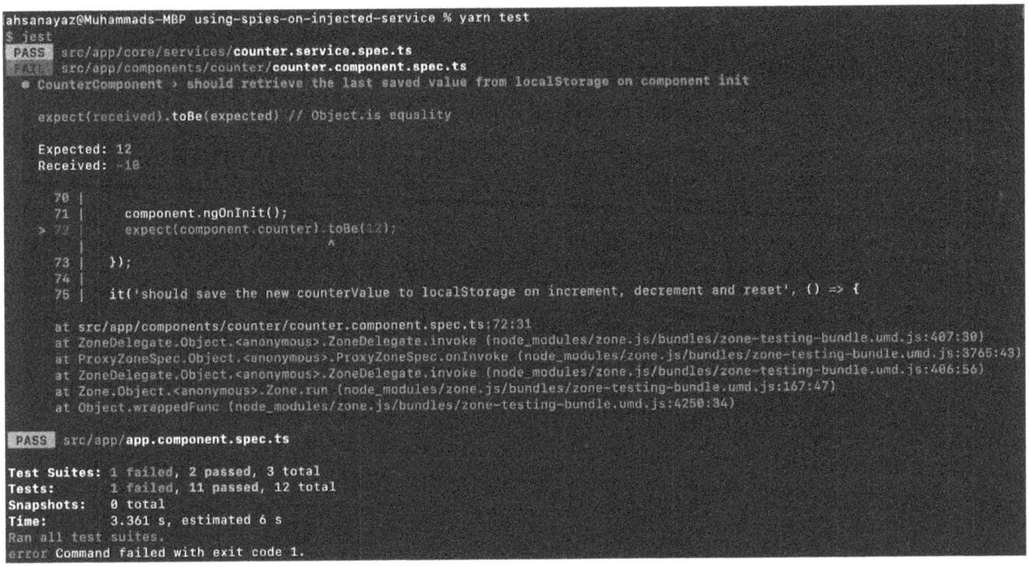

Figure 10.8 – Unit tests failing for the 'using-spies-on-injected-service' project

Now that we have the tests running locally, in the next section, let's go through the steps of the recipe.

How to do it...

The tests we have in the code for `CounterComponent` are incomplete. That's because we're missing the `expect()` blocks and the code to spy on the methods of `CounterService`. Let's get started with the recipe to complete writing the tests using spies on the actual `CounterService`, as follows:

1. First, we need to get an instance of the actual injected service in our tests. So, we'll create a variable and get the injected service in a `beforeEach()` method. Update the `counter.component.spec.ts` file as follows:

    ```
    ...
    describe('CounterComponent', () => {
        let component: CounterComponent;
        let fixture: ComponentFixture<CounterComponent>;
        let counterService: CounterService;

        beforeEach(async () => {...});

        beforeEach(() => {
    ```

```
      fixture = TestBed.createComponent(CounterComponent);
      component = fixture.componentInstance;
      fixture.detectChanges();
      counterService = TestBed.inject(CounterService);
    });
    ...
  });
```

2. Now, we'll write our first `expect()` block for the service. For the test that says `'should call the localStorage.getItem method on component init'`, add the following `spyOn()` and `expect()` blocks:

```
...
describe('CounterComponent', () => {
  ...
  it('should call the localStorage.getItem method on
  component init', () => {
    spyOn(counterService, 'getFromStorage');
    component.ngOnInit();
    expect(counterService.getFromStorage).
    toHaveBeenCalled();
  });
  ...
});
```

If you run `npm run test` again, you should still see one test failing but the rest of them passing.

3. Now, let's fix the failing test. That is `'should retrieve the last saved value from localStorage on component init'`. In this case, we need to spy on the `getFromStorage()` method of `CounterService` to return the expected value of `12`. To do so, update the test file, as follows:

```
...
describe('CounterComponent', () => {
  ...
  it('should retrieve the last saved value from
  localStorage on component init', () => {
    spyOn(counterService, 'getFromStorage').and.
    returnValue(12);
```

```
      component.ngOnInit();
      expect(component.counter).toBe(12);
    });
    ...
  });
```

4. Finally, let's fix our last test where we expect the `increment()`, `decrement()`, and `reset()` methods to call the `saveToStorage()` method of `CounterService`. Update the test as follows:

```
    ...
    describe('CounterComponent', () => {
      ...
      it('should save the new counterValue to localStorage
      on increment, decrement and reset', () => {
        spyOn(counterService, 'saveToStorage');
        component.counter = 0;
        component.increment();
        expect(counterService.saveToStorage).
        toHaveBeenCalledWith(1);
        component.counter = 20;
        component.decrement();
        expect(counterService.saveToStorage).
        toHaveBeenCalledWith(19);
        component.reset();
        expect(counterService.saveToStorage).
        toHaveBeenCalledWith(0);
      });
    });
```

Awesome! With this change, you should see all 12 tests passing. Let's take a look at the next section to understand how it works.

How it works...

This recipe contained a lot of knowledge from the previous recipes of this chapter. However, the key highlight is the `TestBed.inject()` method. Essentially, this magical method gets the instance of the provided service – `CounterService` – to us. This is the instance of the service that is bound with the instance of `CounterComponent`. Since we have access to the same instance of the service that is being used by the component's instance, we can spy on it directly and expect it to be called – or even mock the returned values.

See also

- An introduction to Angular TestBed (`https://angular.io/guide/testing-services#angular-testbed`)

Mocking child components and directives using the ng-mocks package

Unit tests mostly revolve around testing components in isolation. However, what if your component depends completely on another component or directive to work properly? In such cases, you usually provide a mock implementation for the component, but that is a lot of work. However, with the `ng-mocks` package, it is super easy. In this recipe, we'll learn an advanced example of how to use `ng-mocks` for a parent component that depends on a child component to work properly.

Getting ready

The project that we are going to work with resides in `chapter10/start_here/` `mocking-components-with-ng-mocks`, which is inside the cloned repository. Perform the following steps:

1. Open the project in Visual Studio Code.

2. Open the Terminal and run `npm install` to install the dependencies of the project.

3. Once done, run `ng serve -o`.

This should open the app in a new browser tab. You should see something similar to the following screenshot:

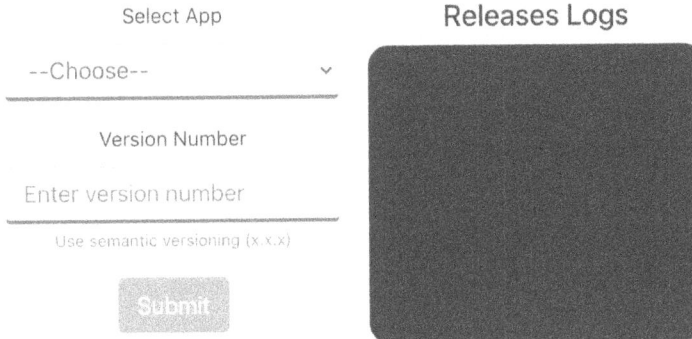

Figure 10.9 – The mocking-components-with-ng-mocks app running on http://localhost:4200

Now that we have the app running locally, in the next section, let's go through the steps of the recipe.

How to do it...

If you run the `yarn test` command or the `npm run test` command, you'll see that not all of our tests pass. Additionally, there are a bunch of errors on the console, as follows:

Figure 10.10 – An unknown elements error during unit tests

Let's go through the recipe to make sure that our tests pass correctly without any errors using the ng-mocks package:

1. First, let's install the ng-mocks package within our project. To do this, run the following command from your project root in the Terminal:

```
npm install ng-mocks --save
# or
yarn add ng-mocks
```

2. Now, we'll try to fix the tests for AppComponent. To only run specific tests based on a string regex, we can use the -t parameter with the jest command. Run the following command to only run the tests for AppComponent:

```
npm run test -- -t 'AppComponent'
#or
yarn test -- -t 'AppComponent'
```

Now you can see that we only run the tests for AppComponent, and they fail as follows:

```
components-with-ng-mocks
> jest "-t" "AppComponent"

PASS  src/app/app.component.spec.ts
  ● Console

    console.error
      NG0304: 'app-version-control' is not a known element:
      1. If 'app-version-control' is an Angular component, then verify that it is part of this module
      .
      2. If 'app-version-control' is a Web Component then add 'CUSTOM_ELEMENTS_SCHEMA' to the '@NgMod
      ule.schemas' of this component to suppress this message.

      at logUnknownElementError (../packages/core/src/render3/instructions/element.ts:220:15)
      at elementStartFirstCreatePass (../packages/core/src/render3/instructions/element.ts:41:16)
      at ɵɵelementStart (../packages/core/src/render3/instructions/element.ts:87:7)
      at ɵɵelement (../packages/core/src/render3/instructions/element.ts:180:3)
      at AppComponent_Template (ng:/AppComponent.js:16:9)
      at executeTemplate (../packages/core/src/render3/instructions/shared.ts:511:5)
      at renderView (../packages/core/src/render3/instructions/shared.ts:301:7)
      at renderComponent (../packages/core/src/render3/instructions/shared.ts:1765:3)
```

Figure 10.11 – Error – 'app-version-control' is not a known element

3. To fix the error shown in *Figure 10.11*, we'll import `VersionControlComponent` into the `TestBed` definition inside the `app.component.spec.ts` file. This is so that our test environment also knows the missing `VersionControlComponent`. To do this, modify the mentioned file as follows:

    ```
    . . .
    import { VersionControlComponent } from './components/
    version-control/version-control.component';
    . . .
    describe('AppComponent', () => {
      beforeEach(waitForAsync(() => {
        TestBed.configureTestingModule({
          imports: [RouterTestingModule],
          declarations: [AppComponent,
          VersionControlComponent],
        }).compileComponents();
      }));
      . . .
    });
    ```

    If you rerun the tests for `AppComponent`, you'll see some fresher and newer errors. Surprise! Well, that's what happens with dependencies. We'll discuss the details in more detail in the *How it works...* section. However, to fix this, let's follow the next steps.

4. Instead of providing the `VersionControlComponent` directly, we need to mock it since we don't really care about it for the tests for `AppComponent`. To do this, update the `app.component.spec.ts` file as follows:

    ```
    . . .
    import { MockComponent } from 'ng-mocks';
    . . .
    describe('AppComponent', () => {
      beforeEach(waitForAsync(() => {
        TestBed.configureTestingModule({
          imports: [RouterTestingModule],
          declarations: [AppComponent,
          MockComponent(VersionControlComponent)],
        }).compileComponents();
      }));
    ```

```
        . . .
    });
```

Boom! Problem solved. Run the tests again, just for the `AppComponent`, and you should see them all pass as follows:

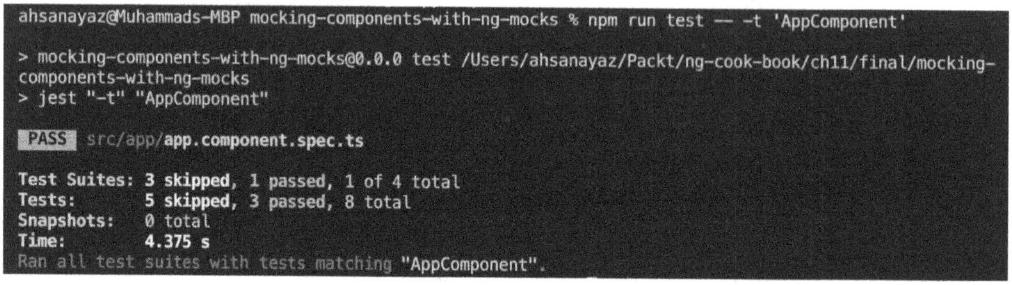

```
ahsanayaz@Muhammads-MBP mocking-components-with-ng-mocks % npm run test -- -t 'AppComponent'

> mocking-components-with-ng-mocks@0.0.0 test /Users/ahsanayaz/Packt/ng-cook-book/ch11/final/mocking-
components-with-ng-mocks
> jest "-t" "AppComponent"

 PASS  src/app/app.component.spec.ts

Test Suites: 3 skipped, 1 passed, 1 of 4 total
Tests:       5 skipped, 3 passed, 8 total
Snapshots:   0 total
Time:        4.375 s
Ran all test suites with tests matching "AppComponent".
```

Figure 10.12 – Passing all of the tests for AppComponent

5. Now, let's talk about the tests for `VersionControlComponent`. This depends on the `ReleaseFormComponent` as well as the `ReleaseLogsComponent`. Let's mock them like a pro this time, using the `MockBuilder` and `MockRender` methods, so we can get rid of the errors during the tests. After the update, the `version-control.component.spec.ts` file should appear as follows:

```
import { MockBuilder, MockedComponentFixture, MockRender
} from 'ng-mocks';
import { ReleaseFormComponent } from '../release-form/
release-form.component';
import { ReleaseLogsComponent } from '../release-logs/
release-logs.component';

import { VersionControlComponent } from './version-
control.component';

describe('VersionControlComponent', () => {
  let component: VersionControlComponent;
  let fixture: MockedComponentFixture
  <VersionControlComponent>;
  beforeEach(() => {
    return MockBuilder(VersionControlComponent)
      .mock(ReleaseFormComponent)
      .mock(ReleaseLogsComponent);
```

```
    });
    beforeEach(() => {
      fixture = MockRender(VersionControlComponent);
      component = fixture.point.componentInstance;
    });
    it('should create', () => {...});
  });
```

If you run npm run test now, you should see all of the tests passing. In the next steps, let's actually write some interesting tests.

6. VersionControlComponent uses ReleaseLogsComponent as a child. Additionally, it provides the releaseLogs property as @Input() to ReleaseLogsComponent via the [logs] attribute. We can actually check whether the input's value is set correctly. To do so, update the version-control.component.spec.ts file, as follows:

```
import {
  MockBuilder,
  MockedComponentFixture,
  MockRender,
  ngMocks,
} from 'ng-mocks';
import { Apps } from 'src/app/constants/apps';
...
describe('VersionControlComponent', () => {
  ...
  it('should set the [logs] @Input for the
  ReleaseLogsComponent', () => {
    const releaseLogsComponent = ngMocks.
    find<ReleaseLogsComponent>(
      'app-release-logs'
    ).componentInstance;
    const logsStub = [{ app: Apps.DRIVE, version:
    '2.2.2', message: '' }];
    component.releaseLogs = [...logsStub];
    fixture.detectChanges();
    expect(releaseLogsComponent.logs.length).toBe(1);
    expect(releaseLogsComponent.logs).toEqual([...
```

```
logsStub]);
    });
});
```

7. Now we'll make sure that when we have a new log created via
   `ReleaseFormComponent`, we show this new log by adding it to the
   `releaseLogs` array in `VersionControlComponent`. Then, we'll also pass that
   as `@Input logs` to `ReleaseLogsComponent`. Add the following tests to the
   `version-control.component.spec.ts` file:

```
...
describe('VersionControlComponent', () => {
    ...
    it('should add the new log when it is created via
    ReleaseFormComponent', () => {
        const releaseFormsComponent = ngMocks.
        find<ReleaseFormComponent>('app-release-form').
        componentInstance;
        const releaseLogsComponent = ngMocks.
        find<ReleaseLogsComponent>('app-release-logs').
        componentInstance;
        const newLogStub = { app: Apps.DRIVE, version:
        '2.2.2', message: '' };
        component.releaseLogs = []; // no logs initially
        releaseFormsComponent.newReleaseLog.emit(newLogStub);
        // add a new log
        fixture.detectChanges(); // detect changes
        expect(component.releaseLogs).toEqual([newLogStub]);
        // VersionControlComponent logs
        expect(releaseLogsComponent.logs).
        toEqual([newLogStub]); // ReleaseLogsComponent logs
    });
});
```

Boom! We have implemented some interesting tests by using the ng-mocks package.
I absolutely love it every time I use it. Now that we've finished the recipe, in the next
section, let's take a look at how it all works.

How it works...

There are a couple of interesting things that we have covered in this recipe. First of all, to avoid any errors on the console complaining about unknown components, we use the `MockComponent` method from the `ng-mocks` package, to declare the components we're dependent on, as mocks. That is absolutely the simplest thing we achieve with the `ng-mocks` package. However, we do move on to an advanced situation, which I will admit is sort of an unconventional approach; that is testing the `@Input` and `@Output` emitters of the child components in the parent component in order to test an entire flow. This is what we do for the tests of `VersionControlComponent`.

Notice that we remove the usage of the `@angular/core/testing` package completely from the `version-control.component.spec.ts` file. This is because we're no longer using `TestBed` to create the test environment. Instead, we use the `MockBuilder` method from the `ng-mocks` package to build the test environment for our `VersionControlComponent`. Then, we use the `.mock()` method to mock each child component that we want to work with inside the tests later on. The `.mock()` method is not only used to mock components, but it can also be used to mock services, directives, pipes, and more. Please refer to the next section for further resources to read.

Then, in the `'should add the new log when it is created via ReleaseFormComponent'` test, pay attention to the `ngMocks.find()` method, which we use to find the relevant component and get its instance. Its use is relatively similar to what we would do in `TestBed`, as follows:

```
fixture.debugElement.query(
    By.css('app-release-form')
).componentInstance
```

However, using `ngMocks.find()` is better suited, as it has better support for types. Once we get a hold of the instance of `ReleaseFormComponent`, we use the `@Output` named `newReleaseLog` to create a new log using the `.emit()` method. Then, we do a quick `fixture.detectChanges()` to trigger the Angular change detection. We also check the `VersionControl.releaseLogs` array to determine whether our new release log has been added to the array. Afterward, we also check the `ReleaseLogsComponent.logs` property to make sure that the child component has updated the `logs` array via `@Input`.

> **Important note**
>
> Notice that we don't use a spy on the `VersionControlComponent.`
> `addNewReleaseLog` method. That is because if we do so, that function
> will become a Jest spy function. Therefore, it'll lose its functionality inside. In
> return, it'll never add the new log to the `releaseLogs` array, and none of
> our tests will pass. You can try it out for fun.

See also

- The ng-mocks `.mock` method (`https://ng-mocks.sudo.eu/api/`
  `MockBuilder#mock`)

- The ng-mocks official documentation (`https://ng-mocks.sudo.eu`)

Creating even easier component tests with Angular CDK component harnesses

When writing tests for components, there might be scenarios where you'd actually want
to interact with the DOM elements. Now, this can already be achieved by using the
`fixture.debugElement.query` method to find the element using a selector and
then triggering events on it. However, that means maintaining it for different platforms,
knowing the identifiers of all the selectors, and then exposing all of that in the tests. And
this is even worse if we're talking about an Angular library. It certainly isn't necessary
for each developer who interacts with my library to know all the element selectors in
order to write the tests. Only the author of the library should know that much to respect
encapsulation. Luckily, we have the component harnesses from the Angular CDK team,
which were released with Angular 9 along with the IVY compiler. And they've led by
example, by providing component harnesses for the Angular material components as well.
In this recipe, you'll learn how to create your own component harnesses.

Getting ready

The project that we are going to work with resides in `chapter10/start_here/`
`tests-using-cdk-harness`, which is inside the cloned repository. Perform the
following steps:

1. Open the project in Visual Studio Code.

2. Open the Terminal and run `npm install` to install the dependencies of the
   project.

3. Once done, run `ng serve -o`.

This should open the app in a new browser tab. You should see something similar to the following screenshot:

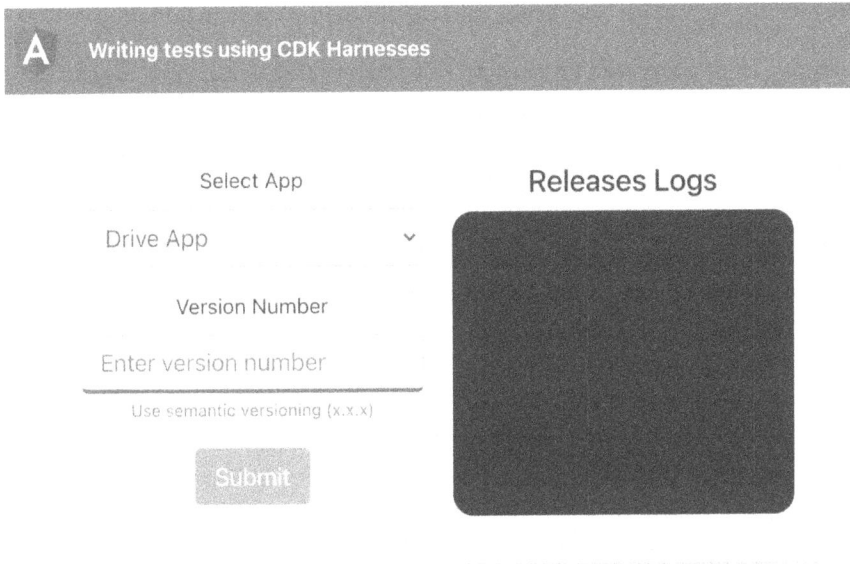

Figure 10.13 – The tests-using-cdk-harness app running on http://localhost:4200

Now that you have the app running, let's move on to the next section to follow the recipe.

How to do it...

We have our favorite Angular version control app that allows us to create release logs. And we have the tests written already, including tests that interact with the DOM element to validate a few use cases. Let's follow the recipe to use component harnesses instead, and discover how easy it becomes to use in the actual tests:

1. First, open a new Terminal window/tab and ensure you're inside the `chapter10/start_here/tests-using-cdk-harness` folder. Once inside, run the following command to install the Angular CDK:

    ```
    npm install --save @angular/cdk@12.0.0
    ```

2. You have to restart your Angular server. So, rerun the `ng serve` command.

3. First, we'll create a **component harness** for the `ReleaseFormComponent`. Let's create a new file inside the `release-form` folder, and name it `release-form.component.harness.ts`. Then, add the following code inside it:

```
import { ComponentHarness } from '@angular/cdk/testing';

export class ReleaseFormComponentHarness extends
ComponentHarness {
  static hostSelector = 'app-release-form';
  protected getSubmitButton = this.
  locatorFor('button[type=submit]');
  protected getAppNameInput = this.
  locatorFor(`#appName`);
  protected getAppVersionInput = this.
  locatorFor(`#versionNumber`);
  protected getVersionErrorEl = async () => {
    const alerts = await this.locatorForAll('.alert.
    alert-danger')();
    return alerts[1];
  };
}
```

4. Now we need to set up the harness environment for our tests for `VersionControlComponent`. For this, we'll use `HarnessLoader` and `TestbedHarnessEnvironment` from the Angular CDK. Update the `version-control.component.spec.ts` file as follows:

```
...
import { HarnessLoader } from '@angular/cdk/testing';
import { TestbedHarnessEnvironment } from '@angular/cdk/
testing/testbed';

describe('VersionControlComponent', () => {
  let component: VersionControlComponent;
  let fixture: ComponentFixture<VersionControlComponent>;
  let harnessLoader: HarnessLoader;

  ...
  beforeEach(() => {
```

```
    fixture = TestBed.
    createComponent (VersionControlComponent);
    component = fixture.componentInstance;
    fixture.detectChanges();
    harnessLoader = TestbedHarnessEnvironment.
    loader(fixture);
  });

  ...
});
```

5. Now, let's write some methods in our `ReleaseFormComponentHarness` class to get the relevant information. We'll use these methods in the later steps. Update the `release-form.component.harness.ts` file as follows:

```
...
export class ReleaseFormComponentHarness extends
ComponentHarness {
  ...
  async getSelectedAppName() {
    const appSelectInput = await this.getAppNameInput();
    return appSelectInput.getProperty('value');
  }

  async clickSubmit() {
    const submitBtn = await this.getSubmitButton();
    return await submitBtn.click();
  }

  async setNewAppVersion(version: string) {
    const versionInput = await this.getAppVersionInput();
    return await versionInput.sendKeys(version);
  }

  async isVersionErrorShown() {
    const versionErrorEl = await this.
    getVersionErrorEl();
    const versionErrorText = await versionErrorEl.text();
    return (
```

```
        versionErrorText.trim() === 'Please write an
        appropriate version number'
      );
    }
  }
```

6. Next, we'll work on our first test, named `'should have the first app selected for the new release log'`, with the component harness. Update the `version-control.component.spec.ts` file as follows:

```
...
import { ReleaseFormComponentHarness } from '../release-
form/release-form.component.harness';

describe('VersionControlComponent', () => {
  ...
  it('should have the first app selected for the new
  release log', async () => {
    const rfHarness = await harnessLoader.getHarness(
      ReleaseFormComponentHarness
    );
    const appSelect = await rfHarness.
    getSelectedAppName();
    expect(appSelect).toBe(Apps.DRIVE);
  });
  ...
});
```

Now if you run npm run test, you should see all of the tests passing, which means our first test with the component harness works. Woohoo!

7. Now, we'll work on our second test, that is, for `'should show error on wrong version number input'`. Update the test in the `version-control.component.spec.ts` file, as follows:

```
...
describe('VersionControlComponent', () => {
  ...
  it('should show error on wrong version number input',
  async () => {
```

```
  const rfHarness = await harnessLoader.getHarness(
    ReleaseFormComponentHarness
  );
  await rfHarness.setNewAppVersion('abcd');
  const isErrorshown = await rfHarness.
  isVersionErrorShown();
  expect(isErrorshown).toBeTruthy();
});

...

});
```

Boom! Note that we just reduced the lines of code for this test from nine statements to only four statements. Isn't that amazing? I believe it is awesome and much cleaner, to be honest.

8. For the final test, we also need a component harness for `ReleaseLogsComponent`. Let's quickly create it. Add a new file inside the `release-logs` folder, named `release-logs.component.harness.ts`, and add the following code:

```
import { ComponentHarness } from '@angular/cdk/testing';

export class ReleaseLogsComponentHarness extends
ComponentHarness {
  static hostSelector = 'app-release-logs';
  protected getLogsElements = this.locatorForAll
  ('.logs__item');

  async getLogsLength() {
    const logsElements = await this.getLogsElements();
    return logsElements.length;
  }

  async getLatestLog() {
    const logsElements = await this.getLogsElements();
    return await logsElements[0].text();
  }

  async validateLatestLog(version, app) {
```

```
        const latestLogText = await this.getLatestLog();
        return (
          latestLogText.trim() === `Version ${version}
          released for app ${app}`
        );
    }
}
```

9. Finally, let's modify our final tests in the `version-control.component.spec.ts` file as follows:

```
...
import { ReleaseFormComponentHarness } from '../release-
form/release-form.component.harness';
import { ReleaseLogsComponentHarness } from '../release-
logs/release-logs.component.harness';

describe('VersionControlComponent', () => {
  ...
  it('should show the new log in the list after adding
  submitting a new log', async () => {
    const rfHarness = await harnessLoader.getHarness(
      ReleaseFormComponentHarness
    );
    const rLogsHarness = await harnessLoader.getHarness(
      ReleaseLogsComponentHarness
    );
    let logsLength = await rLogsHarness.getLogsLength();
    expect(logsLength).toBe(0); // no logs initially
    const APP = Apps.DRIVE;
    const VERSION = '2.3.6';
    await rfHarness.setNewAppVersion(VERSION);
    await rfHarness.clickSubmit();
    logsLength = await rLogsHarness.getLogsLength();
    expect(logsLength).toBe(1);
    const isNewLogAdded = await rLogsHarness.
    validateLatestLog(VERSION, APP);
```

```
        expect(isNewLogAdded).toBe(true);
    });
});
```

Voila! That's some amazing testing right there using the Angular CDK component harnesses. If you run the tests now, you should see all of the tests passing. Now that you've finished the recipe, please refer to the next section to learn how this works.

How it works...

All right! That was a cool recipe, which I enjoyed working on myself. The key factor of this recipe is the `@angular/cdk/testing` package. If you have worked with e2e tests using Protractor before, this is a similar concept to the `Pages` in Protractor. First, we create a component harness for both the `ReleaseLogsComponent` and the `ReleaseFormComponent`.

Notice that we import the `ComponentHarness` class from `@angular/cdk/testing` for both component harnesses. Then, we extend our custom classes called `ReleaseFormComponentHarness` and `ReleaseLogsComponentHarness` from the `ComponentHarness` class. Essentially, this is the correct way to author component harnesses. Did you notice the static property called `hostSelector`? We need this property for every component harness class that we create. And the value is always the selector of the target element/component. This ensures that when we load this harness into the test environment, the environment is able to find the host element in the DOM – for which we're creating the component harness. In our component harness class, we use the `this.locatorFor()` method to find elements within the host component. The `locateFor()` method takes a single argument as the `css selector` of the element to be found and returns an `AsyncFactoryFn`. This means the returned value is a function that we can use at a later time to get the required elements.

In the `ReleaseFormComponentHarness` class, we find the submit button, the app name input, and the version number input using the `protected` methods' `getSubmitButton`, `getAppNameInput`, and `getAppVersionInput`, respectively, which are all of the `AsyncFactoryFn` type, as mentioned earlier. We have these methods set as `protected` because we don't want the people writing the unit tests to access or care about the information of the DOM elements. This makes it much easier for everyone to write tests without worrying about the internal implementation of accessing the DOM.

Notice that the `getVersionErrorEl()` method is slightly different. It is not actually of the `AsyncFactoryFn` type. Instead, it is a regular `async` function that first calls the `locatorForAll` method to get all the elements with the `alert` class and the `alert-danger` class, which results in the error messages. Then, it selects the second alert element, which is for the app version number input.

One important thing to mention here is that when we call the `locatorFor()` method or the `locatorForAll()` method, we get back a `Promise` with the `TestElement` item or a `Promise` with a list of `TestElement` items, respectively. Each `TestElement` item has a bunch of handy methods such as `.click()`, `.sendKeys()`, `.focus()`, `.blur()`, `.getProperty()`, `.text()`, and more. And these methods are what we're interested in since we use them behind the scenes to interact with the DOM elements.

Now, let's talk about configuring the test environment. In the `version-control.component.spec.ts` file, we set up the environment to use component harnesses for both `ReleaseLogsComponent` and `ReleaseFormComponent`. The `TestbedHarnessEnvironment` element is the key element here. We use the `.loader()` method of the `TestbedHarnessEnvironment` class by providing our `fixture` as an argument. Note that the fixture is what we get in the test environment using the `TestBed.createComponent(VersionControlComponent)` statement. Because we provide this fixture to the `TestbedHarnessEnvironment.loader()` method, we get back an element of the `HarnessLoader` statement, which can now load component harnesses for the other components – that is, for `ReleaseLogsComponent` and `ReleaseFormComponent`.

Notice that in the tests, we use the `harnessLoader.getHarness()` method by providing the harness class as an argument. This enables the test environment to find the DOM element associated with the `hostSelector` property of the harness class. Additionally, we get back the instance of the component harness that we can use further in the test.

See also

- Finding components in the DOM with component harnesses (`https://material.angular.io/cdk/test-harnesses/overview#finding-elements-in-the-components-dom`)

- *API for component harness authors* (`https://material.angular.io/cdk/test-harnesses/overview#api-for-component-harness-authors`)

Unit testing components with Observables

If you're building an Angular application, it is very likely that you'll work with Observables inside the app at some point. For instance, you could be fetching data from a third-party API or perhaps just managing the state. In either case, it becomes slightly difficult to test applications that have Observables in action. In this recipe, we're going to learn how to test unit tests with Observables.

Getting ready

The project for this recipe resides in `chapter10/start_here/unit-testing-observables`. Perform the following steps:

1. Open the project in Visual Studio Code.

2. Open the Terminal and run `npm install` to install the dependencies of the project.

3. Once done, run `ng serve -o`.

This should open the app in a new browser tab. You should see something similar to the following screenshot:

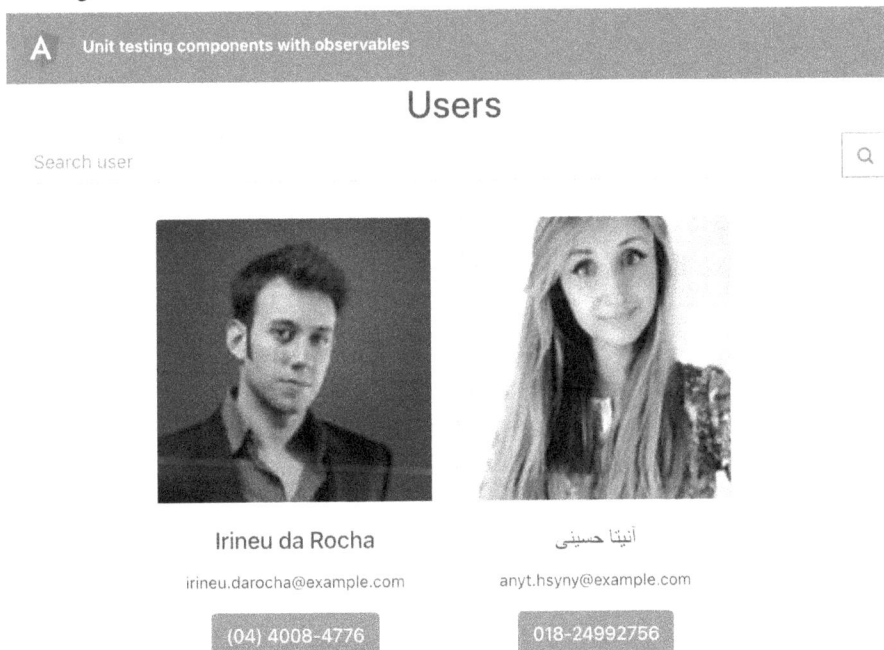

Figure 10.14 – The unit-testing-observables app running on http://localhost:4200

Now that we have the app running locally, in the next section, let's take a look at the steps of the recipe.

How to do it...

We'll start by writing test cases, which technically involve the usage of Observables. Essentially, we have to mock the methods using Observables, and we have to use the `fakeAsync` and `tick()` methods provided by Angular to reach our goal of writing good unit tests with Observables. Let's get started:

1. First and foremost, we'll write a test to see what happens when we use an `expect()` clause in a test that involves a function containing an Observable. Update the `users.component.spec.ts` file by adding a test, which checks whether we get the users from the server when the component initiates:

```
import { HttpClientModule } from '@angular/common/http';
import {
  ComponentFixture,
  fakeAsync,
  TestBed,
  tick,
} from '@angular/core/testing';
...

describe('UsersComponent', () => {
  ...
  it('should get users back from the API component init',
  fakeAsync(() => {
    component.ngOnInit();
    tick(500);
    expect(component.users.length).toBeGreaterThan(0);
  }));
});
```

Now, as soon as you run the `npm run test` command, you'll see that the test fails with the following message:

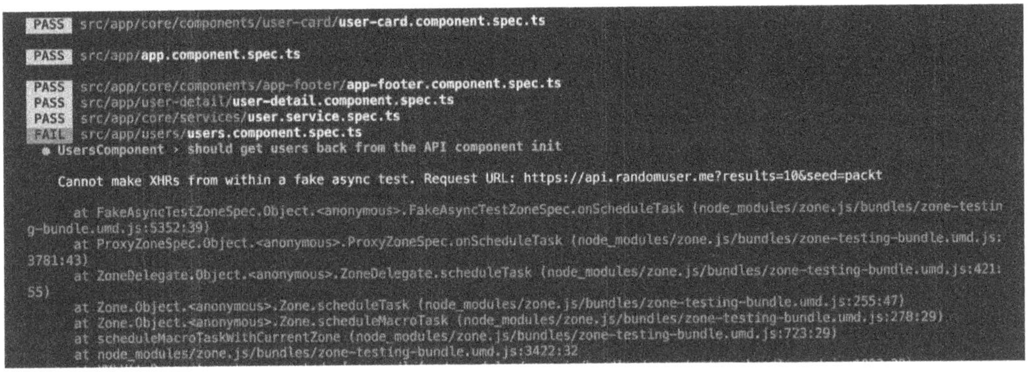

Figure 10.15 – Error – Cannot make XHRs from within a fake async test

What this means is that we can't make real HTTP calls from the `fakeAsync` tests, which is what happens after the `ngOnInit()` method is called.

2. The proper way to test this is to mock `UserService`. Luckily, we've already done this as we have the `UserServiceMock` class in the project. We need to provide it as a `useClass` property for `UserService` in `TestBed` and update our test slightly. Let's modify the `users.component.spec.ts` file as follows:

```
...
import {
  DUMMY_USERS,
  UserServiceMock,
} from 'src/__mocks__/services/user.service.mock';
...
describe('UsersComponent', () => {
  ...
  beforeEach(async () => {
    await TestBed.configureTestingModule({
      declarations: [UsersComponent, UserCardComponent],
      providers: [
        {
          provide: UserService,
          useClass: UserServiceMock,
        },
      ],
      imports: [HttpClientModule, ReactiveFormsModule,
      RouterTestingModule],
```

```
    }).compileComponents();
  });

  ...

  it('should get users back from the API component init',
  fakeAsync(() => {
    component.ngOnInit();
    tick(500);
    expect(component.users.length).toBe(2);
    expect(component.users).toEqual(DUMMY_USERS);
  }));
});
```

Now, if you run the tests again, they should pass. We'll cover this in more detail in the *How it works...* section later.

3. Let's add another test for a scenario in which we want to search users. We'll set the value for the username form control and search users using UserService, or more technically, UserServiceMock. Then, we will expect the results to be appropriate. Add a test in the users.component.spec.ts file as follows:

```
...
describe('UsersComponent', () => {

  ...

  it('should get the searched users from the API upon
  searching', fakeAsync(() => {
    component.searchForm.get('username').
    setValue('hall');
    // the second record in our DUMMY_USERS array has
    the name Mrs Indie Hall
    const expectedUsersList = [DUMMY_USERS[1]];
    component.searchUsers();
    tick(500);
    expect(component.users.length).toBe(1);
    expect(component.users).toEqual(expectedUsersList);
  }));
});
```

4. Now we'll write a test for `UserDetailComponent`. We need to test that our `UserDetailComponent` gets the appropriate user from the server when the component is initiated and that we get similar users as well. Update the `user-detail.component.spec.ts` file by adding a test, as follows:

```
...
import {..., fakeAsync, tick, } from '@angular/core/
testing';
...
import { UserServiceMock } from 'src/__mocks__/services/
user.service.mock';
describe('UserDetailComponent', () => {
    ...
    beforeEach(
        waitForAsync(() => {
            TestBed.configureTestingModule({
                declarations: [...],
                imports: [HttpClientModule, RouterTestingModule],
                providers: [
                    {
                        provide: UserService,
                        useClass: UserServiceMock,
                    },
                ],
            }).compileComponents();
        })
    );
    ...
    it('should get the user based on routeParams on page
load', fakeAsync(() => {
        component.ngOnInit();
        tick(500);
        expect(component.user).toBeTruthy();
    }));
});
```

The new test should be failing at the moment. We will fix it in the next steps.

5. To debug, we can quickly add a `console.log()` to the `params` that we get from subscribing to the `route.paramMap` Observable in the `ngOnInit()` method. Modify the `user-detail.component.ts` file, and then run the tests again:

```
...
@Component({...})
export class UserDetailComponent implements OnInit,
OnDestroy {
  ...
  ngOnInit() {
    this.isComponentAlive = true;
    this.route.paramMap
      .pipe(
        takeWhile(() => !!this.isComponentAlive),
        flatMap((params) => {
          this.user = null;
          console.log('params', params);
          ...
          return this.userService.getUser(userId).
          pipe(...);
        })
      )
      .subscribe((similarUsers: IUser[]) => {...});
  }
  ...
}
```

Now when you run the tests, you can see the error, as follows:

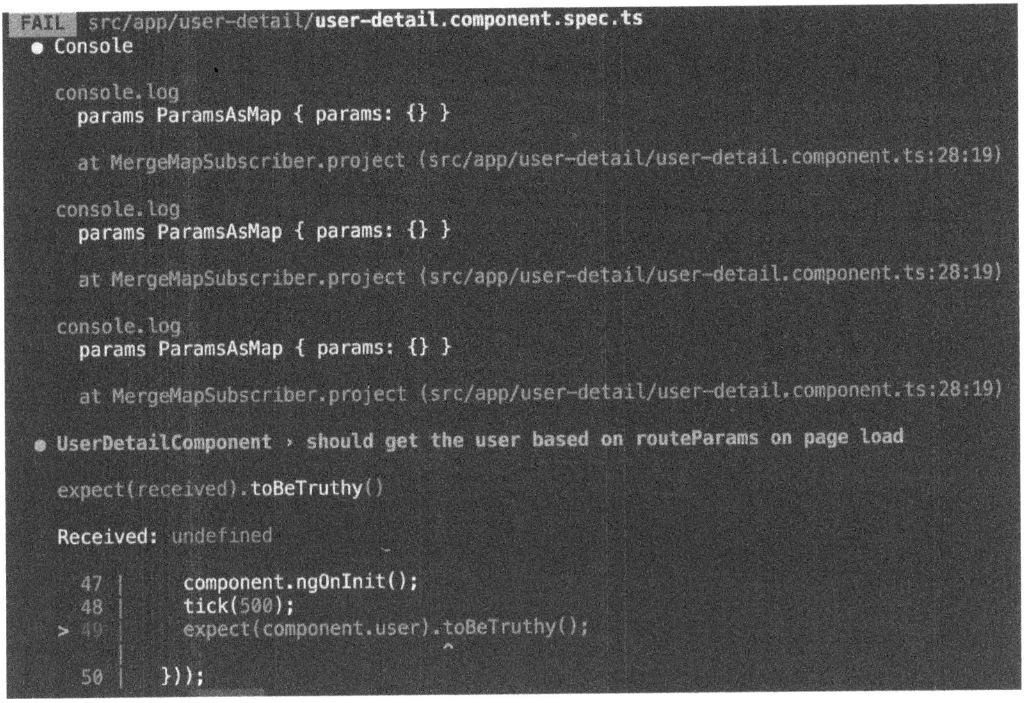

```
FAIL  src/app/user-detail/user-detail.component.spec.ts
 ● Console

   console.log
     params ParamsAsMap { params: {} }

     at MergeMapSubscriber.project (src/app/user-detail/user-detail.component.ts:28:19)

   console.log
     params ParamsAsMap { params: {} }

     at MergeMapSubscriber.project (src/app/user-detail/user-detail.component.ts:28:19)

   console.log
     params ParamsAsMap { params: {} }

     at MergeMapSubscriber.project (src/app/user-detail/user-detail.component.ts:28:19)

 ● UserDetailComponent › should get the user based on routeParams on page load

   expect(received).toBeTruthy()

   Received: undefined

     47 |        component.ngOnInit();
     48 |        tick(500);
   > 49 |        expect(component.user).toBeTruthy();
        |                               ^
     50 |      }));
```

Figure 10.16 – Error – empty params and missing uuid

6. As you can see in *Figure 10.16*, we don't have the uuid in the Params object.
    This is because it is not a real routing process for a real user. So, we need to mock
    the ActivatedRoute service that is used in UserDetailComponent to get
    the desired result. Let's create a new file inside the __mocks__ folder, named
    activated-route.mock.ts, and add the following code to it:

```
import { convertToParamMap, ParamMap, Params } from '@
angular/router';
import { ReplaySubject } from 'rxjs';


/**
 * An ActivateRoute test double with a `paramMap`
 observable.
 * Use the `setParamMap()` method to add the next
 `paramMap` value.
 */
export class ActivatedRouteMock {
    // Use a ReplaySubject to share previous values with
    subscribers
```

```
    // and pump new values into the `paramMap` observable
    private subject = new ReplaySubject<ParamMap>();

    constructor(initialParams?: Params) {
      this.setParamMap(initialParams);
    }

    /** The mock paramMap observable */
    readonly paramMap = this.subject.asObservable();

    /** Set the paramMap observables's next value */
    setParamMap(params?: Params) {
      this.subject.next(convertToParamMap(params));
    }
  }
}
```

7. Now we'll use this mock in our tests for UserDetailComponent. Update the
   user-detail.component.spec.ts file, as follows:

```
...
import { ActivatedRouteMock } from 'src/__mocks__/
activated-route.mock';
import {
  DUMMY_USERS,
  UserServiceMock,
} from 'src/__mocks__/services/user.service.mock';
...
describe('UserDetailComponent', () => {
  ...
  let activatedRoute;
  beforeEach(
    waitForAsync(() => {
      TestBed.configureTestingModule({
        ...
        providers: [
          {...},
          {
```

```
                  provide: ActivatedRoute,
                  useValue: new ActivatedRouteMock(),
              },
          ],
      }).compileComponents();
  })
 );

 beforeEach(() => {
   . . .
   fixture.detectChanges();
   activatedRoute = TestBed.inject(ActivatedRoute);
 });
 . . .
});
```

8. Now that we have injected the mock into the test environment, let's modify our test to get the second user from the DUMMY_USERS array. Update the tests file as follows:

```
 . . .
describe('UserDetailComponent', () => {
   . . .
  it('should get the user based on routeParams on page
  load', fakeAsync(() => {
    component.ngOnInit();
    activatedRoute.setParamMap({ uuid: DUMMY_USERS[1].
    login.uuid });
    tick(500);
    expect(component.user).toEqual(DUMMY_USERS[1]);
  }));
});
```

9. Now we'll write a test that allows us to get similar users when
   `UserDetailComponent` is loaded. Remember that according to our current
   business logic, similar users are all users except the current user on the page,
   which is saved in the `user` property. Let's add the test in the `user-detail.`
   `component.spec.ts` file, as follows:

```
...
describe('UserDetailComponent', () => {
  ...
  it('should get similar user based on routeParams uuid
  on page load', fakeAsync(() => {
    component.ngOnInit();
    activatedRoute.setParamMap({ uuid: DUMMY_USERS[1].
    login.uuid }); // the second user's uuid
    const expectedSimilarUsers = [DUMMY_USERS[0]]; //
    the first user
    tick(500);
    expect(component.similarUsers).
    toEqual(expectedSimilarUsers);
  }));
});
```

If you run the tests, you should see them all pass as follows:

```
console.log
  params ParamsAsMap {
    params: { uuid: 'd2775083-57a8-4034-983b-844cbd58aba1' }
  }

  at MergeMapSubscriber.project (src/app/user-detail/user-detail.component.ts:28:19)

PASS  src/app/users/users.component.spec.ts

Test Suites: 7 passed, 7 total
Tests:       12 passed, 12 total
Snapshots:   0 total
Time:        5.156 s
Ran all test suites.
:+  Done in 6.80s.
```

Figure 10.17 – All of the tests are passing with mocked Observables

Great! You now know how to work with Observables when writing unit tests for
components. Although there's still a lot to learn about testing Observables in Angular,
the purpose of this recipe was to keep everything simple and sweet.

Now that you have finished the recipe, please refer to the next section to understand how
it works.

How it works...

We start our recipe by using the `fakeAsync()` and `tick()` methods from the `'@angular/core/testing'` package. Notice that we wrap our tests' callback method using the `fakeAsync()` method. The method wrapped in the `fakeAsync()` method is executed in something called a `fakeAsync` zone. This is contrary to how it works in the actual Angular application, which runs inside `ngZone`.

> **Important note**
>
> In order to work with the `fakeAsync` zone, we need to import the `zone.js/dist/zone-testing` library in our test environment. This is usually done in the `src/test.ts` file when you create an Angular project. However, since we migrated to Jest, we removed that file.

"Okay. How does it work then, Ahsan?" Well, I'm glad you asked. While setting up for Jest, we use the `jest-preset-angular` package. This package ultimately requires all the necessary files for the `fakeAsync` tests, as follows:

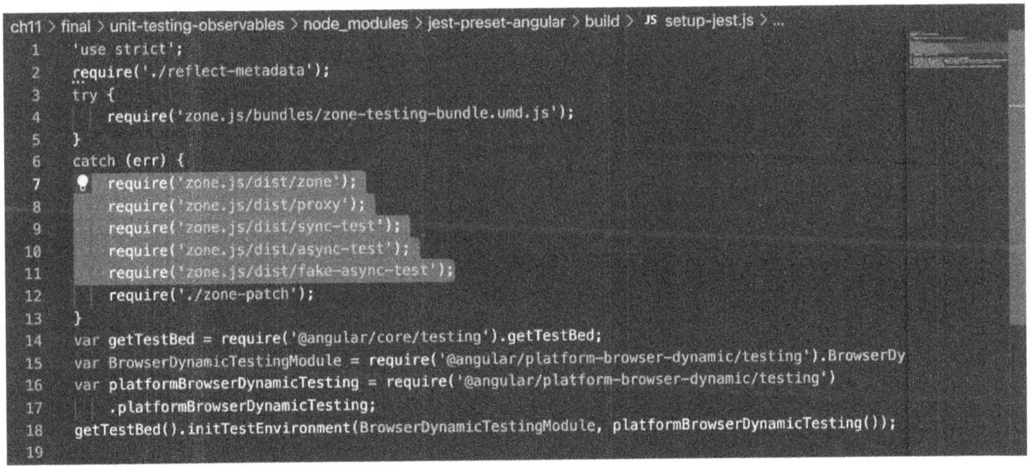

```js
ch11 > final > unit-testing-observables > node_modules > jest-preset-angular > build > JS setup-jest.js > ...
1   'use strict';
2   require('./reflect-metadata');
3   try {
4       require('zone.js/bundles/zone-testing-bundle.umd.js');
5   }
6   catch (err) {
7       require('zone.js/dist/zone');
8       require('zone.js/dist/proxy');
9       require('zone.js/dist/sync-test');
10      require('zone.js/dist/async-test');
11      require('zone.js/dist/fake-async-test');
12      require('./zone-patch');
13  }
14  var getTestBed = require('@angular/core/testing').getTestBed;
15  var BrowserDynamicTestingModule = require('@angular/platform-browser-dynamic/testing').BrowserDy
16  var platformBrowserDynamicTesting = require('@angular/platform-browser-dynamic/testing')
17      .platformBrowserDynamicTesting;
18  getTestBed().initTestEnvironment(BrowserDynamicTestingModule, platformBrowserDynamicTesting());
19
```

Figure 10.18 – The jest-preset-angular package importing the required zone.js files

Essentially, the `tick()` method simulates the passage of time in this virtual `fakeAsync` zone until all of the asynchronous tasks are finished. It takes a parameter as milliseconds, which either reflects how many milliseconds have passed or how much the virtual clock has advanced. In our case, we use `500` milliseconds as the value for the `tick()` method.

Notice that we're mocking `UserService` for the tests for `UsersComponent`. Specifically for `'should get users back from the API component init'`, we call the `component.ngOnInit()` method in the test and then call the `tick()` method. In the meantime, the `ngOnInit()` method calls the `searchUsers()` method, which calls the `UserServiceMock.searchUsers()` method since we've provided it as the `useClass` property in our test environment for `UserService`. Finally, that returns the value of the `DUMMY_USERS` array that we have defined in the `user.service.mock.ts` file. The other test for the `UsersComponent`, that is, `'should get the searched users from the API upon searching'`, is quite similar as well.

In terms of the tests for `UserDetailComponent`, we do something different, that is, we also have to mock the `activatedRoute` service. Why? Well, that is because the `UserDetailComponent` is a page that can be navigated with a `uuid` and because its path is defined as `'/users/:uuid'` in the `app-routing.module.ts` file. Therefore, we need to populate this `uuid` parameter in our tests to work with the `DUMMY_USERS` array. For this, we use the `ActivatedRouteMock` class inside the `__mocks__` folder. Notice that it has a `setParamMap()` method. This allows us to specify the `uuid` parameter in our tests. Then, when the actual code subscribes to the `this.route.paramMap` Observable, our set `uuid` parameter can be found there.

For the `'should get the user based on routeParams on page load'` test, we set the second user's `uuid` from the `DUMMY_USERS` array as the `uuid` route parameter's value. Then, we use the `tick()` method, after which we expect the `user` property to have the second user from the `DUMMY_USERS` array as the value. The other test in the file is also quite similar and self-explanatory. Please refer to the next section for more useful links regarding unit testing scenarios.

See also

- Angular testing component scenarios (`https://docs.angular.lat/guide/testing-components-scenarios`)

- Testing routed Angular components with `RouterTestingModule` (`https://dev.to/this-is-angular/testing-angular-routing-components-with-the-routertestingmodule-4cj0`)

Unit testing Angular Pipes

In my personal opinion, pipes are the easiest components to test in an Angular application. Why? Well, this is because they're (supposed to be) pure functions that return the same result based on the same set of inputs. In this recipe, we'll write some tests for a really simple pipe in an Angular application.

Getting ready

The project that we are going to work with resides in `chapter10/start_here/unit-testing-pipes`, which is inside the cloned repository. Perform the following steps:

1. Open the project in Visual Studio Code.

2. Open the Terminal and run `npm install` to install the dependencies of the project.

3. Once done, run `ng serve -o`.

This should open the app in a new browser tab. You should see something similar to the following screenshot:

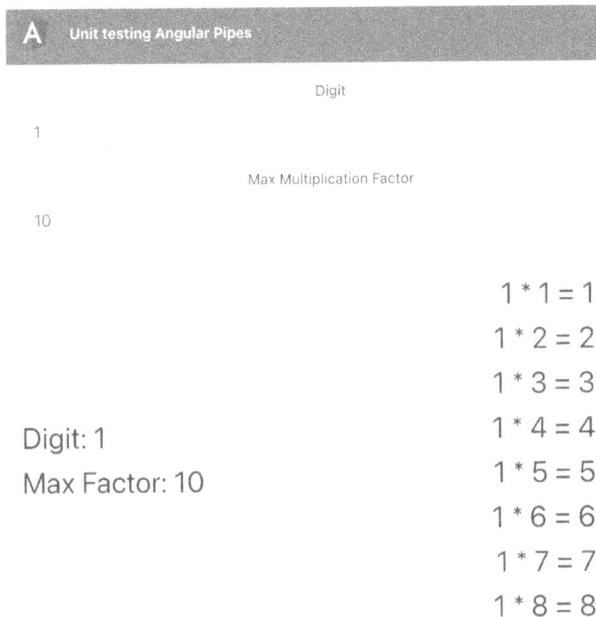

A   **Unit testing Angular Pipes**

Digit

1

Max Multiplication Factor

10

Digit: 1
Max Factor: 10

1 * 1 = 1
1 * 2 = 2
1 * 3 = 3
1 * 4 = 4
1 * 5 = 5
1 * 6 = 6
1 * 7 = 7
1 * 8 = 8

Figure 10.19 – The unit-testing-pipes app running on http://localhost:4200

Now that we have the app running locally, in the next section, let's go through the steps of the recipe.

How to do it...

Here, we have a simple recipe that takes two inputs – the digit and the max factor value. Based on these inputs, we show a multiplication table. We already have the `MultTablePipe` that is working fine according to our business logic. We'll now write some unit tests to validate our inputs and expected outputs, as follows:

1. Let's write our first test for `MultTablePipe`. We'll make sure it returns an empty array when we have an invalid value for the `digit` input. Update the `mult-table.pipe.spec.ts` file, as follows:

```
. . .
describe('MultTablePipe', () => {
  . . .
  it('should return an empty array if the value of digit
  is not valid', () => {
    const digit = 0;
    const limit = 10;
    const outputArray = pipe.transform(null, digit,
    limit);
    expect(outputArray).toEqual([]);
  });
});
```

2. Let's write another test to validate the `limit` input so that we also return an empty array if it is invalid:

```
. . .
describe('MultTablePipe', () => {
  . . .
  it('should return an empty array if the value of limit
  is not valid', () => {
    const digit = 10;
    const limit = 0;
    const outputArray = pipe.transform(null, digit,
    limit);
    expect(outputArray).toEqual([]);
  });
});
```

3. Now we'll write a test to validate the output of the pipe's transform method when both the `digit` and `limit` inputs are valid. In this scenario, we should get back the array containing the multiplication table. Write another test as follows:

```
...
describe('MultTablePipe', () => {
  ...
  it('should return the correct multiplication table when
    both digit and limit inputs are valid', () => {
    const digit = 10;
    const limit = 2;
    const expectedArray = ['10 * 1 = 10', '10 * 2 = 20'];
    const outputArray = pipe.transform(null, digit,
    limit);
    expect(outputArray).toEqual(expectedArray);
  });
});
```

4. Right now, within the app, we have the possibility to provide decimal digits for the `limit` input. For instance, we can write 2.5 as the max factor in the input. To handle this, we use a `Math.floor()` in `MultTablePipe` to round it down to the lower number. Let's write a test to make sure this works:

```
...
describe('MultTablePipe', () => {
  ...
  it('should round of the limit if it is provided in
    decimals', () => {
    const digit = 10;
    const limit = 3.5;
    const expectedArray = ['10 * 1 = 10', '10 * 2 = 20',
    '10 * 3 = 30']; // rounded off to 3 factors instead
    of 3.5
    const outputArray = pipe.transform(null, digit,
    limit);
    expect(outputArray).toEqual(expectedArray);
  });
});
```

Easy peasy! Writing tests for Angular pipes is so straightforward that I love it. We could call this the power of pure functions. Now that you've finished the recipe, please refer to the next section for more informative links.

See also

- Testing Angular pipes official documentation (`https://angular.io/guide/testing-pipes`)

- *Test Angular Pipes With Services* (`https://levelup.gitconnected.com/test-angular-pipes-with-services-4cf77e34e576`)

11
E2E Tests in Angular with Cypress

An app having a couple of **end-to-end** (E2E) tests surely promises more reliability than an app having no tests at all, and in today's world, with emerging businesses and complex applications, it becomes essential at some point to have E2E tests written to capture the entire flow of an application. Cypress is one of the best tools out there today when it comes to E2E tests for web applications. In this chapter, you'll learn how to test your E2E flows in an Angular app with Cypress. Here are the recipes we're going to cover in this chapter:

- Writing your first Cypress test
- Validating if a **Document Object Model** (DOM) element is visible on the view
- Testing form inputs and submission
- Waiting for **XMLHttpRequests** (XHRs) to finish
- Using Cypress bundled packages
- Using Cypress fixtures to provide mock data

Technical requirements

For the recipes in this chapter, make sure you have **Git** and **Node.js** installed on your machine. You also need to have the `@angular/cli` package installed, which you can do with `npm install -g @angular/cli` from your terminal. The code for this chapter can be found at `https://github.com/PacktPublishing/Angular-Cookbook/tree/master/chapter11`.

Writing your first Cypress test

If you have been writing E2E tests already, you might have been doing this using Protractor. Working with Cypress is a completely different experience, though. In this recipe, you'll set up Cypress with an existing Angular application and will write your first E2E test with Cypress.

Getting ready

The project that we are going to work with resides in `chapter11/start_here/angular-cypress-starter`, inside the cloned repository:

1. Open the project in **Visual Studio Code** (**VS Code**).

2. Open the terminal and run `npm install` to install the dependencies of the project.

Now that we have the project opened locally, let's see the steps of the recipe in the next section.

How to do it...

The app we're working with is a simple counter application. It has a minimum and maximum values and some buttons that can increment, decrement, and reset the counter's value. We'll start by configuring Cypress for our application and will then move toward writing the test:

1. First, open a new terminal window/tab and make sure you're inside the `chapter11/start_here/angular-cypress-starter` folder. Once inside, run the following command to install `Cypress` and `concurrently` in our project:

    ```
    npm install -d cypress concurrently
    ```

2. Now, open your `package.json` file and add the following script inside the `scripts` object, as follows:

```
{
    "name": "angular-cypress-starter",
    "version": "0.0.0",
    "scripts": {
        ...
        "e2e": "ng e2e",
        "start:cypress": "cypress open",
        "cypress:test": "concurrently 'npm run start' 'npm run start:cypress'"

    },
    ...
}
```

3. Let's run the `cypress:test` command to simultaneously start the `http://localhost:4200` Angular server and to start Cypress tests as well, as follows:

```
npm run cypress:test
```

You should also see that Cypress creates a folder named `cypress` and some example tests inside it by default. Cypress also creates a `cypress.json` file to be able to provide some configuration. We will not remove these default tests but will instead ignore them in the next step.

4. Ignore the default/example tests by modifying the `cypress.json` file, as follows:

```
{
    "baseUrl": "http://localhost:4200",
    "ignoreTestFiles": "**/examples/*",
    "viewportHeight": 760,
    "viewportWidth": 1080
}
```

5. If you look again at the Cypress window now, you should see that we don't have any
    integration tests, as follows:

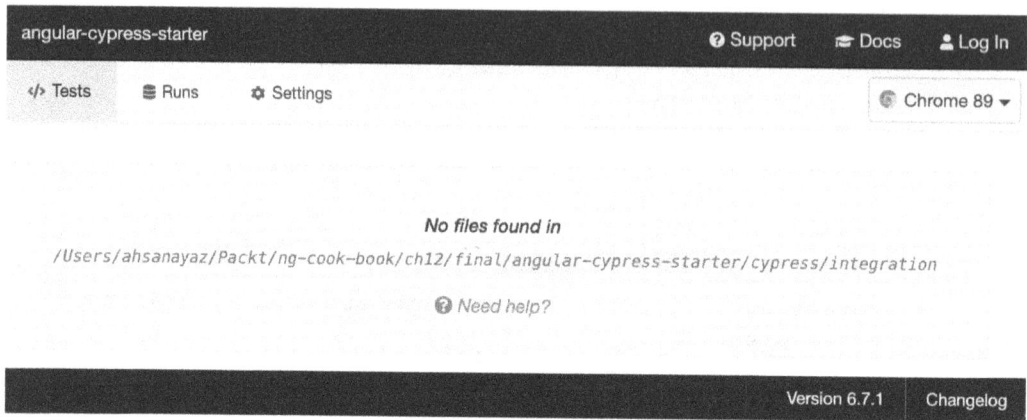

Figure 11.1 – No integration tests to execute

6. Let's create our first test now. We'll just check whether the browser title of our
    app is **Writing your first Cypress test**. Create a new file inside the cypress/
    integration folder named app.spec.js, and paste the following code inside:

```
/// <reference types="cypress" />

context('App', () => {
  beforeEach(() => {
    cy.visit('/');
  });

    it('should have the title "Writing your first Cypress
    test "', () => {
      // https://on.cypress.io/title
      cy.title().should('eq', 'Writing your first Cypress
      test');
    });
});
```

7. If you look again at the Cypress window, you should see a new app.spec.js file
    listed, as follows:

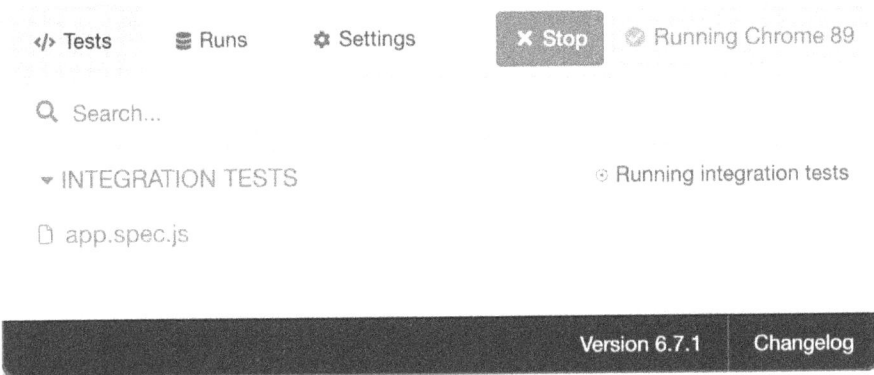

Figure 11.2 – The new app.spec.js test file being shown

8. Tap the `app.spec.js` file in the window shown in *Figure 11.2*, and you should see the Cypress tests passing for the tests written in the file.

Kaboom! Within a few steps, we have now set up Cypress for our Angular application and have written our first test. You should see the Cypress window, as follows:

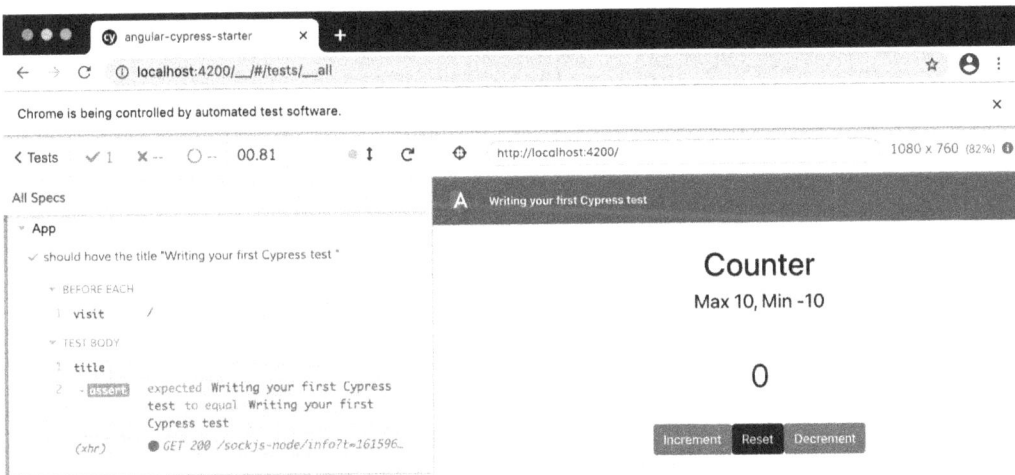

Figure 11.3 – Our first Cypress test passes

Easy! Right? Now that you know how to configure Cypress for an Angular app, see the next section to understand how it works.

How it works...

Cypress can be integrated with absolutely any framework and web development project. One interesting fact is that Cypress uses Mocha as the test runner behind the scenes. The tooling for Cypress watches for code changes so that you don't have to recompile the tests time and time again. Cypress also adds a shell around the application being tested to capture logs and access DOM elements during the tests, and some functionality for debugging tests.

At the very top of our `app.spec.js` file, we use the `context()` method that defines the test suite, basically defining the context of the tests about to be written inside. Then, we use a `beforeEach()` method to specify what should happen before each test is executed. Since each test starts with no data, we first have to make sure that Cypress navigates to our application's `http://localhost:4200` **Uniform Resource Locator (URL)**. The reason we just specify `cy.visit('/')` and it still works is that we have already specified the `baseUrl` property in the `cypress.json` file. Therefore, we just have to provide relative URLs in our tests.

Finally, we use the `it()` method to specify the titles for our first test, and then we use the `cy.title()` method, which is a handy helper, to fetch the text value of the **Title** of our **HyperText Markup Language (HTML)** page currently being rendered. We use the `'eq'` operator to check its value against the `'Writing your first Cypress test'` string, and it all works!

See also

- `cy.title()` documentation (`https://docs.cypress.io/api/commands/title.html#Syntax`)

- Cypress documentation—*Writing Your First Test* (`https://docs.cypress.io/guides/getting-started/writing-your-first-test.html`)

Validating if a DOM element is visible on the view

In the previous recipe, we learned how to install and configure Cypress in an Angular app. There might be different cases in your application where you'd want to see if an element is visible on the DOM or not. In this recipe, we'll write some tests to identify if any elements are visible on the DOM.

Getting ready

The project for this recipe resides in `chapter11/start_here/cypress-dom-element-visibility`:

1. Open the project in VS Code.

2. Open the terminal and run `npm install` to install the dependencies of the project.

3. Once done, run `npm run cypress:test`.

This should run the app at `https://localhost:4200` and should open the Cypress window, as follows:

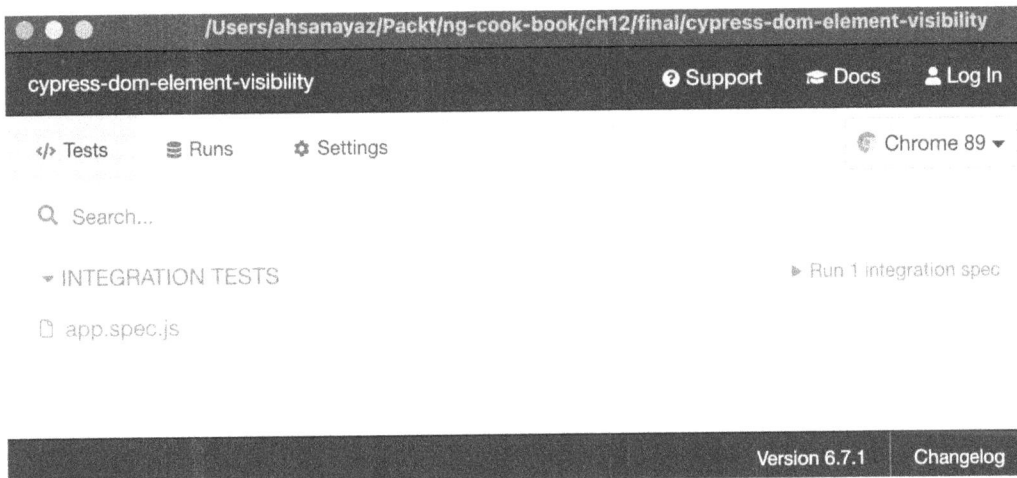

Figure 11.4 – Cypress tests running for the cypress-dom-element-visibility app

Now that we have the app and the Cypress tests running locally, let's see the steps of the recipe in the next section.

How to do it...

We have the same old counter app from the previous recipe. However, some things have changed. We now have a button at the top that toggles the visibility of the the counter component (CounterComponent). Also, we have to hover over the counter card to actually see the **Increment**, **Decrement**, and **Reset** action buttons. Let's start writing some tests to check the visibility of the the counter component (CounterComponent) and for the actions:

1. Let's write a test to check the visibility of the the counter component (CounterComponent) when we have clicked the **Toggle Counter Visibility** button to show it. We'll check it by asserting the visibility of the elements having .counter__heading and .counter classes. Update the cypress/integration/app.spec.js file, as follows:

```
...
context('App', () => {
  ...
  it('should show the counter component when the "Toggle
  Counter Visibility" button is clicked', () => {
    cy.get('.counter__heading').should('have.length', 0);
    cy.get('.counter').should('have.length', 0);
    cy.contains('Toggle Counter Visibility').click();
    cy.get('.counter__heading').should('be.visible');
    cy.get('.counter').should('be.visible');
  });
});
```

2. Now, we'll write a test to check if our action buttons (**Increment**, **Decrement**, and **Reset**) show up when we hover over the counter component. Update the app.spec.js file, as follows:

```
...
context('App', () => {
  ...
  it('should show the action buttons on hovering the
  counter card', () => {
    cy.contains('Toggle Counter Visibility').click();
    cy.get('.counter').trigger('mouseover');
    cy.get('.counter__actions__action').
```

```
        should('have.length', 3);
        cy.contains('Increment').should('be.visible');
        cy.contains('Decrement').should('be.visible');
        cy.contains('Reset').should('be.visible');
    });
});
```

If you look at the Cypress window now, you should see the test failing, as follows:

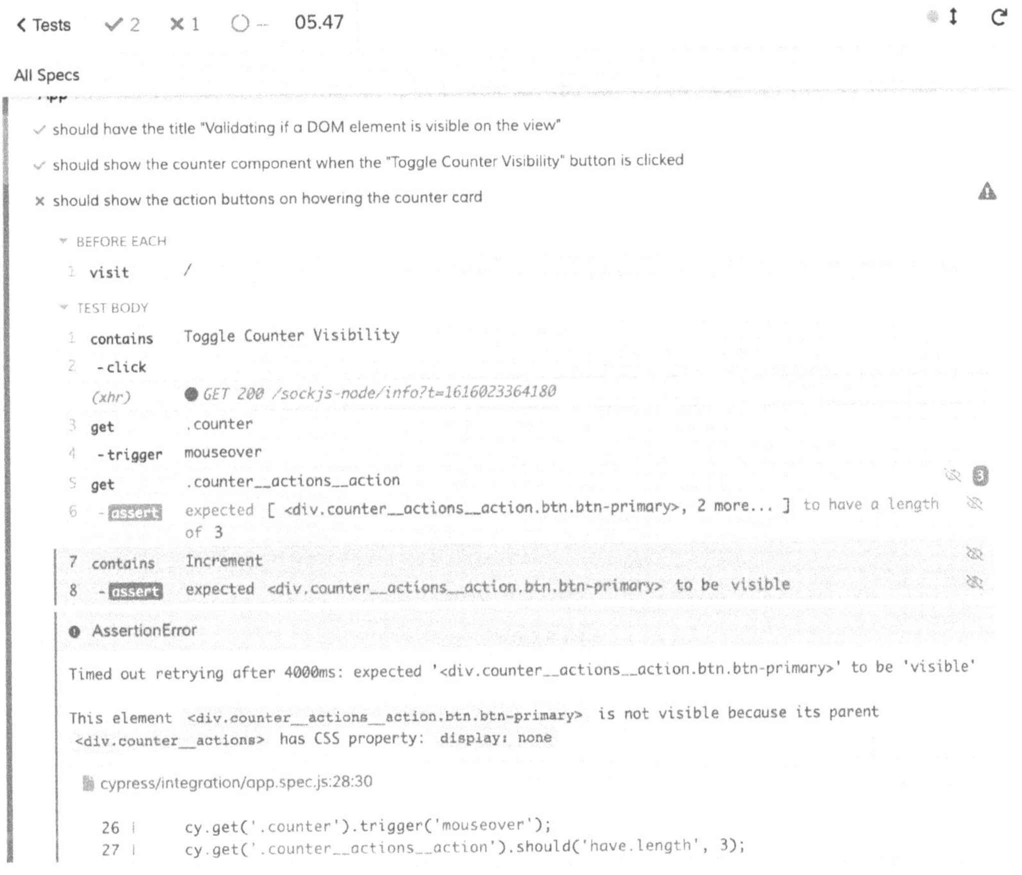

< Tests   ✓ 2   ✗ 1   ○ --   05.47   ● ↕ ↻

All Specs

✓ should have the title "Validating if a DOM element is visible on the view"

✓ should show the counter component when the "Toggle Counter Visibility" button is clicked

✗ should show the action buttons on hovering the counter card   ⚠

```
    ▼ BEFORE EACH
    1  visit        /
    ▼ TEST BODY
    1  contains     Toggle Counter Visibility
    2  - click
      (xhr)         ● GET 200 /sockjs-node/info?t=1616023364180
    3  get          .counter
    4  - trigger    mouseover
    5  get          .counter__actions__action
    6  - assert     expected [ <div.counter__actions__action.btn.btn-primary>, 2 more... ] to have a length
                    of 3
    7  contains     Increment
    8  - assert     expected <div.counter__actions__action.btn.btn-primary> to be visible
```

● AssertionError

Timed out retrying after 4000ms: expected '<div.counter__actions__action.btn.btn-primary>' to be 'visible'

This element <div.counter__actions__action.btn.btn-primary> is not visible because its parent
<div.counter__actions> has CSS property: display: none

📄 cypress/integration/app.spec.js:28:30

```
    26 |    cy.get('.counter').trigger('mouseover');
    27 |    cy.get('.counter__actions__action').should('have.length', 3);
```

Figure 11.5 – Unable to get action buttons on hovering

The reason for the test's failure is that Cypress doesn't currently provide a **Cascading Style Sheets (CSS)** hover effect. In order to work around this, we'll install a package in the next step.

3. Stop the running Cypress and Angular app and then install the `cypress-real-events` package, as follows:

```
npm install --save-dev cypress-real-events
```

4. Now, open the `cypress/support/index.js` file and update it, as follows:

```
. . .

// Import commands.js using ES2015 syntax:
import './commands';
import 'cypress-real-events/support';

. . .
```

5. Now, update the `app.spec.js` file to use the `.realHover()` method from the package on the `.counter` element, as follows:

```
/// <reference types="cypress" />
/// <reference types="cypress-real-events" />
context('App', () => {
  . . .
  it('should show the action buttons on hovering the
  counter card', () => {
    cy.contains('Toggle Counter Visibility').click();
    cy.get('.counter').realHover();
    cy.get('.counter__actions__action').
    should('have.length', 3);

    . . .
  });
});
```

6. Now, run the `cypress:test` command again using `npm run cypress:test`. Once the app is running and the Cypress window is opened, you should see all the tests passing, as follows:

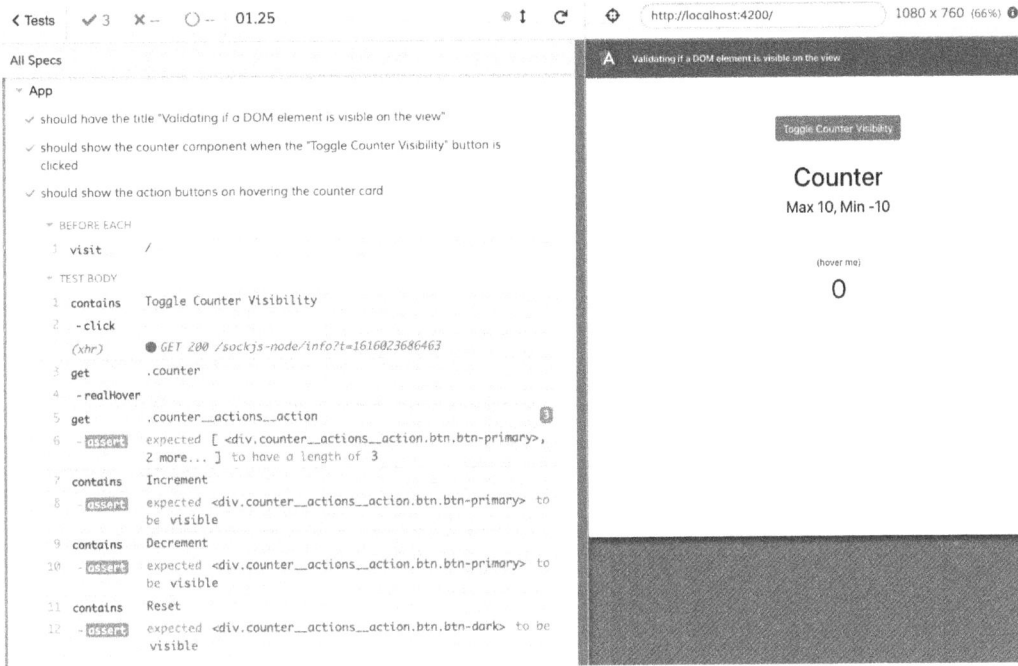

Figure 11.6 – All tests passing after using the cypress-real-events package

Awesomesauce! You've just learned how to check the visibility of DOM elements in different scenarios. These are, of course, not the only options available to identify and interact with DOM elements. Now that you've finished the recipe, see the next section to understand how it works.

How it works...

At the beginning of the recipe, in our first test we use the `.should('have.length', 0)` assertion. When we use the `'have.length'` assertion, Cypress checks the `length` property of the DOM elements found using the `cy.get()` method. Another assertion that we use is `.should('be.visible')`, which checks if an element is visible on the DOM. This assertion will pass as long as the element is visible on the screen—that is, none of the parent elements are hidden.

In the later test, we try to hover over the element with the `'.counter'` selector, using `cy.get('.counter').trigger('mouseover');`. This fails our test. Why? Because all the hover workarounds in Cypress eventually lead to triggering the JavaScript events and not affecting the CSS pseudo selectors, and since we have our action buttons (with the `'.counter__actions__action'` selector) shown on the `:hover` (CSS) of the element with the `'.counter'` selector, our tests fail because in the tests our action buttons are not actually shown. To tackle the issue, we use the `cypress-real-events` package, which has the `.realHover()` method that affects the pseudo selectors and eventually shows our action buttons.

See also

- Cypress official documentation on the visibility of items (`https://docs.cypress.io/guides/core-concepts/interacting-with-elements.html#Visibility`)

- `cypress-real-events` project repository (`https://github.com/dmtrKovalenko/cypress-real-events`)

Testing form inputs and submission

If you're building a web app, there's a high chance that you're going to have at least one form in it, and when it comes to forms we need to make sure that we have the right **user experience (UX)** and the right business logic in place. What better way to make sure everything works as expected than writing E2E tests for them? In this recipe, we're going to test a login form using Cypress.

Getting ready

The project for this recipe resides in `chapter11/start_here/cy-testing-forms`:

1. Open the project in VS Code.

2. Open the terminal and run `npm install` to install the dependencies of the project.

3. Once done, run `npm run cypress:test`.

This should open a new Cypress window. Tap the `app.spec.ts` file and you should see the tests, as follows:

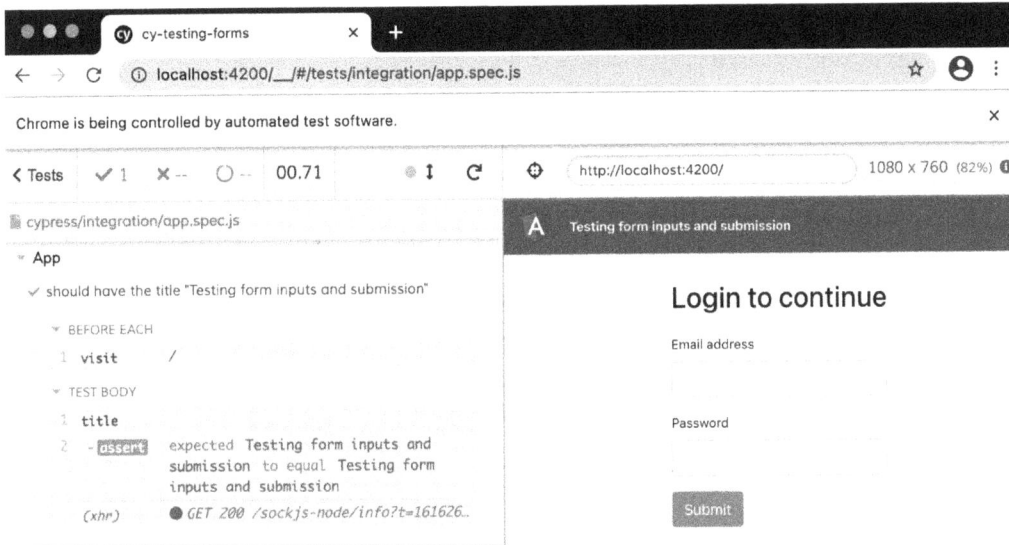

Figure 11.7 – Cypress tests running for the app cy-testing-forms

Now that we have the Cypress tests running, let's see the steps of the recipe in the next section.

How to do it...

We have to make sure that we see a **Success** alert when the form is successfully submitted. We also need to make sure that we see relevant errors if any of the inputs have an invalid value. Let's get started:

1. Let's create a new file inside the `cypress/integration` folder, named `login.spec.js`.

2. We'll first make sure that our form cannot be submitted unless we have valid form inputs. To do that, let's make sure that the **Submit** button is disabled when there are either no input values or invalid values. Open the `login.spec.js` file and add a test, as follows:

```
/// <reference types="cypress" />

context('Login', () => {
  beforeEach(() => {
    cy.visit('/');
  });
```

```
it('should have the button disabled if the form inputs
are not valid', () => {
    // https://on.cypress.io/title
    // No input values
    cy.contains('Submit').should('be.disabled');
    cy.get('#passwordInput').type('password123');
    cy.contains('Submit').should('be.disabled');


    cy.get('#emailInput').type('ahsanayaz@gmail.com');
    cy.get('#passwordInput').clear();
    cy.contains('Submit').should('be.disabled');
  });
});
```

Now, open the `login.spec.js` file in the Cypress window and you should see the tests passing, as follows:

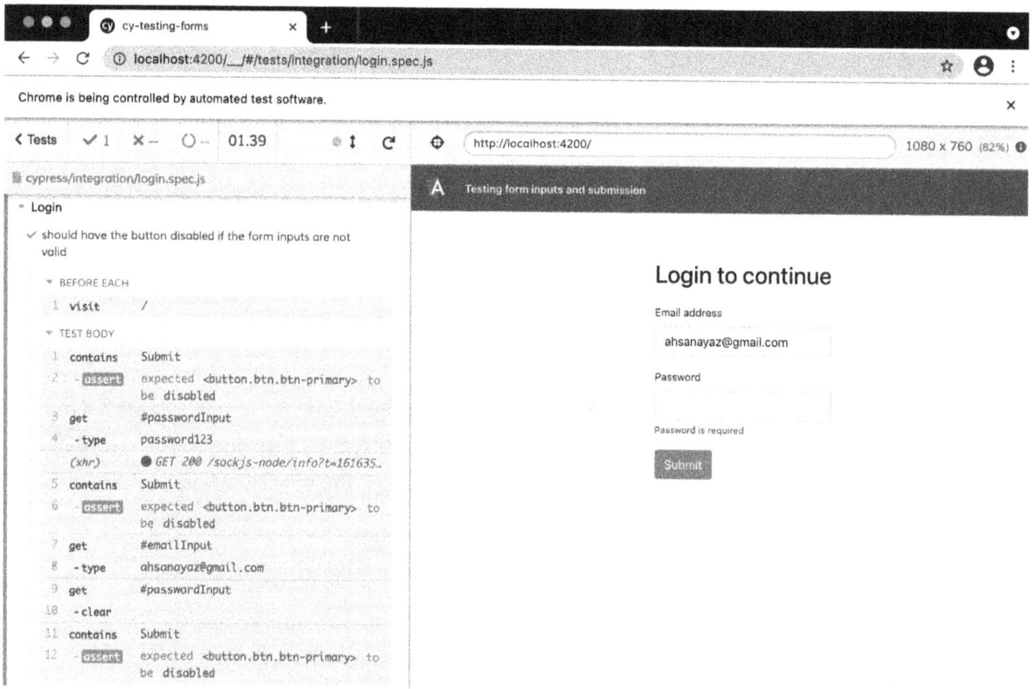

Figure 11.8 – Checking if the Submit button is disabled when there is invalid input

3. Let's add another test that validates that we see a success alert on submitting the right values for the inputs. Add another test in the `login.spec.js` file, as follows:

```
...
context('Login', () => {
    ...
    it('should submit the form with the correct values and
    show the success alert', () => {
        cy.get('#emailInput')
            .type('ahsan.ayaz@domain.com')
            .get('#passwordInput')
            .type('password123');
        cy.contains('Submit').click();
        cy.get('.alert.alert-success').should('be.visible');
    });
});
```

4. We'll add another test now to make sure the success alert hides on tapping the **Close** button. Since we're using the same logic/code for the successful login, we'll create a function to reuse it. Let's modify the `login.spec.js` file, as follows:

```
...
context('Login', () => {
    ...
    it('should submit the form with the correct values and
    show the success alert', () => {
        successfulLogin();
        cy.get('.alert.alert-success').should('be.visible');
    });

    it('should hide the success alert on clicking close
    button', () => {
        successfulLogin();
        cy.get('.alert.alert-success').find('.btn-close').
        click();
        cy.get('.alert.alert-success').should((domList) => {
            expect(domList.length).to.equal(0);
        });
    });
});
```

```
});

function successfulLogin() {
  cy.get('#emailInput')
    .type('ahsan.ayaz@domain.com')
    .get('#passwordInput')
    .type('password123');
  cy.contains('Submit').click();
}
```

5. The success alert should also hide when the input changes. To check that as well, let's add another test, as follows:

```
...
context('Login', () => {
  ...
  it('should hide the success alert on changing the
  input', () => {
    successfulLogin();
    cy.get('#emailInput').clear().
    type('mohsin.ayaz@domain.com');
    cy.get('.alert.alert-success').should((domList) => {
      expect(domList.length).to.equal(0);
    });
  });
});
```

6. Finally, let's write a test to make sure we show error messages on invalid inputs. Add another test in the `logic.spec.js` file, as follows:

```
...
context('Login', () => {
  ...
  it('should show the (required) input errors on invalid
  inputs', () => {
    ['#emailHelp', '#passwordHelp'].map((selector) => {
      cy.get(selector).should((domList) =>
      expect(domList.length).to.equal(0));
    });
```

```
        cy.get('#emailInput').type(
        'mohsin.ayaz@domain.com').clear().blur();
        cy.get('#emailHelp').should('be.visible');
        cy.get('#passwordInput').type(
        'password123').clear().blur();
        cy.get('#passwordHelp').should('be.visible');
    });
});
```

If you look at the **Tests** window now, you should see all the tests passing, as follows:

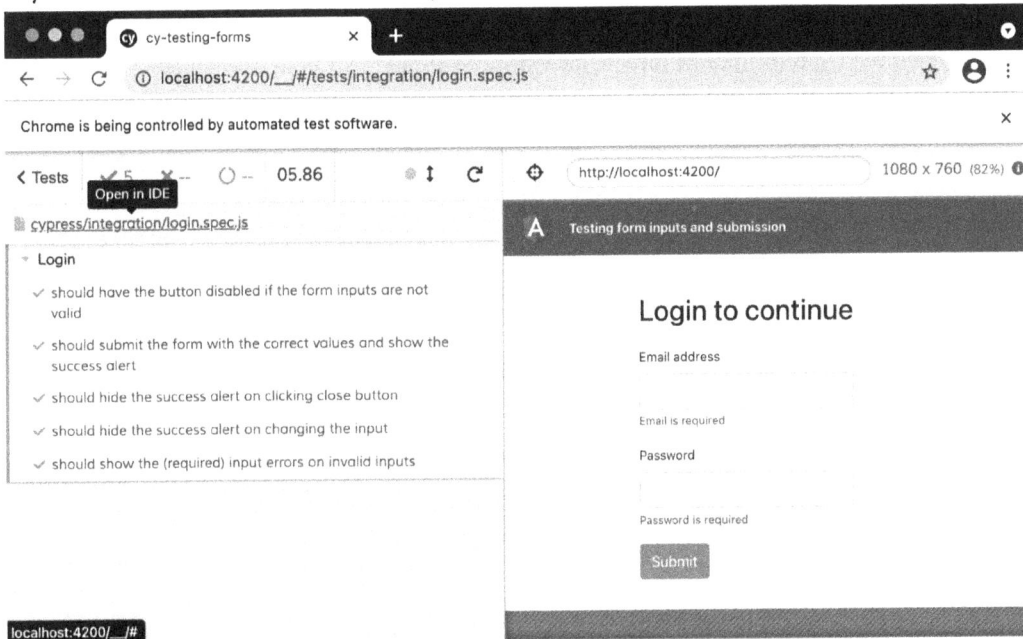

Figure 11.9 – All tests passing for the Login page

Awesome! You now know how to use Cypress to test forms with some interesting use cases and assertions. See the next section to understand how it all works.

How it works...

Since our app's logic has a rule that the **Submit** button should be disabled until both the email and password inputs have valid values, we check if the button is disabled in our tests. We do this by using the 'be.disabled' assertion on the **Submit** button, as follows:

```
cy.contains('Submit').should('be.disabled');
```

We then use .type() method chaining on the cy.get() selector to type in both inputs one by one, and check if the button is disabled either when we have an invalid value for any input or no value entered at all.

To perform a successful login, we execute the following code:

```
cy.get('#emailInput')
    .type('ahsan.ayaz@domain.com')
    .get('#passwordInput')
    .type('password123');
  cy.contains('Submit').click();
```

Notice that we get each input and type valid values in them, and then we call the .click() method on the **Submit** button. We then check if the success alert exists using the '.alert.alert-success' selector and the should('be.visible') assertion.

In cases where we want to check that the success alert has been dismissed on clicking the **Close** button on the alert or when any of the inputs change, we can't just use the should('not.be.visible') assertion. This is because Cypress in this case would expect the alert to be in the DOM but just not be visible, whereas in our case (in our Angular app), the element doesn't even exist in the DOM, so Cypress fails to get it. Therefore, we use the following code to check that the success alert doesn't even exist:

```
cy.get('.alert.alert-success').should((domList) => {
    expect(domList.length).to.equal(0);
});
```

One final interesting thing is when we want to check if error messages for each input show when we type something in either of the inputs and clear the input. In this case, we use the following code:

```
cy.get('#emailInput').type('mohsin.ayaz@domain.com').clear().
blur();
cy.get('#emailHelp').should('be.visible');
cy.get('#passwordInput').type('password123').clear().blur();
cy.get('#passwordHelp').should('be.visible');
```

The reason we use the `.blur()` method is because when Cypress just clears the input the Angular change detection doesn't take place, which results in the error messages not showing on the view immediately. Since Angular's change detection does monkey-patching on the browser events, we trigger a `.blur()` event on both the inputs to trigger the change detection mechanism. As a result, our error messages show properly.

See also

- Cypress recipes: Form interactions (`https://github.com/cypress-io/cypress-example-recipes/tree/master/examples/testing-dom__form-interactions`)

- Cypress recipes: Login form (`https://github.com/cypress-io/cypress-example-recipes/tree/master/examples/logging-in__html-web-forms`)

Waiting for XHRs to finish

Testing **user interface** (**UI**) transitions is the essence of E2E testing. While it is important to test the predicted outcome of an action right away, there might be cases where the outcome actually has a dependency. For instance, if a user fills out the **Login** form, we can't show the success toast until we have a successful response from the backend server, hence we can't test whether the success toast is shown right away. In this recipe, you're going to learn how to wait for a specific XHR call to be completed before performing an assertion.

Getting ready

The project for this recipe resides in `chapter11/start_here/waiting-for-xhr`.

1. Open the project in VS Code.

2. Open the terminal and run `npm install` to install the dependencies of the project.

3. Once done, run `npm run cypress:test`.

This should open a new Cypress window. Tap the `user.spec.ts` file and you should see the tests, as follows:

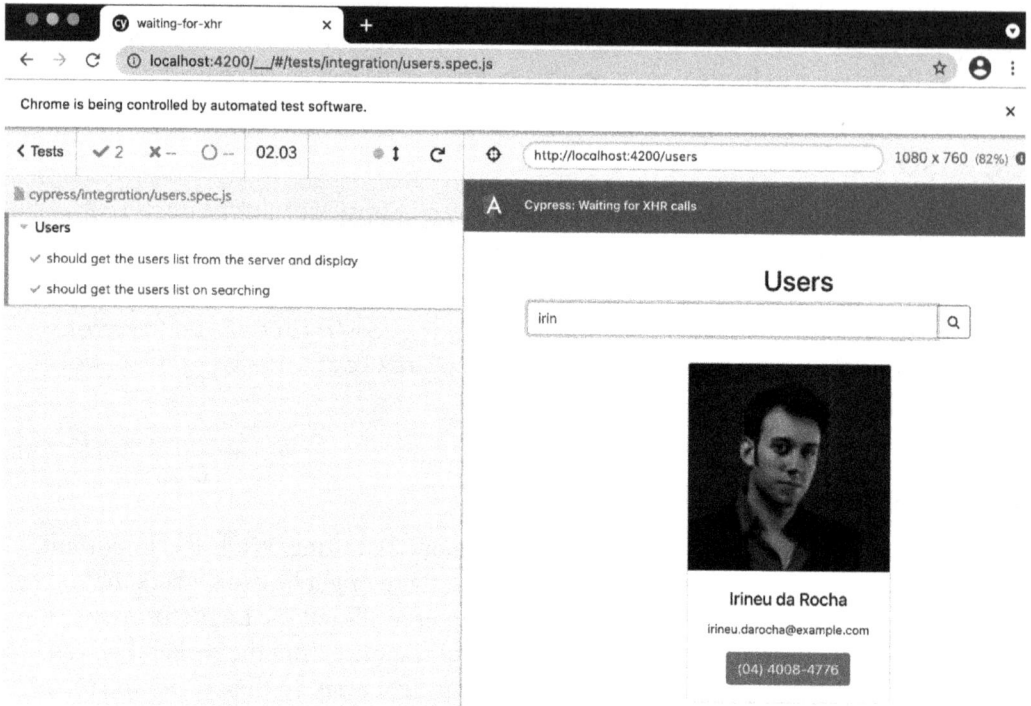

Figure 11.10 – Cypress tests running for the waiting-for-xhr app

Now that we have the Cypress tests running, let's see the steps of the recipe in the next section.

How to do it...

All the tests right now work fine, even though we have XHR calls involved in getting the data. So, what is this recipe about exactly? Well, Cypress has a timeout of 4,000 **milliseconds (ms)** (4 seconds), during which it tries the assertion again and again until the assertion passes. What if our XHR takes more than 4,000 ms? Let's try it out in the recipe:

1. First of all, we need to simulate the scenario where the desired result occurs after 4,000 ms. We'll use the `debounceTime` operator from `rxjs` for this, with a delay of 5,000 ms. Let's apply it on the `valueChanges` Observable of the `searchForm` property in the `users.component.ts` file, as follows:

```
. . .
import { debounceTime, takeWhile } from 'rxjs/operators';

@Component({...})
export class UsersComponent implements OnInit {
  . . .
  ngOnInit() {
    . . .
    this.searchForm
      .get('username')
      .valueChanges.pipe(
        takeWhile(() => !!this.componentAlive),
        debounceTime(5000)
      )
      .subscribe(() => {
        this.searchUsers();
      });
  }
  . . .
}
```

If you now check the Cypress tests, you should see a test failing, as follows:

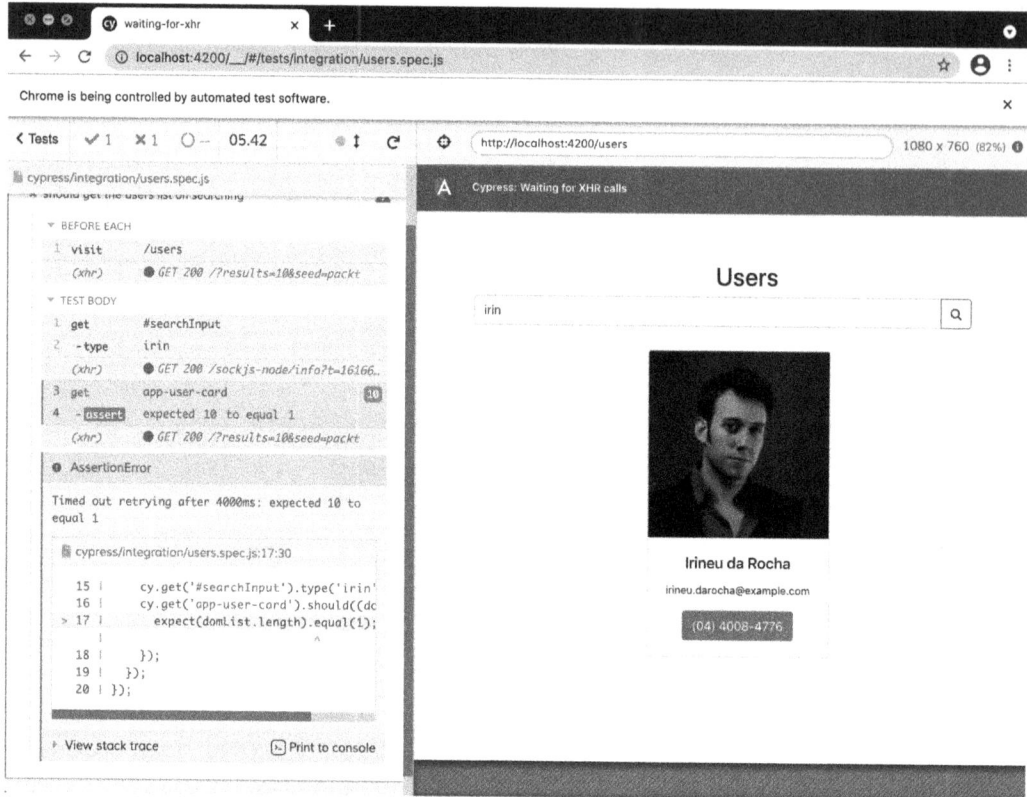

Figure 11.11 – Assertion failing for the test for searching a particular user

2. We can now try to fix this, so it doesn't matter how long the XHR takes—we'll always wait for it to be completed before doing an assertion. Let's intercept the XHR call and create an alias for it so that we can use it later to wait for the XHR call. Update the users.spec.js file, as follows:

```
. . .
context('Users', () => {
    . . .
    it('should get the users list on searching', () => {
        cy.intercept('https://api.randomuser.me/*')
        .as('searchUsers');
        cy.get('#searchInput').type('irin');
```

```
            cy.get('app-user-card').should((domList) => {
              expect(domList.length).equal(1);
            });
          });
        });
```

3. Now, let's use the alias to wait for the XHR call to complete before the assertion.
    Update the `users.spec.js` file, as follows:

```
    ...
    context('Users', () => {
      ...
      it('should get the users list on searching', () => {
        cy.intercept('https://api.randomuser.me/*')
        .as('searchUsers');
        cy.get('#searchInput').type('irin');
        cy.wait('@searchUsers');
        cy.get('app-user-card').should((domList) => {
          expect(domList.length).equal(1);
        });
      });
    });
```

If you check the Cypress tests now for `user.spec.js`, you should see all of them pass, as follows:

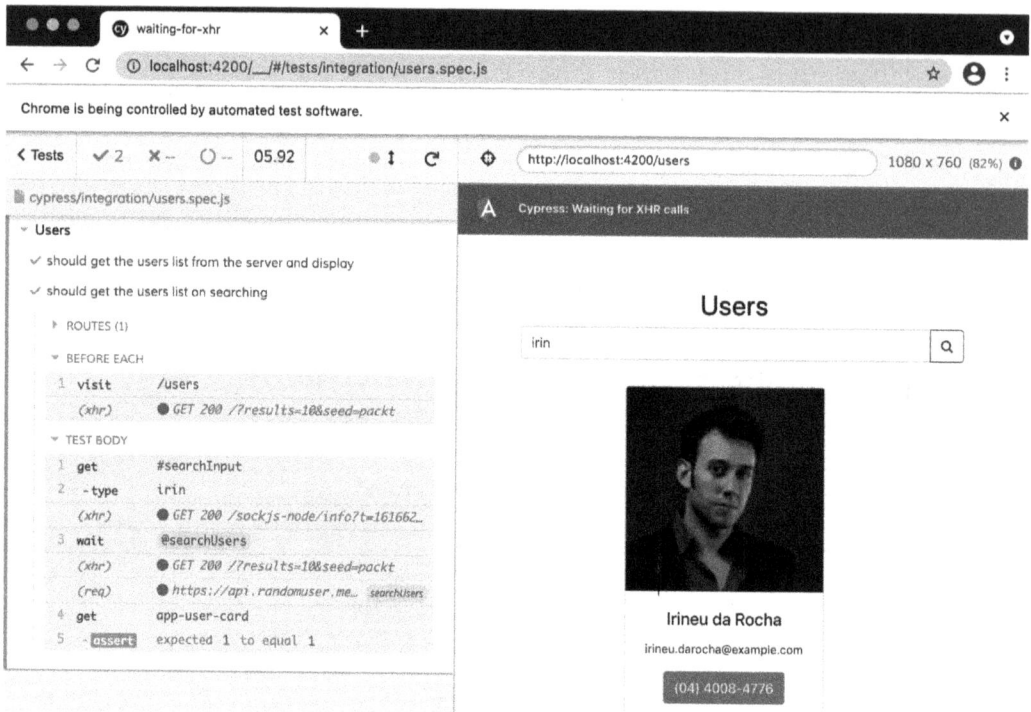

Figure 11.12 – Test waiting for the XHR call to be completed before the assertion

Great!! You now know how to implement E2E tests with Cypress that include waiting for a particular XHR call to finish before an assertion. To understand all the magic behind the recipe, see the next section.

How it works...

In the recipe, we use something called variable aliasing. We first use the `cy.intercept()` method so that Cypress can listen to the network call. Note that we use a wildcard for the URL by using `https://api.randomuser.me/*` as the parameter, and then we use a `.as('searchUsers')` statement to give an alias for this interception.

Then, we use the `cy.wait('@searchUsers');` statement, using the `searchUsers` alias to inform Cypress that it has to wait until the aliased interception happens—that is, until the network call is made, regardless of how long it takes. This makes our tests pass, even though the regular 4,000 ms Cypress timeout has already passed before actually getting the network call. Magic, isn't it?

Well, I hope you liked this recipe—see the next section to view a link for further reading.

See also

- Waiting in Cypress (`https://docs.cypress.io/guides/guides/network-requests#Waiting`)

Using Cypress bundled packages

Cypress provides a bunch of bundled tools and packages that we can use in our tests to make things easier, not because writing tests with Cypress is otherwise hard, but because these libraries are used by many developers already and so they're familiar with them. In this recipe, we're going to look at the bundled `jQuery, Lodash, and Minimatch` libraries to test some of our use cases.

Getting ready

The project that we are going to work with resides in `chapter11/start_here/using-cypress-bundled-packages`, inside the cloned repository:

1. Open the project in VS Code.
2. Open the terminal and run `npm install` to install the dependencies of the project.
3. Once done, run `npm run cypress:test`.

This should open a new Cypress window. Tap the `users.spec.ts` file and you should see the tests, as follows:

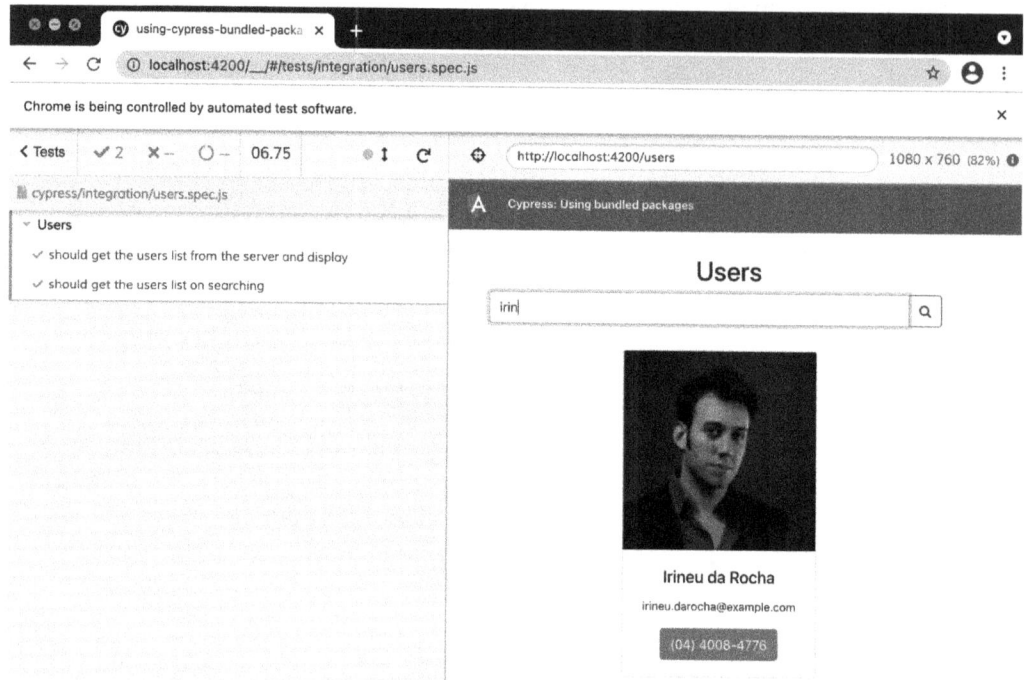

Figure 11.13 – using-cypress-bundled-packages tests running with Cypress

Now that we have the Cypress tests running, let's see the steps of the recipe in the next section.

How to do it...

For this recipe, we have the `users` list and a search app that fetches some users from an **application programming interface (API)** endpoint. We're going to assert a few conditions for the DOM, validate the response from the API, and will assert the URL changes as well. Let's get started:

1. First of all, we'll try out the bundled `jQuery` library along with Cypress. We can access this using `Cypress.$`. Let's add another test and log out some DOM elements. Update the `users.spec.js` file, as follows:

    ```
    . . .
    context('Users', () => {
        . . .
    ```

```
    it('should have the search button disabled when there
    is no input', () => {
      const submitButton = Cypress.$('#userSearchSubmit');
      console.log(submitButton);
    });
  });
```

If you look at the tests now and specifically the console, you should see the log, as follows:

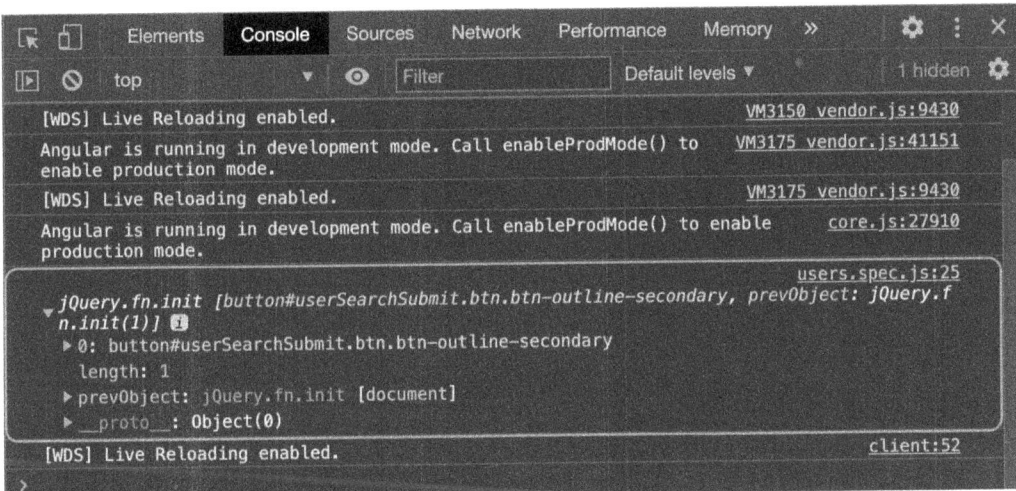

Figure 11.14 – Search button logged using jQuery via Cypress.$

2. Now, let's try to log the user cards that we see after the HTTP call. Add another query and log in to the same test, as follows:

```
...
context('Users', () => {
  ...
  it('should have the search button disabled when there
  is no input', () => {
    const submitButton = Cypress.$('#userSearchSubmit');
    console.log(submitButton);
    const appUserCards = Cypress.$('app-user-card');
    console.log(appUserCards);
  });
});
```

If you see the test and the logs again on the console in the Cypress window, you will see that the `Cypress.$('app-user-card')` query doesn't return any DOM elements. This is because when the query is run, the HTTP call isn't completed. So, should we wait for the HTTP call to finish? Let's try that.

3. Let's add a `cy.wait(5000)` to wait for 5 seconds, during which the HTTP call should have been completed, and let's put an assertion with the `cy.wrap()` method as well to check that the **Search** button is disabled when there's no value provided for the search input. Update the test, as follows:

```
...
context('Users', () => {
  ...
  it('should have the search button disabled when there
  is no input', () => {
    const submitButton = Cypress.$('#userSearchSubmit');
    cy.wrap(submitButton).should('have.attr',
    'disabled');
    cy.get('#searchInput').type('irin');
    cy.wait(5000);
    const appUserCards = Cypress.$('app-user-card');
    console.log(appUserCards);
    cy.wrap(submitButton).should('not.have.attr',
    'disabled');
  });
});
```

If you see the Cypress test and the console, you will see that we still get no DOM elements for the `<app-user-card>` elements:

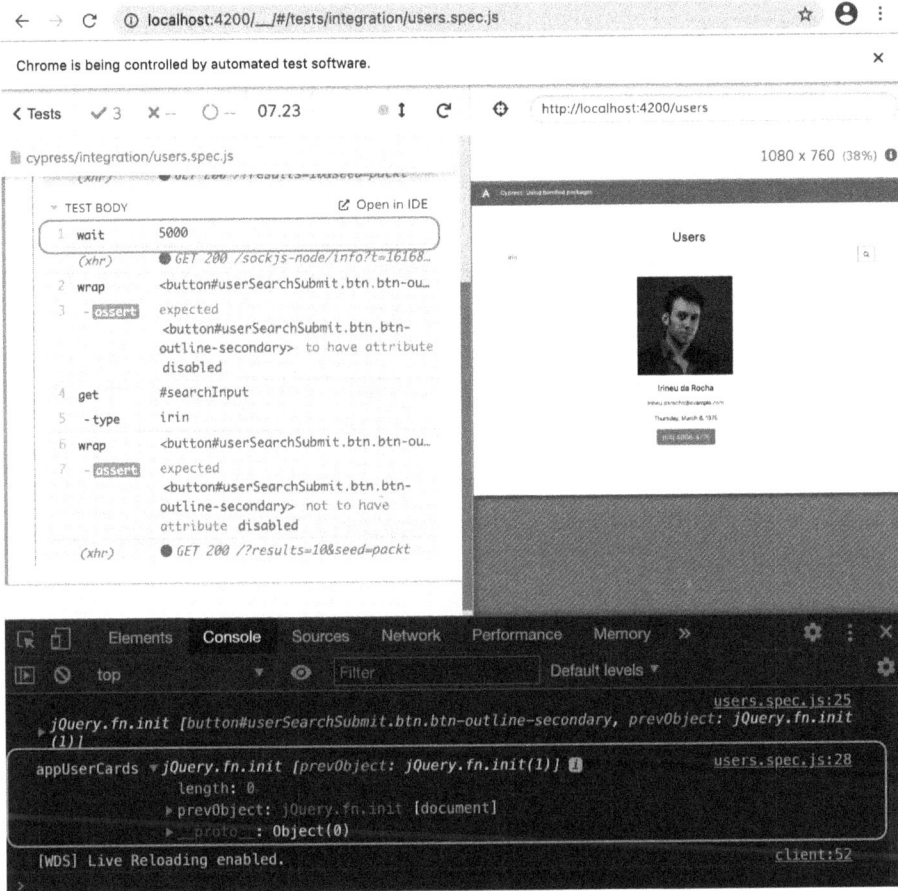

Figure 11.15 – No user cards found using Cypress.$ even after using cy.wait (5000)

We'll discuss in the *How it works...* section why this happens. For now, understand that you should only use `Cypress.$` for elements that are present in the DOM right from when the page is loaded.

4. Let's clean up our test by removing the `cy.wait()` method and the console logs. It should then look like this:

```
...
context('Users', () => {
    ...
    it('should have the search button disabled when there
    is no input', () => {
        const submitButton = Cypress.$('#userSearchSubmit');
        cy.wrap(submitButton).should('have.attr',
```

```
'disabled');
    cy.get('#searchInput').type('irin');
    cy.wrap(submitButton).should('not.have.attr',
    'disabled');
  });
});
```

5. We'll now add a test to verify that we get the same users from the Random User API for the same seed string. We already have the `API_USERS.js` file that contains the expected result. Let's use the bundled `lodash` library in our next test to assert the matching values for the first name, the last name, and the email of the returned users, as follows:

```
...
import API_USERS from '../constants/API_USERS';
context('Users', () => {
  ...
  it('should return the same users as the seed data
  every time', async () => {
    const { _ } = Cypress;
    const response = await cy.request(
      'https://api.randomuser.me/?
      results=10&seed=packt'
    );
    const propsToCompare = ['name.first', 'name.last',
    'email'];
    const results = _.get(response, 'body.results');
    _.each(results, (user, index) => {
      const apiUser = API_USERS[index];
      _.each(propsToCompare, (prop) => {
        const userPropVal = _.get(user, prop);
        const apiUserPropVal = _.get(apiUser, prop);
        return expect(userPropVal).
        to.equal(apiUserPropVal);
      });
    });
  });
});
```

If you see the test now in Cypress, it should be passing, as follows:

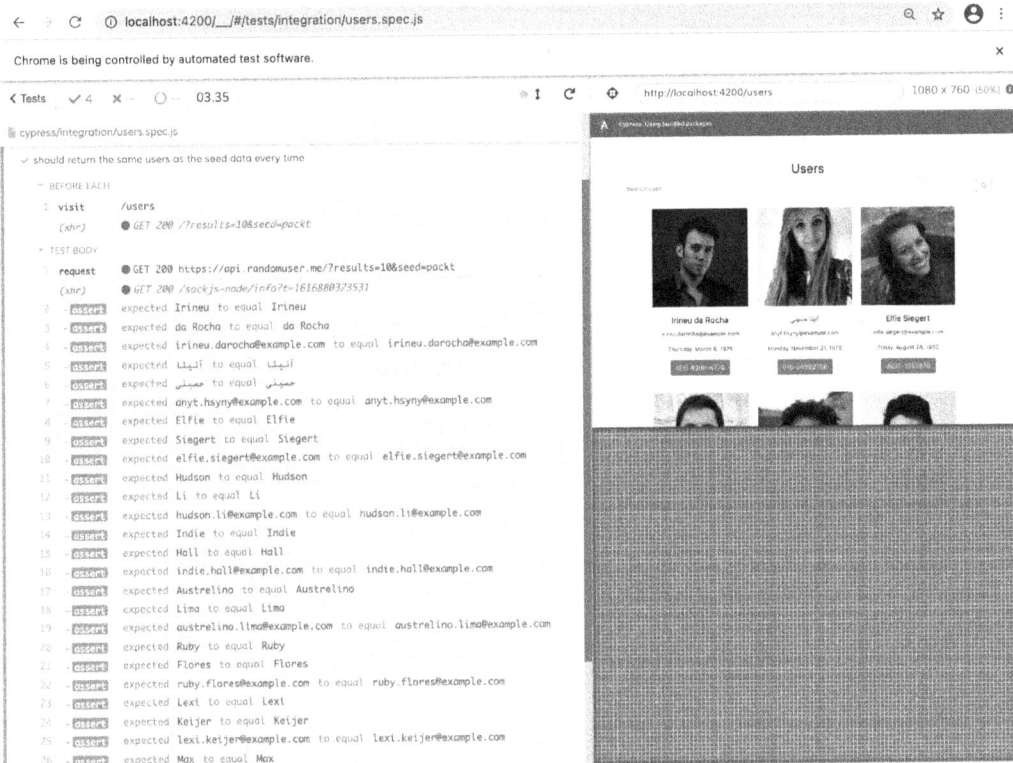

Figure 11.16 – Test passing with the usage of lodash via Cypress

6. We're now going to use the `moment.js` package that is bundled with Cypress as well. Let's assert that the user cards show the formatted date correctly, using `moment.js`. Write another test in the `users.spec.js` file, as follows:

```
...
context('Users', () => {
    ...
    it('should show the formatted date of birth on the
    user card', () => {
        const { _, moment } = Cypress;
        const apiUserDate = _.get(API_USERS[0], 'dob.date');
        const apiUserDateFormatted = moment(apiUserDate).
        format(
            'dddd, MMMM D, YYYY'
        );
```

```
cy.get('app-user-card')
  .eq(0)
  .find('#userCardDOB')
  .should((el) => {
    expect(el.text().trim()).
    to.equal(apiUserDateFormatted);
  });
});
});
```

7. The next package we'll explore is the `minimatch` package. When we tap on a user card, it opens the user details. Since we append a timestamp to the URL as a query parameter, we can't compare the URL as an exact match with our assertion. Let's use the `minimatch` package to assert using a pattern instead. Add a new test, as follows:

```
...
context('Users', () => {
  ...
  it('should go to the user details page with the user
uuid', () => {
    const { minimatch } = Cypress;
    cy.get('app-user-card').eq(0).click();
    const expectedURL = `http://localhost:4200/
users/${API_USERS[0].login.uuid}`;
    cy.url().should((url) => {
      const urlMatches = minimatch(url,
      `${expectedURL}*`);
      expect(urlMatches).to.equal(true);
    });
  });
});
```

And boom! We now have all the tests passing using the bundled packages with Cypress. Now that we've finished the recipe, let's see in the next section how it all works.

How it works...

Cypress bundles jQuery with it and we use it via the Cypress.$ property. This allows us to perform everything that the jQuery function permits us to. It automatically checks which page is in the view using the cy.visit() method, and then queries the document using the provided selector.

> **Important note**
> Cypress.$ can only fetch from the document elements that are available immediately on the DOM. This is great for debugging the DOM using the Chrome DevTools in the Cypress test window. However, it is important to understand that it doesn't have any context about the Angular change detection. Also, you can't query any element that isn't visible on the page right from the beginning, as we experienced following the recipe—that is, it doesn't respect waiting for XHR calls for the elements to be visible.

Cypress also bundles lodash and exposes it via the Cypress._ object. In the recipe, we use the _.get() method to get the nested properties from the user object. The _.get() method takes two parameters: the object, and a string that reflects the path for the properties—for example, we use _.get(response, 'body.results');, which essentially returns a value for response.body.results. We also use the _.each() method to iterate over the arrays in the recipe. Note that we can use any lodash method in our Cypress test and not just the aforementioned methods.

We also used the minimatch package, which Cypress exposes via the Cypress.minimatch object. The minimatch package is great for matching and testing glob patterns against strings. We use it to test the URL after navigating to a user's detail page using a pattern.

Finally, we also use the moment.js package that Cypress exposes via the Cypress.moment object. We use it to make sure the date of birth of each user is shown in the expected format on the view. Easy peasy.

See also

- Cypress bundled tools (https://docs.cypress.io/guides/references/bundled-tools)
- Moment.js (https://momentjs.com/)
- jQuery (https://jquery.com/)
- lodash (https://lodash.com)
- Minimatch.js (https://github.com/isaacs/minimatch)

516   E2E Tests in Angular with Cypress

Using Cypress fixtures to provide mock data

When it comes to writing E2E tests, fixtures play a great role in making sure the tests are not flaky. Consider that your tests rely on fetching data from your API server or your tests include snapshot testing, which includes fetching images from a **content delivery network** (**CDN**) or a third-party API. Although they're technically required for the tests to run successfully, it is not important that the server data and the images are fetched from the original source, therefore we can create fixtures for them. In this recipe, we'll create fixtures for the users' data as well as for the images to be shown on the UI.

Getting ready

The project that we are going to work with resides in `chapter11/start_here/using-cypress-fixtures`, inside the cloned repository:

1. Open the project in VS Code.

2. Open the terminal and run `npm install` to install the dependencies of the project.

3. Once done, run `npm run cypress:test`.

This should open a new Cypress window. Tap the `users.spec.ts` file and you should see the tests, as follows:

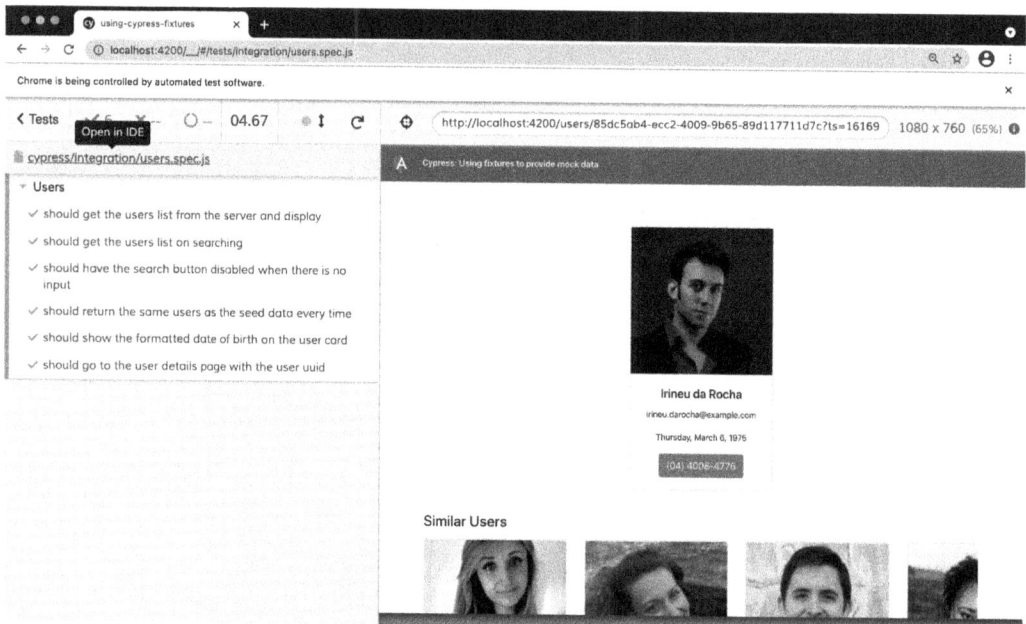

Figure 11.17 – using-cypress-fixtures tests running with Cypress

Now that we have the Cypress tests running, let's see the steps of the recipe in the next section.

How to do it...

We have the same Angular application as in the previous recipe. However, we'll now use Cypress fixtures to provide fixtures for our data and images. Let's get started:

1. We'll first create a fixture for our HTTP call to the `randomuser.me` API. Create a new file under the `cypress/fixtures` folder, named `users.json`. Then, copy the code from the `chapter11/final/using-cypress-fixtures/cypress/fixtures/users.json` file and paste it into the newly created file. It should look like this:

    ```json
    {
        "fixture_version": "1",
        "results": [
          {
            "gender": "male",
            "name": { "title": "Mr", "first": "Irineu",
            "last": "da Rocha" },
            ...
          },
          ...
          {
            "gender": "male",
            "name": { "title": "Mr", "first": "Justin",
            "last": "Grewal" },
            ...
          }
        ]
    }
    ```

2. Now, let's use the fixture in our `users.spec.js` file. We'll use it in the `beforeEach()` life cycle hook since we want to use the fixture for all the tests in the file. This means we'll also remove the existing usage of the `cy.intercept()` method in the file. Update the `users.spec.js` file, as follows:

    ```
    ...
    context('Users', () => {
      beforeEach(() => {
    ```

```
        cy.fixture('users.json')
          .then((response) => {
            cy.intercept('GET', 'https://api.randomuser.
            me/*', response).as(
              'searchUsers'
            );
          })
          .visit('/users');
      });
      ...
      it('should get the users list on searching', () => {
        cy.intercept('
https://api.randomuser.me/*').as('searchUsers'); ← //
REMOVE THIS
        cy.get('#searchInput').type('irin');
        cy.wait('@searchUsers');
        ...
      });
      ...
  });
```

We now need to remove the `constants/API_USERS.js` file from the project since we have the fixture now.

3. We'll create a new variable in which we'll store the value of the `users` array and will use it instead of the `API_USERS` array. Let's modify the `users.spec.js` file further, as follows:

```
...
import API_USERS from '../constants/API_USERS'; ← //
REMOVE THIS
context('Users', () => {
  let API_USERS;
  beforeEach(() => {
    cy.fixture('users.json')
      .then((response) => {
        API_USERS = response.results;
        cy.intercept('GET', 'https://api.randomuser.
        me/*', response).as(
```

```
                    'searchUsers'
            );
        })
        .visit('/users');
    });
});
...
});
```

You'll notice that all of our tests are still passing with the changes done. You can safely remove the `constants/API_USERS.js` file from the project now. Also, you can see the network calls in the Cypress **Tests** window to verify that we're using the fixture instead of the actual API response, as follows:

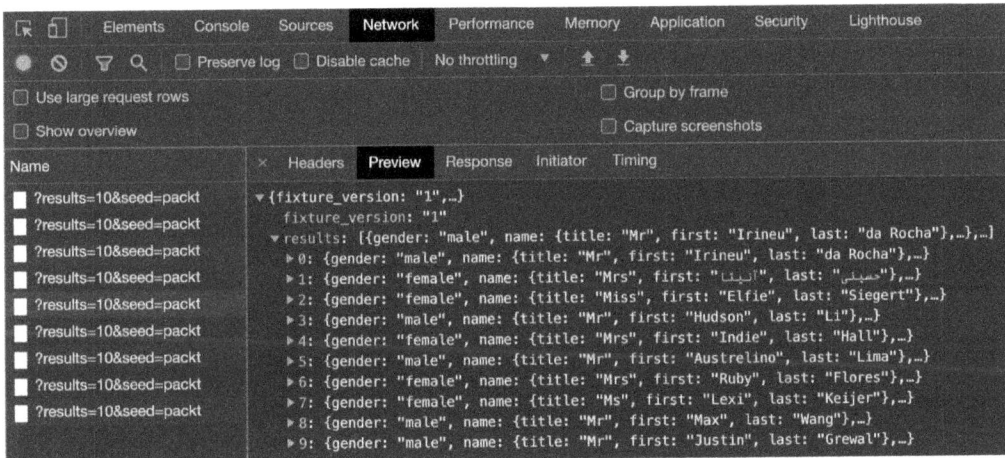

Figure 11.18 – Cypress tests using users.json fixture as XHR response

4. Now, let's try to mock our images to load them from the disk instead of the `randomuser.me` API. For this, we already have the images stored in the `fixtures/images` folder. We just need to use them based on the URL for a particular user. To do so, modify the `users.spec.js` file, as follows:

```
...
context('Users', () => {
    let API_USERS;
    beforeEach(() => {
        cy.fixture('users.json')
            .then((response) => {
```

```
        API_USERS = response.results;

        . . .

        API_USERS.forEach((user) => {
            const url = user.picture.large;

            const imageName = url.substr(url.
            lastIndexOf('/') + 1);

            cy.intercept(url, { fixture:
            `images/${imageName}` });
        });
    .visit('/users');
    });

    . . .

});
```

If you see the tests now, all of them should still be passing, as follows:

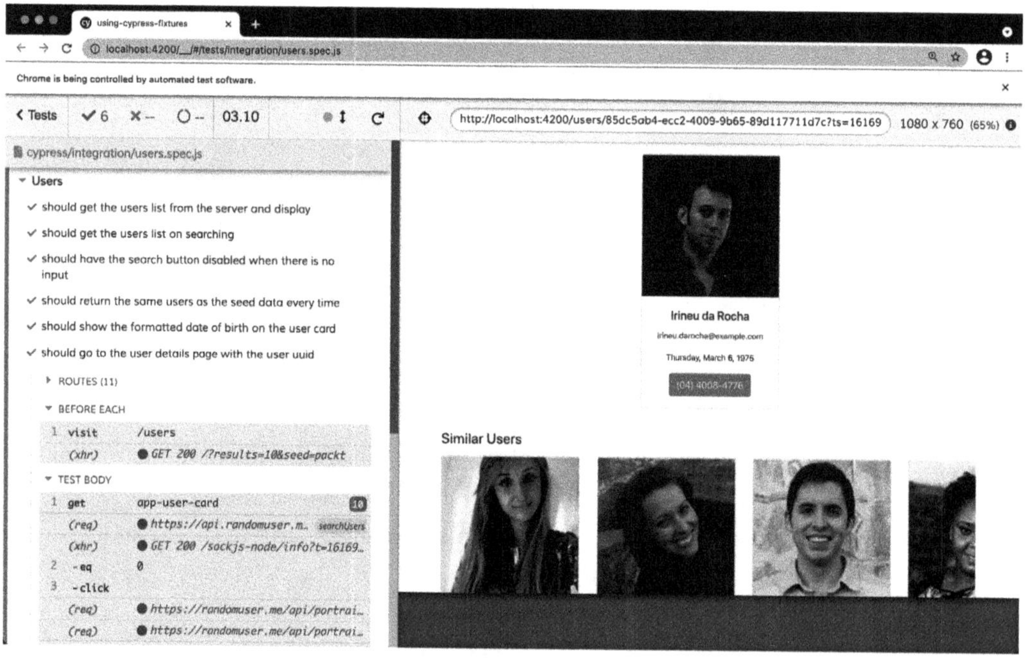

Figure 11.19 – All tests passing after using images fixtures

Looking at the tests, you might be thinking: *"It all seems exactly as before, Ahsan. How do I know we're mocking the images?"* Well, good question. We already have a way to test that.

5. In the `cypress/fixtures/images` folder, we have a file named `9.jpg`, and another test file named `9_test.jpg`. Let's modify the name of the `9.jpg` file to `9_original.jpg` and the `9_test.jpg` file to `9.jpg`. If you see the tests now, you should see a different result for the last test using the replaced file, as follows:

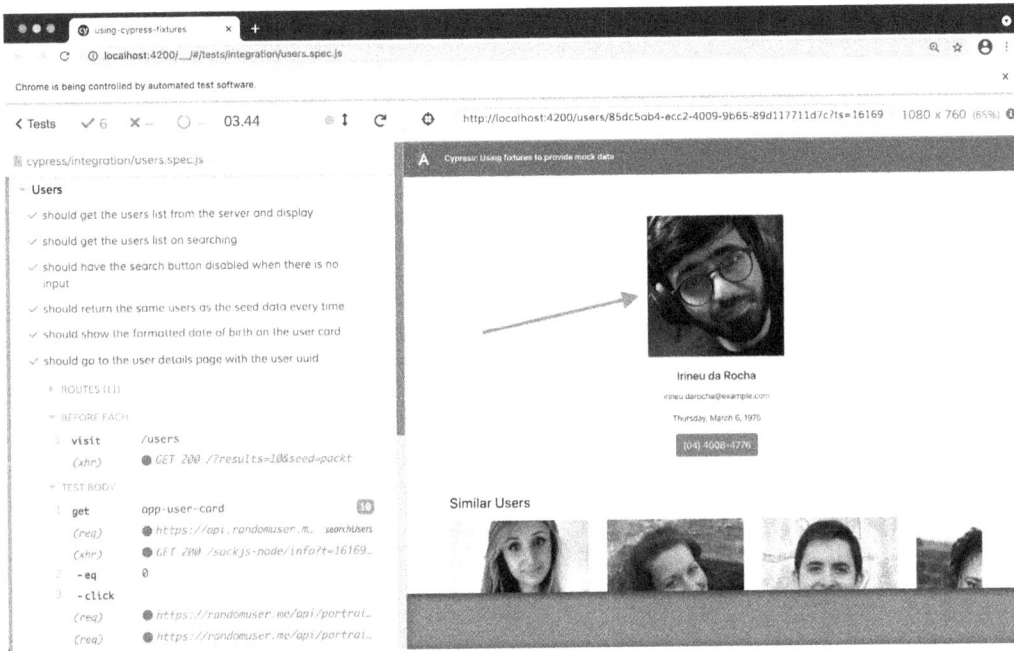

Figure 11.20 – Cypress tests using images from the fixture

Great!!! You now know how to use fixtures in Cypress E2E tests. Now that you've finished the recipe, see the next section on how this works.

How it works...

We use fixtures in a Cypress test using the `cy.fixture()` method, which allows us to use data from a file. In this recipe, we use fixtures for the HTTP call that gets the user data and for the images. But how does it work? Essentially, the `fixture` method has four overloads, as follows:

```
cy.fixture(filePath)
cy.fixture(filePath, encoding)
cy.fixture(filePath, options)
cy.fixture(filePath, encoding, options)
```

The `filePath` parameter takes a string as the file path relative to the `Fixture` folder, which defaults to the `cypress/fixture` path, although we can provide a different `Fixture` folder by defining a `fixturesFolder` property in the `cypress.json` configuration file. Notice that for the HTTP call, we use the `cy.fixture('users.json')` statement, which essentially points to the `cypress/fixture/users.json` file.

First of all, we use the `cy.fixture('users.json')` method before the `cy.visit()` method to ensure that our immediate XHR call that triggers on launching the application uses the fixture. If you change the code otherwise, you'll see that it doesn't work as expected. We then use the `.then()` method to get hold of the data from the `users.json` file. Once we get the data (`response`) object, we use the `cy.intercept()` method using a Minimatch glob pattern to intercept the HTTP call to get the users' data, and we provide this `response` object from the fixture as the response for the HTTP call. As a result, all the calls made to the endpoint matching the `'https://api.randomuser.me/*'` glob use our fixture—that is, the `users.json` file.

We also do one more interesting thing in the recipe, and that is mocking the images to avoid fetching them from their original source. This is super-handy when you use a third-party API and you have to pay for each call made to the API. We already have the fixture images stored in the `cypress/fixture/images` folder. Therefore, we loop over the `API_USERS` array for each user and extract the filename (the `imageName` variable). We then intercept each HTTP call done to fetch the images and use the fixture image instead of the original resource in our tests.

See also

- Cypress fixtures documentation (`https://docs.cypress.io/api/commands/fixture`)

- `cy.intercept()` method documentation (`https://docs.cypress.io/api/commands/intercept`)

12
Performance Optimization in Angular

Performance is always a concern in any product that you build for end users. It is a critical element in increasing the chances of someone using your app for the first time becoming a customer. Now, we can't really improve an app's performance until we identify potential possibilities for improvement and the methods to achieve this. In this chapter, you'll learn some methods to deploy when it comes to improving Angular applications. You'll learn how to analyze, optimize, and improve your Angular app's performance using several techniques. Here are the recipes we're going to cover in this chapter:

- Using `OnPush` change detection to prune component subtrees
- Detaching the change detector from components
- Running `async` events outside Angular with `runOutsideAngular`
- Using `trackBy` for lists with `*ngFor`
- Moving heavy computation to pure pipes

- Using web workers for heavy computation

- Using performance budgets for auditing

- Analyzing bundles with `webpack-bundle-analyzer`

Technical requirements

For the recipes in this chapter, make sure you have **Git** and **Node.js** installed on your machine. You also need to have the `@angular/cli` package installed, which you can do with `npm install -g @angular/cli` from your terminal. The code for this chapter can be found at the following link: `https://github.com/PacktPublishing/Angular-Cookbook/tree/master/chapter12`.

Using OnPush change detection to prune component subtrees

In today's world of modern web applications, performance is one of the key factors for a great **user experience (UX)** and, ultimately, conversions for a business. In this recipe, being the first recipe of this chapter, we're going to discuss the fundamental or the most basic optimization you can do with your components wherever it seems appropriate, and that is by using the `OnPush` change-detection strategy.

Getting ready

The project we are going to work with resides in `Chapter12/start_here/using-onpush-change-detection`, inside the cloned repositor:

1. Open the project in **Visual Studio Code (VS Code)**.

2. Open the terminal and run `npm install` to install the dependencies of the project.

3. Run the `ng serve -o` command to start the Angular app and serve it on the browser. You should see the app, as follows:

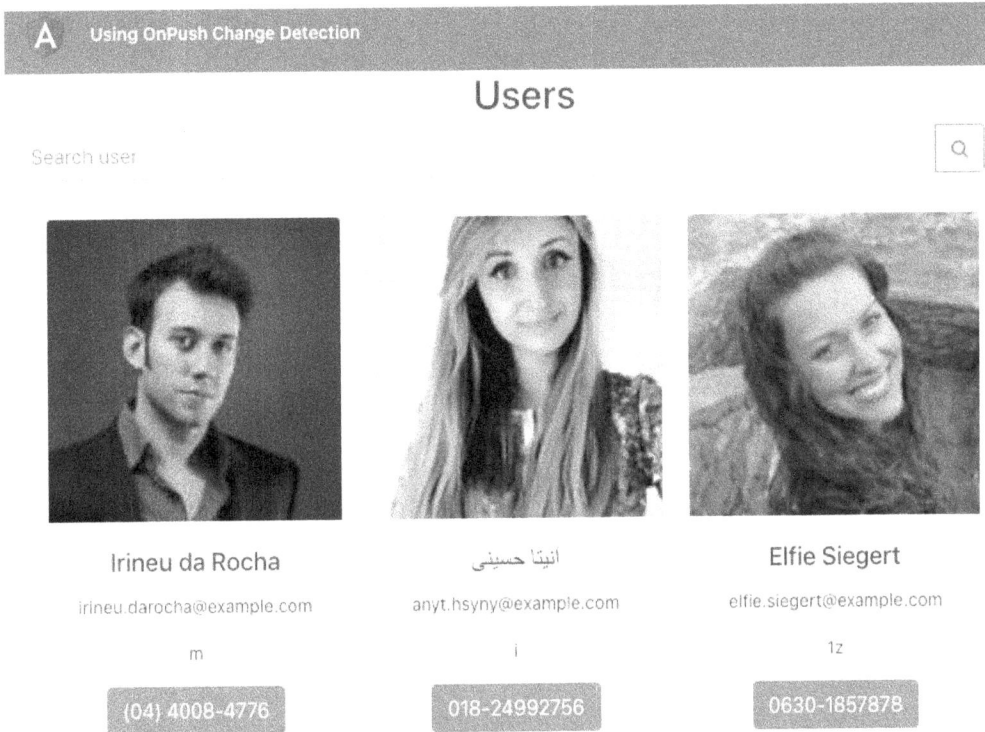

Figure 12.1 – App using OnPush change detection running at http://localhost:4200

Now that we have the project served on the browser, let's see the steps of the recipe in the next section.

How to do it...

The app we're working with has some performance issues, particularly with the `UserCardComponent` class. This is because it is using the `idUsingFactorial()` method to generate a unique ID to show on the card. We're going to experience and understand the performance issue this causes. We will try to fix the issue using the `OnPush` change-detection strategy. Let's get started:

1. First, try to search for a user named `Elfie Siegert` by entering their name in the search box. You'll notice that the app immediately hangs and that it takes a few seconds to show the user. You'll also notice that you don't even see the typed letters in the search box as you type them.

    Let's add some logic to the code. We'll check how many times Angular calls the `idUsingFactorial()` method when the page loads.

2. Modify the `app/core/components/user-card/user-card.component.ts` file, updating it as follows:

```
...
@Component ({...})
export class UserCardComponent implements OnInit {
  ...
  constructor(private router: Router) {}

  ngOnInit(): void {
    if (!window['appLogs']) {
      window['appLogs'] = {};
    }
    if (!window['appLogs'][this.user.email]) {
      window['appLogs'][this.user.email] = 0;
    }
  }
  ...
  idUsingFactorial(num, length = 1) {
    window['appLogs'][this.user.email]++;
    if (num === 1) {...} else {...}
  }
}
```

3. Now, refresh the app and open the Chrome DevTools and, in the **Console** tab, type `appLogs` and press *Enter*. You should see an object, as follows:

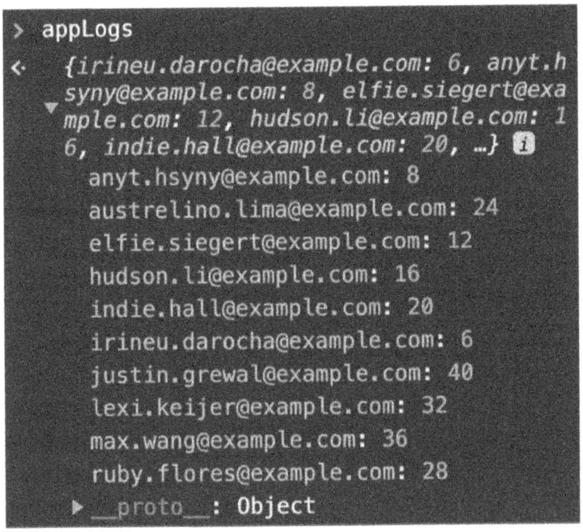

Figure 12.2 – Logs reflecting number of calls to idUsingFactorial() method

4. Now, type the name `Elfie Siegert` again in the search box. Then, type `appLogs` again in the **Console** tab and press *Enter* to see the object again. You'll see that it has some increased numbers. If you didn't make a typo while entering the name, you should see something like this:

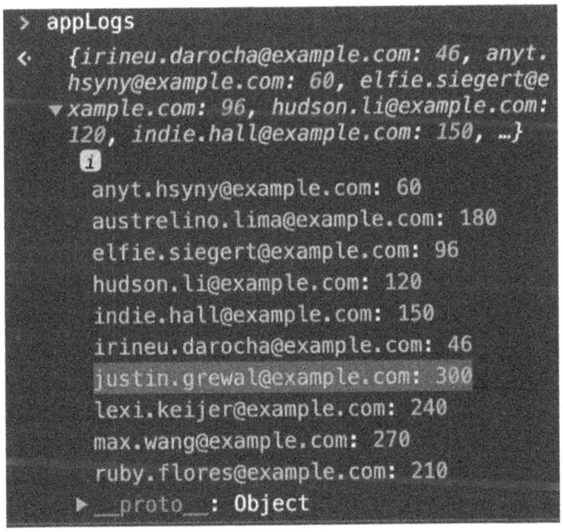

Figure 12.3 – Logs after typing the name Elfie Siegert

Notice the count when calling the `idUsingFactorial()` method for `justin.grewal@example.com`. It has increased from `40` to `300` now, in just a few key presses.

Let's use the `OnPush` change-detection strategy now. This will avoid the Angular change-detection mechanism running on each browser event, which currently causes a performance issue on each key press.

5. Open the `user-card.component.ts` file and update it, as follows:

```
import {
    ChangeDetectionStrategy,
    Component,
    ...
} from '@angular/core';
...
@Component({
    selector: 'app-user-card',
    templateUrl: './user-card.component.html',
    styleUrls: ['./user-card.component.scss'],
    changeDetection: ChangeDetectionStrategy.OnPush,
})
export class UserCardComponent implements OnInit {
    ...
}
```

6. Now, try typing the name `Elfie Siegert` again in the search box. You'll notice that you can now see the typed letters in the search box, and it doesn't hang the app as much. Also, if you look at the `appLogs` object in the **Console** tab, you should see something like this:

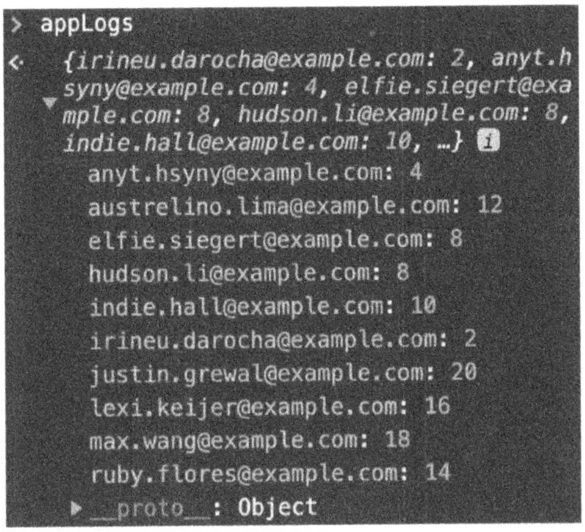

```
>  appLogs
<-   {irineu.darocha@example.com: 2, anyt.h
     syny@example.com: 4, elfie.siegert@exa
   ▼ mple.com: 8, hudson.li@example.com: 8,
     indie.hall@example.com: 10, …} 🛈
       anyt.hsyny@example.com: 4
       austrelino.lima@example.com: 12
       elfie.siegert@example.com: 8
       hudson.li@example.com: 8
       indie.hall@example.com: 10
       irineu.darocha@example.com: 2
       justin.grewal@example.com: 20
       lexi.keijer@example.com: 16
       max.wang@example.com: 18
       ruby.flores@example.com: 14
     ▶ __proto__: Object
```

Figure 12.4 – Logs after typing the name Elfie Siegert with OnPush strategy

Notice that even after refreshing the app, and after typing the name `Elfie Siegert`, we now have a very low number of calls to the `idUsingFactorial()` method. For example, for the `justin.grewal@example.com` email address, we only have **20** hits, instead of the initial **40** hits shown in *Figure 12.2*, and **300** hits, as shown in *Figure 12.3*, after typing.

Great! Within a single step, by using the `OnPush` strategy we were able to improve the overall performance of our `UserCardComponent`. Now you know how to use this strategy, see the next section to understand how it works.

How it works...

Angular by default uses the **Default** change-detection strategy—or technically, it is the `ChangeDetectionStrategy.Default` enum from the `@angular/core` package. Since Angular doesn't know about every component we create, it uses the Default strategy to not encounter any surprises. But as developers, if we know that a component will not change unless one of its `@Input()` variables changes, we can—and we should—use the `OnPush` change-detection strategy for that component. Why? Because it tells Angular to not run change detection until an `@Input()` variable for the component changes. This strategy is an absolute winner for **presentational** components (sometimes called **dumb** components), which are just supposed to show data using `@Input()` variables/ attributes, and emit `@Output()` events on interactions. These presentational components usually do not hold any business logic such as heavy computation, using services to make **HyperText Transfer Protocol** (**HTTP**) calls, and so on. Therefore, it is easier for us to use the `OnPush` strategy in these components because they would only show different data when any of the `@Input()` attributes from the parent component change.

Since we are now using the OnPush strategy on our UserCardComponent, it only triggers change detection when we replace the entire array upon searching. This happens after the **300ms** debounce (*line 28* in the users.component.ts file), so we only do it when the user stops typing. So, essentially, before the optimization, the default change detection was triggering on each keypress being a browser event, and now, it doesn't.

> **Important note**
>
> As you now know that the OnPush strategy only triggers the Angular change-detection mechanism when one or more of the @Input() bindings changes, this means that if we change a property within the component (UserCardComponent), it will not be reflected in the view because the change-detection mechanism won't run in this case, since that property isn't an @Input() binding. You would have to mark the component as dirty so that Angular could check the component and run change detection. You'll do this using the ChangeDetectorRef service—specifically, with the .markForCheck() method.

See also

- Angular ChangeDetectionStrategy official documentation (https://angular.io/api/core/ChangeDetectionStrategy)

- markForCheck() method official documentation (https://angular.io/api/core/ChangeDetectorRef#markforcheck)

Detaching the change detector from components

In the previous recipe, we learned how to use the OnPush strategy in our components to avoid Angular change detection running unless one of the @Input() bindings has changed. There is, however, another way to tell Angular to not run change detection at all, in any instance. This is handy when you want full control on when to run change detection. In this recipe, you'll learn how to completely detach the change detector from an Angular component to gain performance improvements.

Getting ready

The project for this recipe resides in `Chapter12/start_here/detaching-change-detecto`:

1. Open the project in VS Code.

2. Open the terminal and run `npm install` to install the dependencies of the project.

3. Run the `ng serve -o` command to start the Angular app and serve it on the browser. You should see the app, as follows:

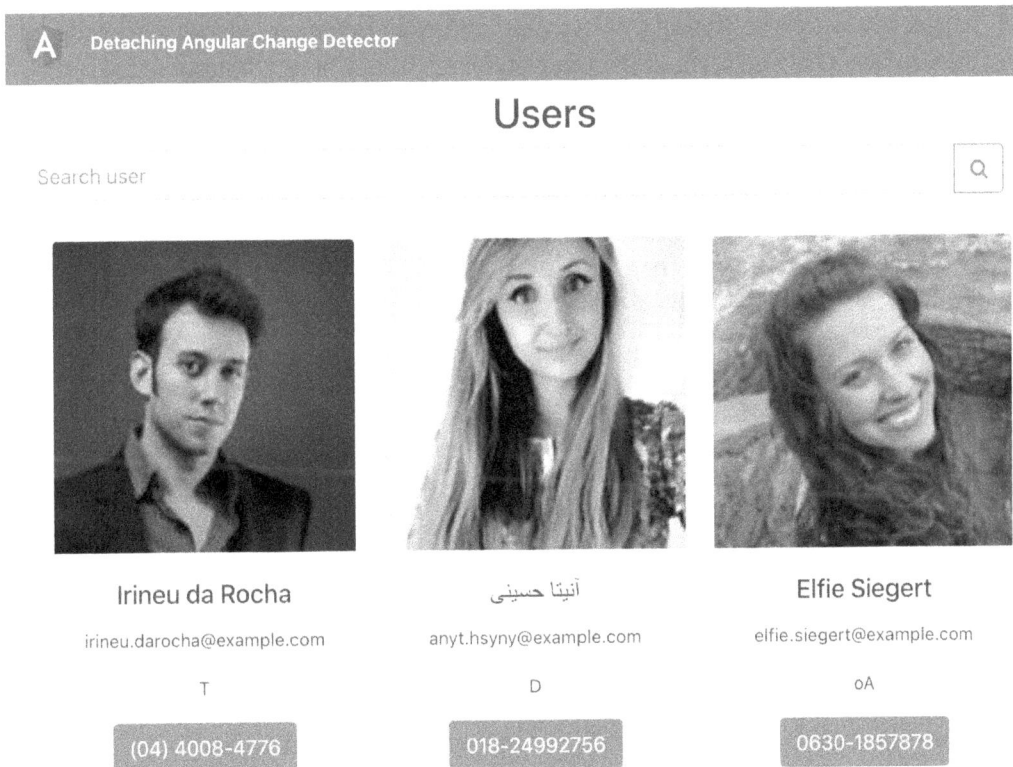

Figure 12.5 – App detaching-change-detector running at http://localhost:4200

Now that we have the project served on the browser, let's see the steps of the recipe in the next section.

How to do it...

We have the same users list application but with a twist. Right now, we have the `UserSearchInputComponent` component that holds the search input box. This is where we type the username to search for it in the users list. On the other hand, we have the `UserCardListComponent` component that has a list of users. We'll first experience the performance issues, and then we'll detach the change detector smartly to gain performance improvements. Let's get starte:

1. Refresh the app in the browser, then just click inside the search input, and then click outside the search input to first trigger a `focus` event on the input and then to trigger a `blur` event. Repeat this two times more, and then, on the console inside the Chrome Dev Tools, check the value of the `appLogs` object. You should see something like this:

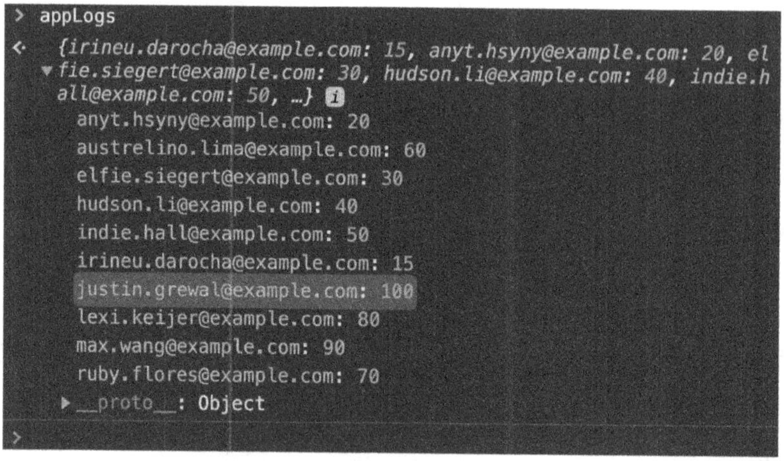

    ```
    >  appLogs
    <·  {irineu.darocha@example.com: 15, anyt.hsyny@example.com: 20, el
        ▼ fie.siegert@example.com: 30, hudson.li@example.com: 40, indie.h
        all@example.com: 50, …} ⓘ
          anyt.hsyny@example.com: 20
          austrelino.lima@example.com: 60
          elfie.siegert@example.com: 30
          hudson.li@example.com: 40
          indie.hall@example.com: 50
          irineu.darocha@example.com: 15
          justin.grewal@example.com: 100
          lexi.keijer@example.com: 80
          max.wang@example.com: 90
          ruby.flores@example.com: 70
        ▶ __proto__: Object
    >
    ```

    Figure 12.6 – Logs after performing focus and blur three times on the search input

    The preceding screenshot shows that the `idUsingFactorial()` method in the `UserCardComponent` class for the `justin.grewal@example.com` user has been called about 100 times, just in the steps we've performed so far.

2. Now, try to search for the `elfie` user by entering the name quickly in the search box.

    You'll notice that the app immediately hangs, and it takes a few seconds to show the user. You'll also notice that you don't even see the letters being typed in the search box as you type them. If you've followed *Step 1* and *Step 2* correctly, you should see an `appLogs` object, as follows:

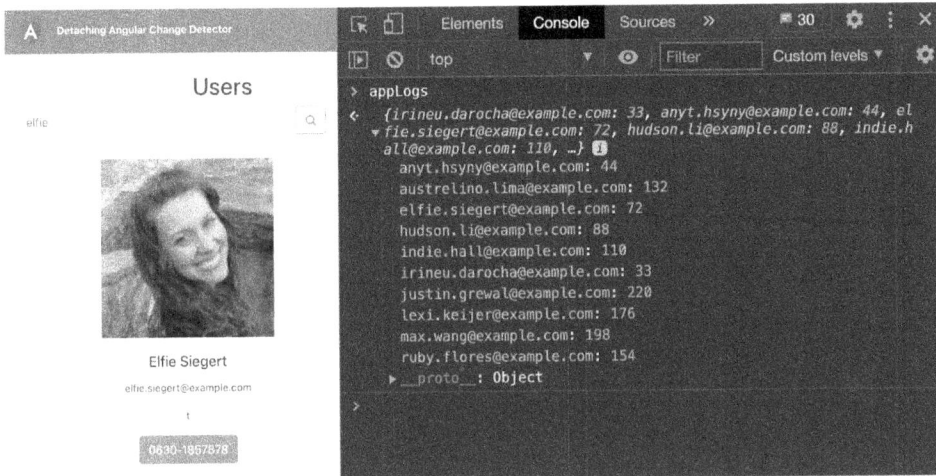

Figure 12.7 – Logs after typing elfie in the input search box

You can see in the preceding screenshot that the `idUsingFactorial()` method for the `justin.grewal@example.com` user has now been called about 220 times.

3. In order to improve performance, we'll use the `ChangeDetectorRef` service in this recipe to completely detach the change detector from the `UsersComponent` component, which is our top component for the **Users** page. Update the `users.component.ts` file, as follows:

```
import { ChangeDetectorRef, Component, OnInit} from '@
angular/core';
...
@Component({...})
export class UsersComponent implements OnInit {
  users: IUser[];
  constructor(
    private userService: UserService,
    private cdRef: ChangeDetectorRef
  ) {}

  ngOnInit() {
    this.cdRef.detach();
    this.searchUsers();
  }
}
```

If you refresh the app now, you'll see… Actually, you won't see anything, and that's fine—we have more steps to follow.

4. Now, since we want to run change detection only when we have searched the users—that is, when the `users` array changes in the `UsersComponent` class—we can use the `detectChanges()` method of the `ChangeDetectorRef` instance. Update the `users.component.ts` file again, as follows:

```
. . .
@Component ({...})
export class UsersComponent implements OnInit {
    . . .
    searchUsers(searchQuery = '') {
        this.userService.searchUsers(
searchQuery).subscribe((users) => {
            this.users = users;
        this.cdRef.detectChanges();
        });
    }
    . . .
}
```

5. Now, try performing the actions again—that is, refresh the page, focus in on the input, focus out, focus in, focus out, focus in, focus out, and then type `elfie` in the search input. Once you've followed the steps, you should see the `appLogs` object, as follows:

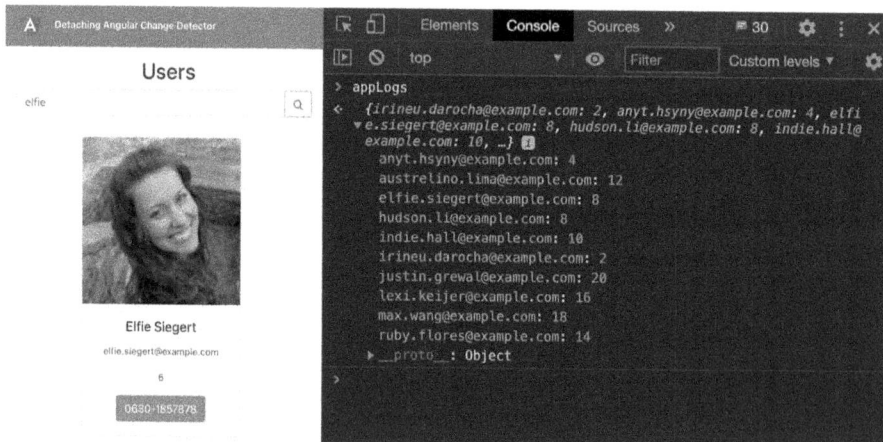

Figure 12.8 – Logs after performing the test steps and using ChangeDetectorRef.detach()

You can see in the preceding screenshot that even after performing all the actions mentioned in *Step 1* and *Step 2*, we have a very low count of the change-detection run cycle.

Awesomesauce! You've just learned how to detach the Angular change detector using the `ChangeDetectorRef` service. Now that you've finished the recipe, see the next section to understand how it works.

How it works...

The `ChangeDetectorRef` service provides a bunch of important methods to control change detection completely. In the recipe, we use the `.detach()` method in the `ngOnInit()` method of the `UsersComponent` class to detach the Angular change-detection mechanism from this component as soon as it is created. As a result, no change detection is triggered on the `UsersComponent` class, nor in any of its children. This is because each Angular component has a change-detection tree in which each component is a node. When we detach a component from the change-detection tree, that component (as a tree node) is detached, and so are its child components (or nodes). By doing this, we end up with absolutely no change detection happening for the `UsersComponent` class. As a result, when we refresh the page nothing is rendered, even after we've got the users from the **application programming interface** (**API**) and have got them assigned to the `users` property inside the `UsersComponent` class. Since we need to show the users on the view, which requires the Angular change-detection mechanism to be triggered, we use the `.detectChanges()` method from the `ChangeDetectorRef` instance, right after we've assigned the users data to the `users` property. As a result, Angular runs the change-detection mechanism, and we get the user cards shown on the view.

This means that in the entire **Users** page (that is, on the `/users` route) the only time the Angular change-detection mechanism would trigger after the `UsersComponent` class has initiated is when we call the `searchUsers()` method, get the data from the API, and assign the result to the `users` property, thus creating a highly controlled change-detection cycle, resulting in much better performance overall.

See also

- `ChangeDetectorRef` official documentation (https://angular.io/api/core/ChangeDetectorRef)

Running async events outside Angular with runOutsideAngular

Angular runs its change-detection mechanism on a couple of things, including—but not limited to—all browser events such as `keyup`, `keydown`, and so on. It also runs change detection on `setTimeout`, `setInterval`, and Ajax HTTP calls. If we had to avoid running change detection on any of these events, we'd have to tell Angular not to trigger change detection on them—for example, if you were using the `setTimeout()` method in your Angular component, it would trigger an Angular change detection each time its callback method was called. In this recipe, you'll learn how to execute code blocks outside of the `ngZone` service, using the `runOutsideAngular()` method.

Getting ready

The project for this recipe resides in `Chapter12/start_here/run-outside-angula`:

1. Open the project in VS Code.

2. Open the terminal and run `npm install` to install the dependencies of the project.

3. Run the `ng serve -o` command to start the Angular app and serve it on the browser. You should see the app, as follows:

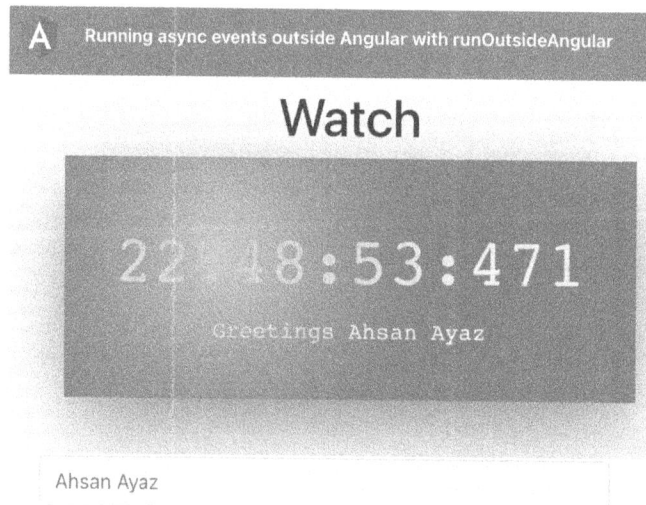

Figure 12.9 – App run-outside-angular running on http://localhost:4200

Now that we have the app running, let's see the steps of the recipe in the next section.

How to do it...

We have an app that shows a watch. However, the change detection right now in the app is not optimal, and we have plenty of room for improvement. We'll try to remove any unnecessary change detection using the `runOutsideAngular` method from `ngZone`. Let's get starte:.

1. The clock values are constantly updating. Thus, we have change detection running for each update cycle. Open the Chrome DevTools and switch to the **Console** tab. Type `appLogs` and press *Enter* to see how many times change detection has run for the `hours`, `minutes`, `seconds`, and `milliseconds` components. It should look like this:

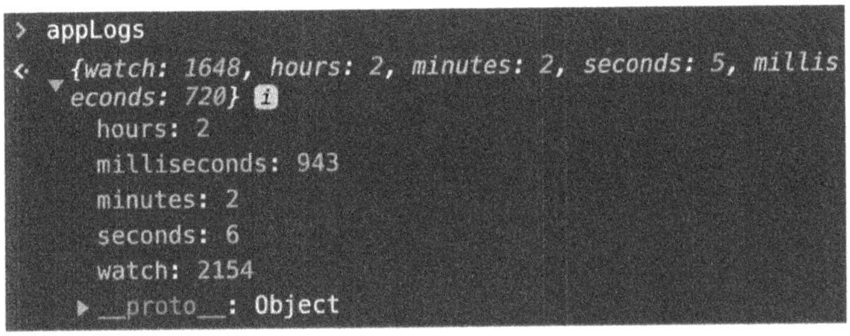

Figure 12.10 – The appLogs object reflecting number of change-detection runs

2. To measure performance, we need to see the numbers within a fixed time period. Let's add some code to turn off the interval timer in 4 seconds from the app's start for the clock. Modify the `watch-box.component.ts` file, as follows:

```
...
@Component({...})
export class WatchBoxComponent implements OnInit {
  ...
  ngOnInit(): void {
    this.intervalTimer = setInterval(() => {
      this.timer();
    }, 1);
    setTimeout(() => {
      clearInterval(this.intervalTimer);
    }, 4000);
  }
  ...
}
```

3.   Refresh the app and wait for 4 seconds for the clock to stop. Then, type `appLogs` multiple times in the **Console** tab, press *Enter*, and see the results. The clock stops but the animation is still running. You should see that change detection for the watch key still increases, as follows:

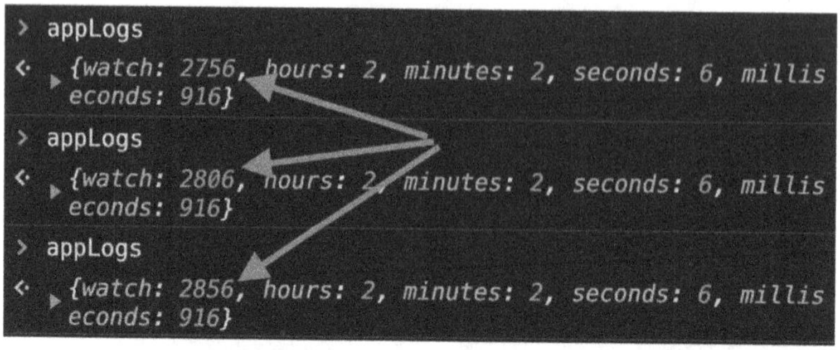

Figure 12.11 – Change detection still running for the watch component

4.   Let's also stop the animation inside the watch after 4 seconds. Update the watch.component.ts file, as follows:

```
. . .
@Component ({...})
export class WatchComponent implements OnInit {
  . . .
  ngOnInit(): void {
    this.intervalTimer = setInterval(() => {
      this.animate();
    }, 30);
    setTimeout(() => {
      clearInterval(this.intervalTimer);
    }, 4000);
  }
  . . .
}
```

Refresh the app and wait for the animation to stop. Have a look at the `appLogs` object in the Chrome DevTools. You should see that change detection stops for the watch key, as follows:

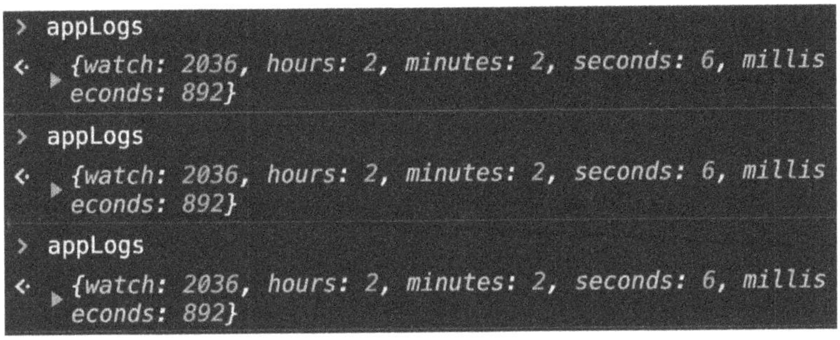

Figure 12.12 – Change detection stops after we stop the animation interval

5. We want the animation to run but without causing additional change-detection runs. This is because we want to make our app more performant. So, let's just stop the clock for now. To do that, update the watch-box.component.ts file, as follows:

```
...
@Component({...})
export class WatchBoxComponent implements OnInit {

  ...

  ngOnInit(): void {
    // this.intervalTimer = setInterval(() => {
    //   this.timer();
    // }, 1);
    // setTimeout(() => {
    //   clearInterval(this.intervalTimer);
    // }, 4000);
  }
}
```

Since we've now stopped the clock, the values for appLogs for the watch key are now only based on the animation for these 4 seconds. You should now see a value between **250** and **260** for the watch key.

6. Let's avoid running change detection on the animation by running the interval outside the ngZone service. We'll use the runOutsideAngular() method for this. Update the watch.component.ts file, as follows:

```
import {
  ...
```

```
    ViewChild,
    NgZone,
} from '@angular/core';
@Component({...})
export class WatchComponent implements OnInit {
    ...
    constructor(private zone: NgZone) {
        ...
    }

    ngOnInit(): void {
        this.zone.runOutsideAngular(() => {
            this.intervalTimer = setInterval(() => {
                this.animate();
            }, 30);
            setTimeout(() => {
                clearInterval(this.intervalTimer);
            }, 2500);
        });
    }
    ...
}
```

Refresh the app and wait for about 5 seconds. If you check the appLogs object now, you should see a decrease in the overall number of change-detection runs for each of the properties, as follows:

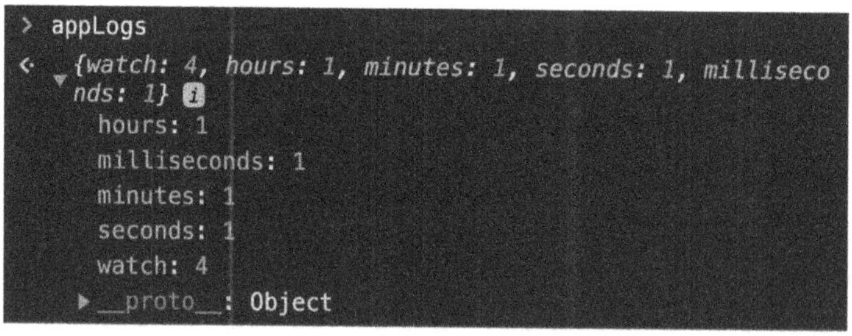

Figure 12.13 – The appLogs object after using runOutsideAngular() in WatchComponent

Yayy! Notice that the value for the watch key in the appLogs object has decreased from about **250** to **4** now. This means that our animation now doesn't contribute to change detection at all.

7. Remove the usage of `clearInterval()` from the animation for the
`WatchComponent` class. As a result, the animation should keep running. Modify
the `watch.component.ts` file, as follows:

```
. . .
@Component({...})
export class WatchComponent implements OnInit {
    . . .

    ngOnInit(): void {
      . . .
      this.ngZone.runOutsideAngular(() => {
        this.intervalTimer = setInterval(() => {
          this.animate();
        }, 30);
        setTimeout(() => { // ← Remove this block
          clearInterval(this.intervalTimer);
        }, 4000);
      });
    }
    . . .
}
```

8. Finally, remove the usage of `clearInterval()` from the `WatchBoxComponent`
class to run the clock. Update the `watch-box.component.ts` file, as follows:

```
import { Component, OnInit } from '@angular/core';

@Component({
  selector: 'app-watch-box',
  templateUrl: './watch-box.component.html',
  styleUrls: ['./watch-box.component.scss'],
})
export class WatchBoxComponent implements OnInit {
  name = '';
  time = {
    hours: 0,
    minutes: 0,
```

```
      seconds: 0,
      milliseconds: 0,
    };
    intervalTimer;
    constructor() {}

    ngOnInit(): void {
      this.intervalTimer = setInterval(() => {
        this.timer();
      }, 1);
      setTimeout(() => { // ← Remove this
        clearInterval(this.intervalTimer);
      }, 4000);
    }
    ...
  }
```

Refresh the app and check the value of the `appLogs` object after a few seconds, multiple times. You should see something like this:

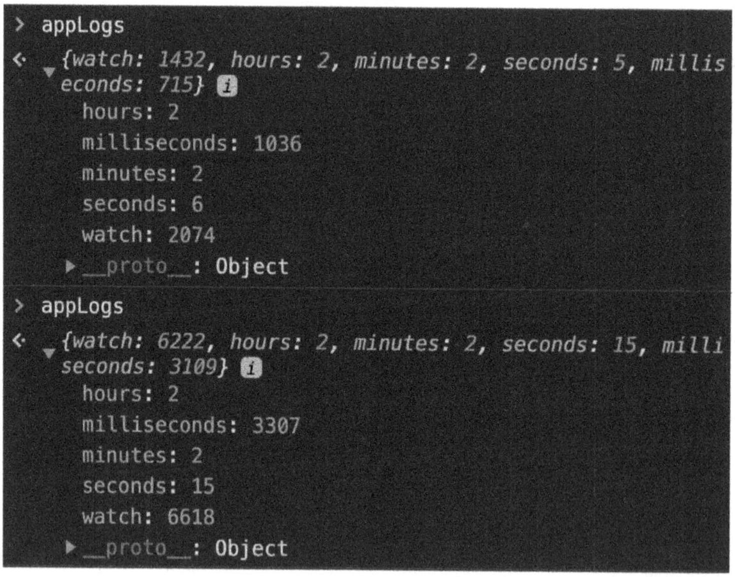

Figure 12.14 – The appLogs object after performance optimization with runOutsideAngular()

Looking at the preceding screenshot, you'd be like: "*Ahsan! What is this? We still have a huge number for the change-detection runs for the watch key. How is this performant exactly?*" Glad you asked. I will tell you the *why* in the *How it works...* section.

9. As a final step, stop the Angular server and run the following command to start the server in production mode:

```
ng serve --prod
```

10. Navigate to `https://localhost:4200` again. Wait for a few seconds and then check the `appLogs` object in the **Console** tab multiple times. You should see the object, as follows:

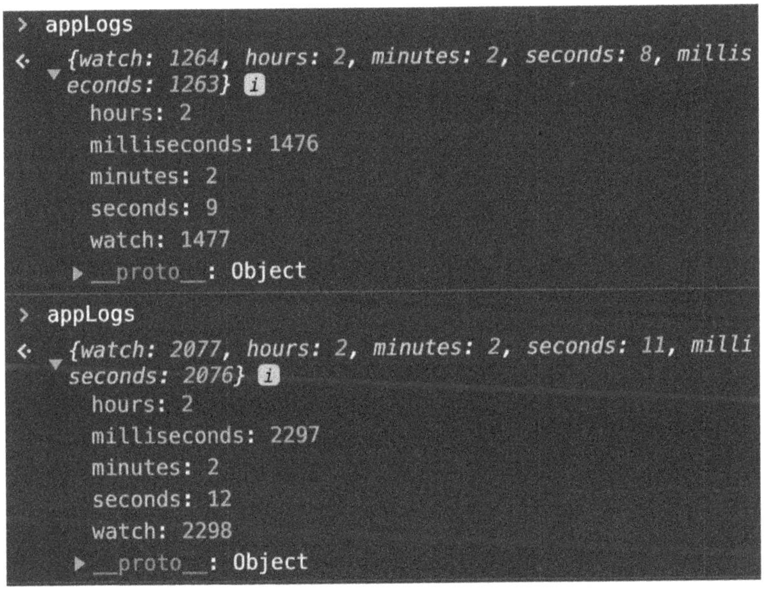

```
> appLogs
<· ▼ {watch: 1264, hours: 2, minutes: 2, seconds: 8, millis
       econds: 1263} ⓘ
       hours: 2
       milliseconds: 1476
       minutes: 2
       seconds: 9
       watch: 1477
     ▶ __proto__: Object
> appLogs
<· ▼ {watch: 2077, hours: 2, minutes: 2, seconds: 11, milli
       seconds: 2076} ⓘ
       hours: 2
       milliseconds: 2297
       minutes: 2
       seconds: 12
       watch: 2298
     ▶ __proto__: Object
```

Figure 12.15 – The appLogs object using the production build

Boom! If you look at the preceding screenshot, you should see that the change-detection run count for the `watch` key is always just one cycle more than the `milliseconds` key. This means that the `WatchComponent` class is almost only re-rendered whenever we have the value of the `@Input()` `milliseconds` binding updated.

Now that you've finished the recipe, see the next section to understand how it all works.

How it works...

In this recipe, we begin by looking at the `appLogs` object, which contains some key-value pairs. The value for each key-value pair represents the number of times Angular ran change detection for a particular component. The `hours`, `milliseconds`, `minutes`, and `seconds` keys represent the `WatchTimeComponent` instance for each of the values shown on the clock. The `watch` key represents the `WatchComponent` instance.

At the beginning of the recipe, we see that the value for the `watch` key is more than twice the value of the `milliseconds` key. Why do we care about the `milliseconds` key at all? Because the `@Input()` attribute binding `milliseconds` changes most frequently in our application—that is, it changes every 1 **millisecond (ms)**. The second most frequently changed values are the `xCoordinate` and `yCoordinates` properties within the `WatchComponent` class, which change every 30 ms. The `xCoordinate` and `yCoordinate` values aren't bound directly to the template (the **HyperText Markup Language (HTML)**) because they change the **Cascading Style Sheets (CSS)** variables of the `stopWatch` view child. This happens inside the `animate()` method, as follows:

```
el.style.setProperty('--x', `${this.xCoordinate}px`);
el.style.setProperty('--y', `${this.yCoordinate}px`);
```

Thus, changing these values shouldn't actually trigger change detection at all. We begin by limiting the clock window, using the `clearInterval()` method in the `WatchBoxComponent` class so that the clock stops within 4 seconds and we can evaluate the numbers. In *Figure 12.11*, we see that even after the clock stops, the change-detection mechanism keeps triggering for the `WatchComponent` class. This increases the count for the `watch` key in the `appLogs` object as time passes. We then stop the animation by using `clearInterval()` in the `WatchComponent` class. This stops the animation after 4 seconds as well. In *Figure 12.12*, we see that the count for the `watch` key stops increasing after the animation stops.

We then try to see the count of change detection only based on the animation. In *Step 6*, we stop the clock. Therefore, we only get a count based on the animation in the `appLogs` object for the `watch` key, which is a value between **250** and **260**.

We then introduce the magic `runOutsideAngular()` method into our code. This method is part of the `NgZone` service. The `NgZone` service is packaged with the `@angular/core` package. The `runOutsideAngular()` method accepts a method as a parameter. This method is executed outside the Angular zone. This means that the `setTimeout()` and `setInterval()` methods used inside the `runOutsideAngular()` method do not trigger the Angular change-detection cycle. You can see in *Figure 12.13* that the count drops to **4** after using the `runOutsideAngular()` method.

We then remove the `clearInterval()` usage from both the `WatchBoxComponent` and the `WatchComponent` classes—that is, to run the clock and the animation again, as we did in the beginning. In *Figure 12.14*, we see that the count for the `watch` key is almost twice the value of the `milliseconds` key. Now, why is that double exactly? This is because in development mode, Angular runs the change-detection mechanism twice. Therefore, in *Step 9* and *Step 10*, we run the application in production mode, and in *Figure 12.15*, we see that the value for the `watch` key is just one greater than the value for the `milliseconds` key, which means that the animation does not trigger any change detection for our application any more. Brilliant, isn't it? If you found this recipe useful, do let me know on my socials.

Now that you understand how it works, see the next section for further reading.

See also

- `NgZone` official documentation (`https://angular.io/api/core/NgZone`)
- Angular `ChangeDetectorRef` official documentation (`https://angular.io/api/core/ChangeDetectorRef`)

Using trackBy for lists with *ngFor

Lists are an essential part of most of the apps we build today. If you're building an Angular app, there's a great chance you will use the `*ngFor` directive at some point. We know that `*ngFor` allows us to loop over arrays or objects generating HTML for each item. However, for large lists, using it may cause performance issues, especially when the source for `*ngFor` is changed completely. In this recipe, we'll learn how we can improve the performance of lists using the `*ngFor` directive with the `trackBy` function. Let's get started.

Getting ready

The project for this recipe resides in `Chapter12/start_here/using-ngfor-trackb`:

1. Open the project in VS Code.
2. Open the terminal and run `npm install` to install the dependencies of the project.

3.   Run the `ng serve -o` command to start the Angular app and serve it on the browser. You should see the app, as follows:

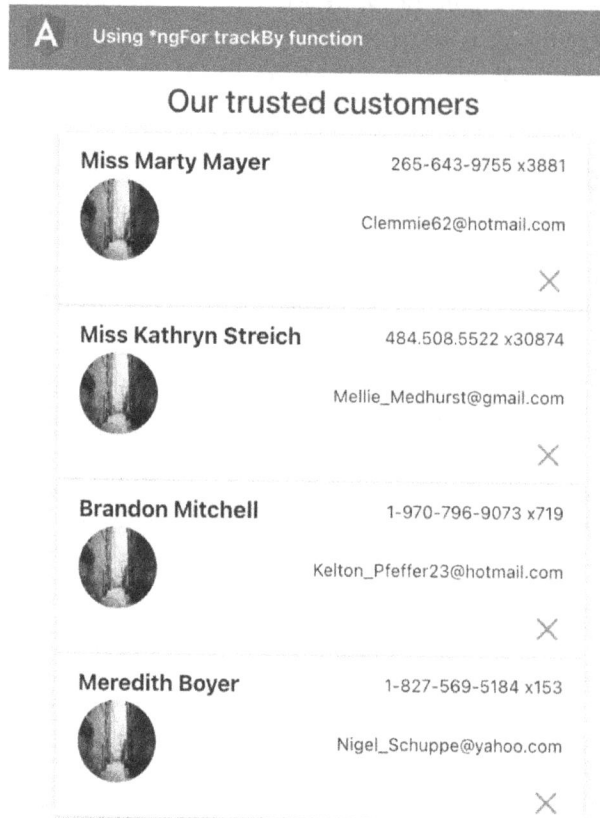

Figure 12.16 – App using-ngfor-trackby running on http://localhost:4200

Now that we have the app running, let's see the steps of the recipe in the next section.

How to do it...

We have an app that has a list of 1,000 users displayed on the view. Since we're not using a virtual scroll and a standard `*ngFor` list, we do face some performance issues at the moment. Notice that when you refresh the app, even after the loader is hidden, you see a blank white box for about 2-3 seconds before the list appears. Let's start the recipe to reproduce the performance issues and to fix them.

1.   First of all, open the Chrome DevTools and look at the **Console** tab. You should see a `ListItemComponent initiated` message logged 1,000 times. This message will be logged any time a list-item component is created/initiated.

2. Now, delete the first item by using the cross button on it. You should see the same message logged again about 999 times now, as shown in the following screenshot. This means we recreate the list-item component for the remaining 999 items:

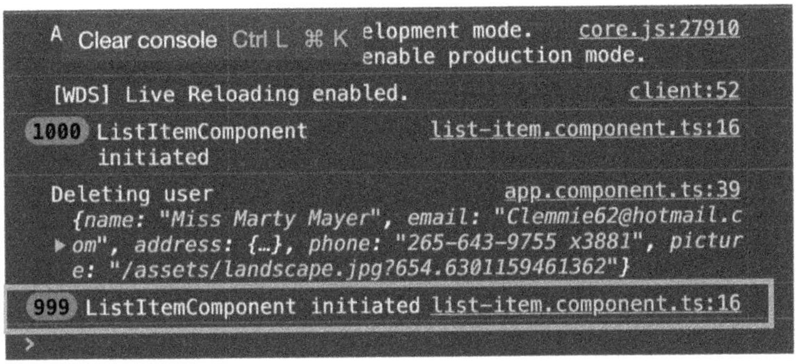

Figure 12.17 – Logs shown again after deleting an item

3. Now, refresh the app and tap on the first list item. You should see the `ListItemComponent initiated` logs again, as shown in the following screenshot. This means we recreate all the list items on an item update. You will notice that the update to the first item's name in the **user interface** (UI) is reflected in about 2-3 seconds:

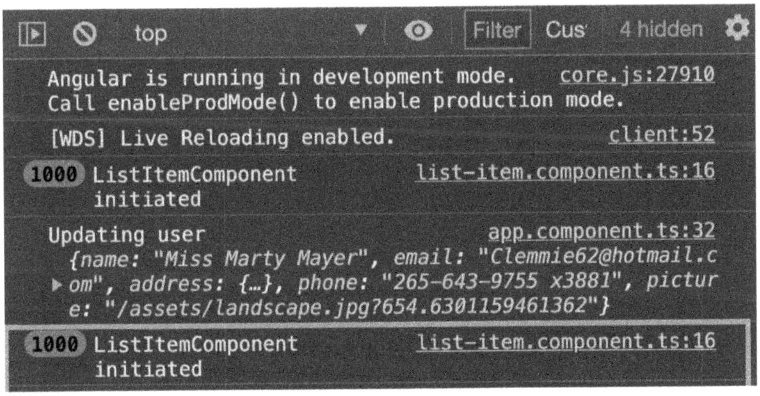

Figure 12.18 – Logs shown again after updating an item

4. Now, let's fix the performance issue by using the `trackBy` function. Open the `the-amazing-list.component.ts` file and update it, as follows:

```
. . .
@Component({...})
export class TheAmazingListComponent implements OnInit {
    . . .
    ngOnInit(): void {}

    trackByFn(_, user: AppUserCard) {
        return user.email;
    }
}
```

5. Now, update the `the-amazing-list.component.html` file to use the `trackByFn()` method we just created, as follows:

```
<h4 class="heading">Our trusted customers</h4>
<div class="list list-group">
    <app-list-item
        *ngFor="let item of listItems; trackBy: trackByFn"
        [item]="item"
        (itemClicked)="itemClicked.emit(item)"
        (itemDeleted)="itemDeleted.emit(item)"
    >
    </app-list-item>
</div>
```

6. Now, refresh the app, and click the first list item to update it. You will notice that the item is updated immediately and we don't log the `ListItemComponent initiated` message again anymore, as shown in the following screenshot:

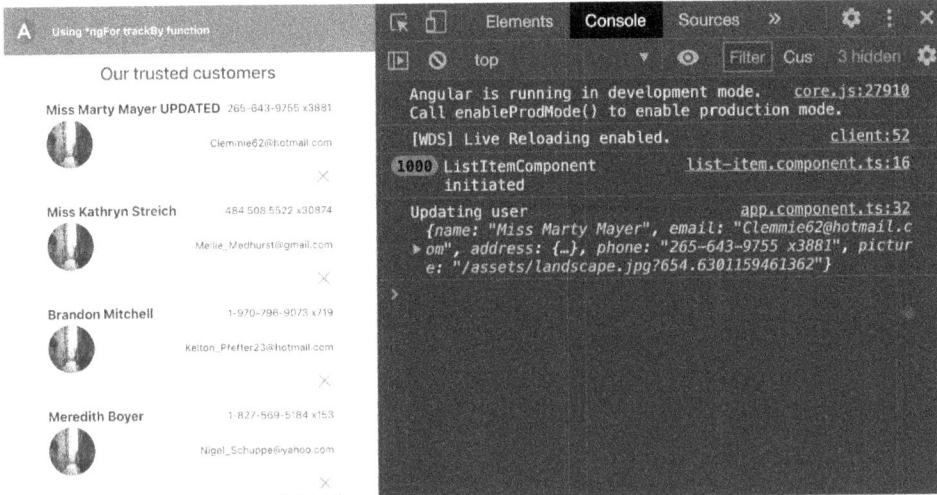

Figure 12.19 – No further logs after updating an item using the trackBy function

7. Delete an item as well now, and you will see we do not log the
    ListItemComponent initiated message again in this case, as well.

Great!! You now know how to use the trackBy function with the *ngFor directive to optimize the performance of lists in Angular. To understand all the magic behind the recipe, see the next section.

How it works...

The *ngFor directive by default assumes that the object itself is its unique identity, which means that if you just change a property in an object used in the *ngFor directive, it won't re-render the template for that object. However, if you provide a new object in its place (different reference in memory), the content for the particular item will re-render. This is what we actually do in this recipe to reproduce the performance-issue content. In the data.service.ts file, we have the following code for the updateUser() method:

```
updateUser(updatedUser: AppUserCard) {
    this.users = this.users.map((user) => {
        if (user.email === updatedUser.email) {
            return {
            ...updatedUser,
        };
        }
        // this tells angular that every object has now
```

```
        a different reference
      return { ...user };
    });
  }
```

Notice that we use the object spread operator ({ ... }) to return a new object for each item in the array. This tells the *ngFor directive to re-render the UI for each item in the listItems array in the TheAmazingListComponent class. Suppose you send a query to the server to find or filter users. The server could return a response that has 100 users. Out of those 100, about 90 were already rendered on the view, and only 10 are different. Angular, however, would re-render the UI for all the list items because of the following potential reasons (but not limited to these):

- The sorting/placement of the users could have changed.
- The length of the users could have changed.

Now, we want to avoid using the object reference as the unique identifier for each list item. For our use case, we know that each user's email is unique, therefore we use the trackBy function to tell Angular to use the user's email as the unique identifier. Now, even if we return a new object for each user after a user update from the updateUser() method (as previously shown), Angular doesn't re-render all the list items. This is because the new objects (users) have the same email and Angular uses it to track them. Pretty cool, right?

Now that you've learned how the recipe works, see the next section to view a link for further reading.

See also

- NgForOf official documentation (https://angular.io/api/common/NgForOf)

Moving heavy computation to pure pipes

In Angular, we have a particular way of writing components. Since Angular is heavily opinionated, we already have a lot of guidelines from the community and the Angular team on what to consider when writing components—for example, making HTTP calls directly from a component is considered a not-so-good practice. Similarly, if we have heavy computation in a component, this is also not considered a good practice. And when the view depends upon a transformed version of the data using a computation constantly, it makes sense to use Angular pipes. In this recipe, you'll learn how to use Angular pure pipes to avoid heavy computation within components.

Getting ready

The project we are going to work with resides in `Chapter12/start_here/using-pure-pipes`, inside the cloned repositor:

1. Open the project in VS Code.

2. Open the terminal and run `npm install` to install the dependencies of the project.

3. Run the `ng serve -o` command to start the Angular app and serve it on the browser. You should see the app, as follows:

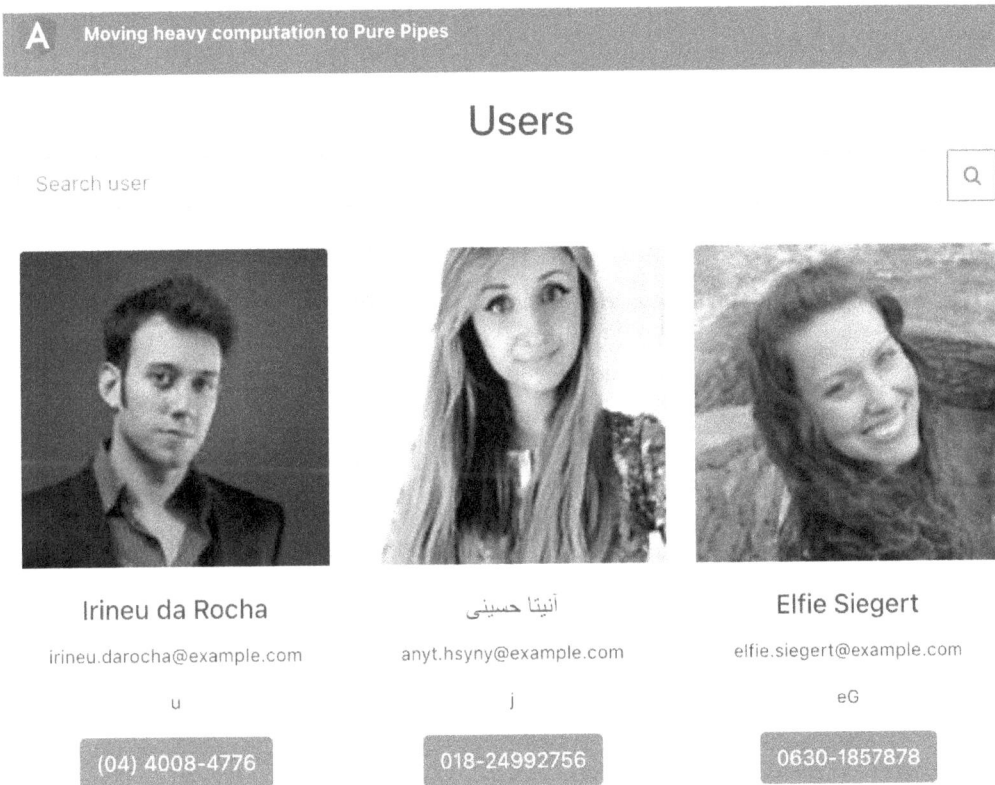

Figure 12.20 – using-pure-pipes app running at http://localhost:4200

Now that we have the project served on the browser, let's see the steps of the recipe in the next section.

How to do it...

The app we're working with has some performance issues, particularly with the `UserCardComponent` class because it uses the `idUsingFactorial()` method to generate a unique ID to show on the card. You'll notice that if you try typing `'irin'` in the search box, the app hangs for a while. We're not able to see the letters being typed instantly in the search box, and it takes a while before the results show. We will fix the issues by moving the computation in the `idUsingFactorial()` method to an Angular (pure) pipe. Let's get starte:

1. Let's create an Angular pipe. We'll move the computation for generating a unique ID for this pipe to later code. In the project root, run the following command in the terminal:

   ```
   ng g pipe core/pipes/unique-id
   ```

2. Now, copy the code for the `createUniqueId()` method from the `user-card.component.ts` file and paste it into the `unique-id.pipe.ts` file. We'll also modify the code a bit, so it should now look like this:

   ```
   ...
   @Pipe({...})
   export class UniqueIdPipe implements PipeTransform {
     characters = 'ABCDEFGHIJKLMNOPQRSTUVWXYZabcdef
     ghijklmnopqrstuvwxyz0123456789';
     createUniqueId(length) {
       var result = '';
       const charactersLength = this.characters.length;
       for (let i = 0; i < length; i++) {
         result += this.characters.charAt(
           Math.floor(Math.random() * charactersLength)
         );
       }
       return result;
     }
     ...
     transform(index: unknown, ...args: unknown[]): unknown
   ```

```
   {
       return null;
     }
   }
```

3. Now, also copy the `idUsingFactorial()` method from the `user-card.component.ts` file to the `unique-id.pipe.ts` file and update the file, as follows:

```
   import { Pipe, PipeTransform } from '@angular/core';

   @Pipe({
     name: 'uniqueId',
   })
   export class UniqueIdPipe implements PipeTransform {
       ...

       idUsingFactorial(num, length = 1) {
         if (num === 1) {
           return this.createUniqueId(length);
         } else {
           const fact = length * (num - 1);
           return this.idUsingFactorial(num - 1, fact);
         }
       }

       transform(index: number): string {
         return this.idUsingFactorial(index);
       }
   }
```

4. Now, update the `user-card.component.html` file to use the `uniqueId` pipe instead of the component's method. The code should look like this:

```
   <div class="user-card">
     <div class="card" *ngIf="user" (click)="cardClicked()">
       <img [src]="user.picture.large" class="card-img-top"
       alt="..." />
       <div class="card-body">
```

```
<h5 class="card-title">{{ user.name.first }}
{{ user.name.last }}</h5>
<p class="card-text">{{ user.email }}</p>
<p class="card-text unique-id" title="{{ index |
uniqueId }}">
    {{ index | uniqueId }}
</p>
<a href="tel: {{ user.phone }}" class="btn
btn-primary">{{
    user.phone
}}</a>
  </div>
 </div>
</div>
```

5. Now, refresh the app and type the name `Elfie Siegert` in the search box. Notice that the UI is not blocked. We're able to see the typed letters immediately as we type them, and the search results are faster as well.

Boom! Now that you know how to optimize performance by moving heavy computation to pure Angular pipes, see the next section to understand how this works.

How it works...

As we know, Angular by default runs change detection on each browser event triggered in the app, and since we're using an `idUsingFactorial()` method in the component template (UI), this function runs each time Angular runs the change-detection mechanism, causing more computation and performance issues. This would also hold true if we used a getter instead of a method. Here, we use a method because each unique ID is dependent on the index and we need to pass the index in the method when calling it.

We can take a step back from the initial implementation and think what the method actually does. It takes an input, does some computation, and returns a value based on the input—a classic example of data transformation, and also an example of where you would use a pure function. Luckily, Angular pure pipes are pure functions, and they do trigger change detection unless the input changes.

In the recipe, we move the computation to a newly created Angular pipe. The pipe's `transform()` method receives the value to which we're applying the pipe, which is the index of each user card in the `users` array. The pipe then uses the `idUsingFactorial()` method and, ultimately, the `createUniqueId()` method to calculate a random unique ID. When we start typing in the search box, the values for the index do not change. This results in no change detection being triggered until we get back a new set of users as output. Therefore, there is no unnecessary computation run as we type the input into the search box, thus optimizing performance and unblocking the UI thread.

See also

- Angular pure and impure pipes official documentation (`https://angular.io/guide/pipes#pure-and-impure-pipes`)

Using web workers for heavy computation

If your Angular application does a lot of computation during an action, there's a great chance that it will block the UI thread. This will cause a lag in rendering the UI because it blocks the main JavaScript thread. Web workers allow us to run heavy computation in the background thread, thus freeing the UI thread as it is not blocked. In this recipe, we're going to use an application that does a heavy computation in the `UserService` class. It creates a unique ID for each user card and saves it into the `localStorage`. However, it loops a couple of thousand times before doing so, which causes our application to hang for a while. In this recipe, we'll move the heavy computation from the components to a web worker and will also add a fallback in case web workers aren't available.

Getting ready

The project we are going to work with resides in `Chapter12/start_here/using-web-workers`, inside the cloned repositor:

1. Open the project in VS Code.
2. Open the terminal and run `npm install` to install the dependencies of the project.

3. Run the ng `serve -o` command to start the Angular app and serve it on the browser. You should see the app, as follows:

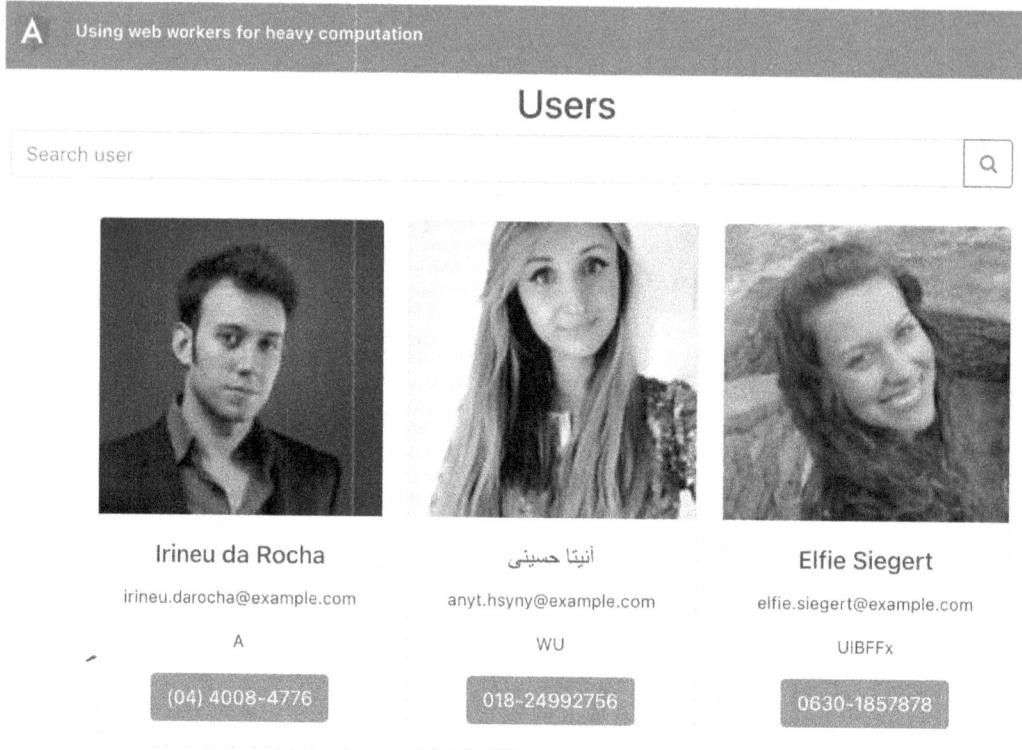

Figure 12.21 – App using-web-workers running at http://localhost:4200

Now that we have the app running, let's see the steps of the recipe in the next section.

How to do it...

Once you open the app, you'll notice that it takes some time before the user cards are rendered. This shows that the UI thread is blocked until we have the computation finished. The culprit is the saveUserUniqueIdsToStorage() method in the UserService class. This generates a unique ID a couple of thousands of times before saving it to the localStorage. Let's start the recipe, to improve the performance of the app. We'll start by implementing the web worke:

1. We'll first create a web worker. Run the following command in the project root:

```
ng generate web-worker core/workers/idGenerator
```

2. Now, copy the `for` loop from the `saveUserUniqueIdsToStorage()` method in the `UserService` class to the newly created `id-generator.worker.ts` file. The code should look like this:

```
/// <reference lib="webworker" />

import createUniqueId from '../constants/create-
unique-id';

addEventListener('message', ({ data }) => {
  console.log('message received IN worker', data);
  const { index, email } = data;
  let uniqueId;
  for (let i = 0, len = (index + 1) * 100000; i < len;
  ++i) {
    uniqueId = createUniqueId(50);
  }
  postMessage({ uniqueId, email });
});
```

3. Now that we have the worker file created, let's create a single instance of a worker to use it in the next steps. Create a new file in the `constants` folder. Name it `get-unique-id-worker.ts` and add the following code inside the file:

```
let UNIQUE_ID_WORKER: Worker = null;

const getUniqueIdWorker = (): Worker => {
  if (typeof Worker !== 'undefined' && UNIQUE_ID_WORKER
  === null) {
    UNIQUE_ID_WORKER = new Worker('../workers/
    id-generator.worker', {
      type: 'module',
    });
  }
  return UNIQUE_ID_WORKER;
};

export default getUniqueIdWorker;
```

4. Now, we'll use the worker in the `user.service.ts` file. Update it, as follows:

```
. . .
import getUniqueIdWorker from '../constants/get-unique-
id-worker';

@Injectable({...})
export class UserService {
    . . .
    worker: Worker = getUniqueIdWorker();
    constructor(private http: HttpClient) {
    this.worker.onmessage = ({ data: { uniqueId, email }
    }) => {
        console.log('received message from worker',
        uniqueId, email);
        const user = this.usersCache.find((user) => user.
        email === email);
        localStorage.setItem(
            `ng_user__${user.email}`,
            JSON.stringify({
                ...user,
                uniqueId,
            })
        );
    };

}
    . . .
}
```

5. We'll update the file again to modify the `saveUserUniqueIdsToStorage()` method. We'll use the worker instead of using the existing code, if we have web workers available in the environment. Update the `user.service.ts` file, as follows:

```
. . .
@Injectable({...})
export class UserService {
    . . .
```

```
saveUserUniqueIdsToStorage(user: IUser, index) {
  let uniqueId;
  const worker: Worker = getUniqueIdWorker();
  if (worker !== null) {
    worker.postMessage({ index, email: user.email });
  } else {
    // fallback
    for(let i = 0, len = (index + 1) * 100000; i<len;
    ++i) {
      uniqueId = createUniqueId(50);
    }
    localStorage.setItem(...);
  }
}
...
}
```

6. Refresh the app and notice how long it takes for the user cards to render. They should appear much faster than before. Also, you should be able to see the following logs reflecting the communication from the app to the web worker, and vice versa:

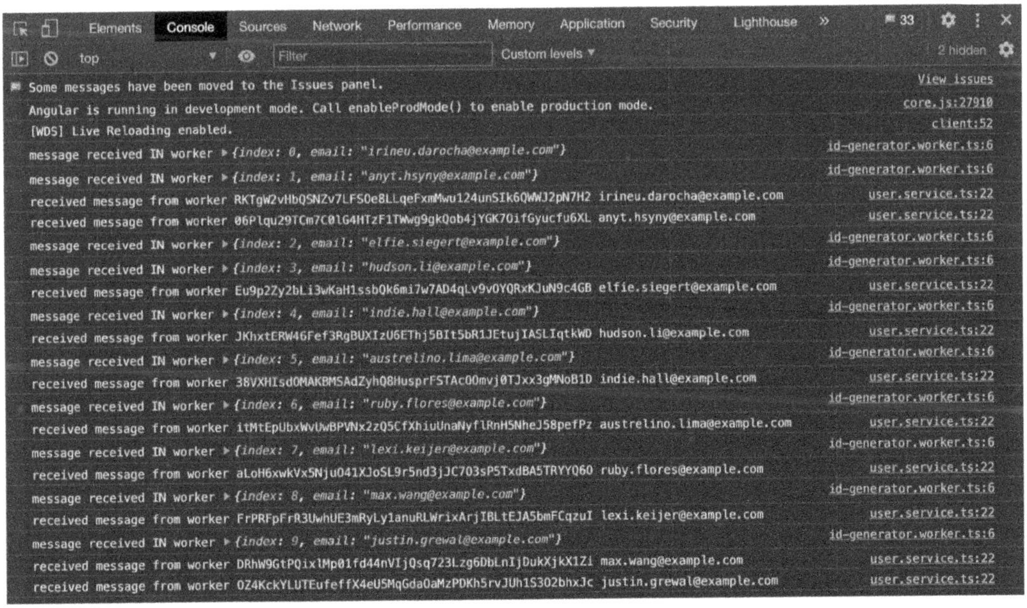

Figure 12.22 – Logs showing messages to and from the app to web workers

Woohoo!!! The power of web workers! And now you know how to use web workers in an Angular app to move heavy computation to them. Since you've finished the recipe, see the next section on how this works.

How it works...

As we discussed in the recipe's description, web workers allow us to run and execute code in a separate thread from the main JavaScript (or UI thread). At the beginning of the recipe, whenever we refresh the app or search for a user, it blocks the UI thread. This is until a unique ID is generated for each card. We begin the recipe by creating a web worker using the Angular **command-line interface (CLI)**. This creates an id-generator.worker.ts file, which contains some boilerplate code to receive messages from the UI thread and to send a message back to it as a response. The CLI command also updates the angular.json file by adding a webWorkerTsConfig property. The value against the webWorkerTsConfig property is the path to the tsconfig.worker.json file, and the CLI command also creates this tsconfig.worker.json file. If you open the tsconfig.worker.json file, you should see the following code:

```
/* To learn more about this file see: https://angular.io/
config/tsconfig. */
{
  "extends": "./tsconfig.json",
  "compilerOptions": {
    "outDir": "./out-tsc/worker",
    "lib": [
      "es2018",
      "webworker"
    ],
    "types": []
  },
  "include": [
    "src/**/*.worker.ts"
  ]
}
```

After creating a web worker file, we create another file named `uniqueIdWorker.ts`. This file exports the `getUniqueIdWorker()` method as the default export. When we call this method, it generates a new `Worker` instance if we don't have a worker generated already. The method uses the `id-generator.worker.ts` file to generate a worker. We also use the `addEventListener()` method inside the worker file to listen to the messages sent from the UI thread (that is, the `UserService` class). We receive the `index` of the user card and the `email` of the user as the data in this message. We then use a `for` loop to generate a unique ID (`uniqueId` variable), and once the loop ends, we use the `postMessage()` method to send the `uniqueId` variable and the `email` back to the UI thread.

Now, in the `UserService` class, we listen to messages from the worker. In the `constructor()` method, we check if web workers are available in the environment by checking the value from the `getUniqueIdWorker()` method, which should be a non-null value. Then, we use the `worker.onmessage` property to assign it a method. This is to listen to the messages from the worker. Since we already know that we get the `uniqueId` variable and the `email` from the worker, we use the `email` to get the appropriate user from the `usersCache` variable. Then, we store the user data with the `uniqueId` variable to the `localStorage` against the user's `email`.

Finally, we update the `saveUserUniqueIdsToStorage()` method to use the worker instance if it is available. Notice that we use the `worker.postMessage()` method to pass the `index` and the `email` of the user. Note also that we are using the previous code as a fallback for cases where we don't have web workers enabled.

See also

- Angular official documentation on web workers (`https://angular.io/guide/web-worker`)
- MDN web worker documentation (`https://developer.mozilla.org/en-US/docs/Web/API/Worker`)

Using performance budgets for auditing

In today's world, most of the population has a pretty good internet connection to use everyday applications, be it a mobile app or a web app, and it is fascinating how much data we ship to our end users as a business. The amount of JavaScript shipped to users has an ever-increasing trend now, and if you're working on a web app, you might want to use performance budgets to make sure the bundle size doesn't exceed a certain limit. With Angular apps, setting the budget sizes is a breeze. In this recipe, you're going to learn how to use the Angular CLI to set up budgets for your Angular apps.

Getting ready

The project for this recipe resides in `Chapter12/start_here/angular-performance-budget`:

1. Open the project in VS Code.

2. Open the terminal and run `npm install` to install the dependencies of the project.

3. Run the `ng build --configuration production` command to build the Angular app in production mode. Notice the output on the console. It should look like this:

```
ahsanayaz@Muhammads-MBP angular-performance-budgets % ng build  --prod
✔ Browser application bundle generation complete.
✔ Copying assets complete.
✔ Index html generation complete.

Initial Chunk Files                  | Names       |       Size
main.636953a84ba78df03f95.js         | main        |  268.05 kB
polyfills.66cbf868de57007a248c.js    | polyfills   |   36.00 kB
runtime.583904e63cb9a5659203.js      | runtime     |    2.31 kB
styles.09e2c710755c8867a460.css      | styles      |    0 bytes

                                     | Initial Total | 306.35 kB

Lazy Chunk Files                     | Names       |       Size
common.18fe967b5b48a6d9b344.js       | common      |    4.56 kB
6.fa8f5359cd76c4c3c3bb.js            | —           |    1.75 kB
7.009a8adad262cdc923ae.js            | —           |  912 bytes
5.a449fd050ca9289dd993.js            | —           |  911 bytes

Build at: 2021-04-12T18:49:47.548Z — Hash: d793cb44c27a566afe6e — Time: 17485ms
```

Figure 12.23 – Build output for production mode, without performance budgets

Notice that the bundle size for the `main.*.js` file is about 260 **kilobytes** (**KB**) at the moment. Now that we have built the app, let's see the steps of the recipe in the next section.

How to do it...

We have an app that is really small in terms of bundle size at the moment. However, this could grow into a huge app with upcoming business requirements. For the sake of this recipe, we'll increase the bundle size deliberately and will then use performance budgets to stop the Angular CLI from building the app for production if the bundle size exceeds the budget. Let's begin the recip:

1. Open the `app.component.ts` file and update it, as follows:

```
...
import * as moment from '../lib/moment';
```

```
import * as THREE from 'three';
@Component({...})
export class AppComponent {

    ...

    constructor(private auth: AuthService, private router:
    Router) {
        const scene = new THREE.Scene();
        console.log(moment().format('MMM Do YYYY'));
    }

    ...

}
```

2. Now, build the app again for production using the ng build
    --configuration production command. You should see that the bundle size
    for the main.*.js file is now 1.12 **megabytes** (**MB**). This is a huge increase in size
    compared to the original 268.05 KB, as you can see in the following screenshot:

```
ahsanayaz@Muhammads-MBP angular-performance-budgets % ng build --prod
✓ Browser application bundle generation complete.
✓ Copying assets complete.
✓ Index html generation complete.

Initial Chunk Files                    | Names          |     Size
main.e48427605c12dd80b439.js           | main           |  1.12 MB
polyfills.66cbf868de57007a248c.js      | polyfills      | 36.00 kB
runtime.583904e63cb9a5659203.js        | runtime        |  2.31 kB
styles.09e2c710755c8867a460.css        | styles         |  0 bytes

                                       | Initial Total  |  1.16 MB

Lazy Chunk Files                       | Names          |     Size
common.18fe967b5b48a6d9b344.js         | common         |  4.56 kB
6.fa8f5359cd76c4c3c3bb.js              | —              |  1.75 kB
7.009a8adad262cdc923ae.js              | —              | 912 bytes
5.a449fd050ca9289dd993.js              | —              | 911 bytes

Build at: 2021-04-12T18:57:53.537Z - Hash: 5ee9352e9b1e5d2f4664 - Time: 35269ms
```

Figure 12.24 – The bundle size for main.*.js increased to 1.11 MB

Let's suppose our business requires us to not ship apps with main bundle sizes more
than 1.0 MB. For this, we can configure our Angular app to throw an error if the
threshold is met.

3. Refresh the app, open the `angular.json file`, and update it. The property
    that we're targeting is `projects.angular-performance-budgets.`
    `architect.build.configurations.production.budgets`. The file
    should look like this:

```
...
{
    "budgets": [
      {
        "type": "initial",
        "maximumWarning": "800kb",
        "maximumError": "1mb"
      },
      {
        "type": "anyComponentStyle",
        "maximumWarning": "6kb",
        "maximumError": "10kb"
      }
    ]
}
...
```

4. Now that we have the budgets in place, let's build the app once again using the `ng`
    `build --configuration production` command. The build should fail and
    you should see both a warning and an error on the console, as follows:

```
ahsanayaz@Muhammads-MBP angular-performance-budgets % ng build --prod
✓ Browser application bundle generation complete.

Warning: budgets: initial exceeded maximum budget. Budget 800.00 kB was not met by 3
83.86 kB with a total of 1.16 MB.

Error: budgets: initial exceeded maximum budget. Budget 1.00 MB was not met by 159.8
6 kB with a total of 1.16 MB.
```

Figure 12.25 – Angular CLI throwing errors and warnings based on performance budgets

5. Let's improve our application by not importing the entire libraries in the `app.`
    `component.ts` file, and use the `date-fns` package instead of `moment.js` to do
    the same thing. Run the following command to install the `date-fns` package:

```
npm install --save date-fns
```

6. Now, update the app.component.ts file, as follows:

```
import { Component } from '@angular/core';
import { Router } from '@angular/router';
import { AuthService } from './services/auth.service';
import { format } from 'date-fns';
import { Scene } from 'three';


@Component({...})
export class AppComponent {

    ...

    constructor(private auth: AuthService, private router:
    Router) {
        console.log(format(new Date(), 'LLL do yyyy'));
        const scene = new Scene();
    }
    ...
}
```

7. Run the ng build --configuration production command again. You should see a decreased bundle size, as follows:

```
ahsanayaz@Muhammads-MBP angular-performance-budgets % ng build --prod
✓ Browser application bundle generation complete.
✓ Copying assets complete.
✓ Index html generation complete.

Initial Chunk Files              | Names       |       Size
main.23532ad8305a49dcab35.js     | main        |   793.42 kB
polyfills.66cbf868de57007a248c.js| polyfills   |    36.00 kB
runtime.583904e63cb9a5659203.js  | runtime     |     2.31 kB
styles.09e2c710755c8867a460.css  | styles      |     0 bytes

                                 | Initial Total |  831.73 kB

Lazy Chunk Files                 | Names       |       Size
common.18fe967b5b48a6d9b344.js   | common      |     4.56 kB
6.fa8f5359cd76c4c3c3bb.js        | —           |     1.75 kB
7.009a8adad262cdc923ae.js        | —           |   912 bytes
5.a449fd050ca9289dd993.js        | —           |   911 bytes

Build at: 2021-04-12T19:23:54.731Z - Hash: 157348e679960d29b796 - Time: 27974ms

Warning: budgets: initial exceeded maximum budget. Budget 800.00 kB was not met by 3
1.73 kB with a total of 831.73 kB.
```

Figure 12.26 – Reduced bundle size after using date-fns and optimized imports

Boom!! You just learned how to use the Angular CLI to define performance budgets. These budgets can be used to throw warnings and errors based on your configuration. Note that the budgets can be modified based on changing business requirements. However, as engineers, we have to be cautious about what we set as performance budgets to not ship JavaScript over a certain limit to the end users.

See also

- Performance budgets with the Angular CLI official documentation (`https://web.dev/performance-budgets-with-the-angular-cli/`)

Analyzing bundles with webpack-bundle-analyzer

In the previous recipe, we looked at configuring budgets for our Angular app, and this is useful because you get to know when the overall bundle size exceeds a certain threshold, although you don't get to know how much each part of the code is actually contributing to the final bundles. This is what we call *analyzing* the bundles, and in this recipe, you will learn how to use `webpack-bundle-analyzer` to audit the bundle sizes and the factors contributing to them.

Getting ready

The project we are going to work with resides in `Chapter12/start_here/using-webpack-bundle-analyzer`, inside the cloned repositor:

1. Open the project in VS Code.

2. Open the terminal and run `npm install` to install the dependencies of the project.

3. Run the `ng serve -o` command to start the Angular app and serve it on the browser. You should see the app, as follows:

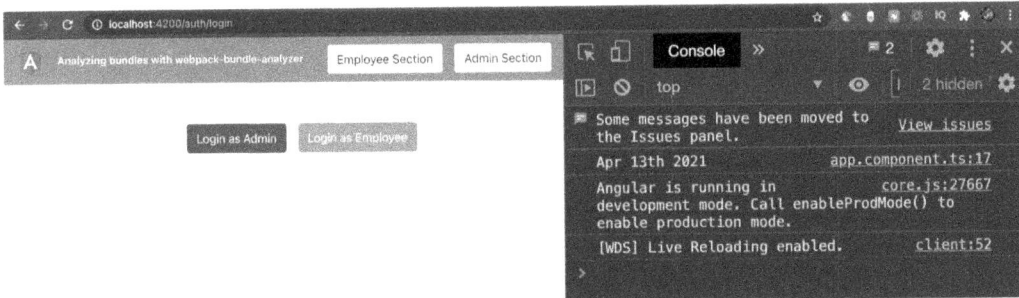

Figure 12.27 – App using-webpack-bundle-analyzer running at http://localhost:4200

4. Now, build the app using the `ng build --configuration production` command to build the Angular app in production mode. You should see the following output:

```
ahsanayaz@Muhammads-MBP using-webpack-bundle-analyzer % ng build --prod
✓ Browser application bundle generation complete.
✓ Copying assets complete.
✓ Index html generation complete.

Initial Chunk Files            | Names       |      Size
main.4919d8b686761d2332f4.js   | main        |    1.12 MB
polyfills.66cbf868de57007a248c.js | polyfills |   36.00 kB
runtime.583904e63cb9a5659203.js | runtime    |    2.31 kB
styles.09e2c710755c8867a460.css | styles     |    0 bytes

                               | Initial Total |  1.16 MB

Lazy Chunk Files               | Names       |      Size
common.18fe967b5b48a6d9b344.js | common      |    4.56 kB
6.fa8f5359cd76c4c3c3bb.js      | —           |    1.75 kB
7.009a8adad262cdc923ae.js      | —           |  912 bytes
5.a449fd050ca9289dd993.js      | —           |  911 bytes

Build at: 2021-04-13T01:40:03.255Z – Hash: 8394a47c42221f2fd6ac – Time: 17011ms
```

Figure 12.28 – The main bundle, having a size of 1.11 MB

Now that we have built the app, let's see the steps of the recipe in the next section.

How to do it...

As you may have noticed, we have a main bundle of size 1.12 MB. This is because we are using the `Three.js` library and the `moment.js` library in our `app.component.ts` file, which imports those libraries, and they end up being in the main bundle. Let's start the recipe to analyze the factors for the bundle size visuall:

1. We'll first install the `webpack-bundle-analyzer` package. Run the following command in the project root:

    ```
    npm install --save-dev webpack-bundle-analyzer
    ```

2. Now, create a script in the package.json file. We'll use this script in the next steps to analyze our final bundles. Update the package.json file, as follows:

```
{
    . . .
    "scripts": {
        "ng": "ng",
        "start": "ng serve",
        "build": "ng build",
        "test": "ng test",
        "lint": "ng lint",
        "e2e": "ng e2e",
        "analyze-bundle": "webpack-bundle-analyzer
        dist/using-webpack-bundle-analyzer/stats.json"
    },
    "private": true,
    "dependencies": {... },
    "devDependencies": {...}
}
```

3. Now, build the production bundle again, but with an argument to generate a stats.json file as well. Run the following command from the project root:

```
ng build --configuration production --stats-json
```

4. Now, run the analyze-bundle script to use the webpack-bundle-analyzer package. Run the following command from the project root:

```
npm run analyze-bundle
```

This will spin up a server with the bundle analysis. You should see a new tab opened in your default browser, and it should look like this:

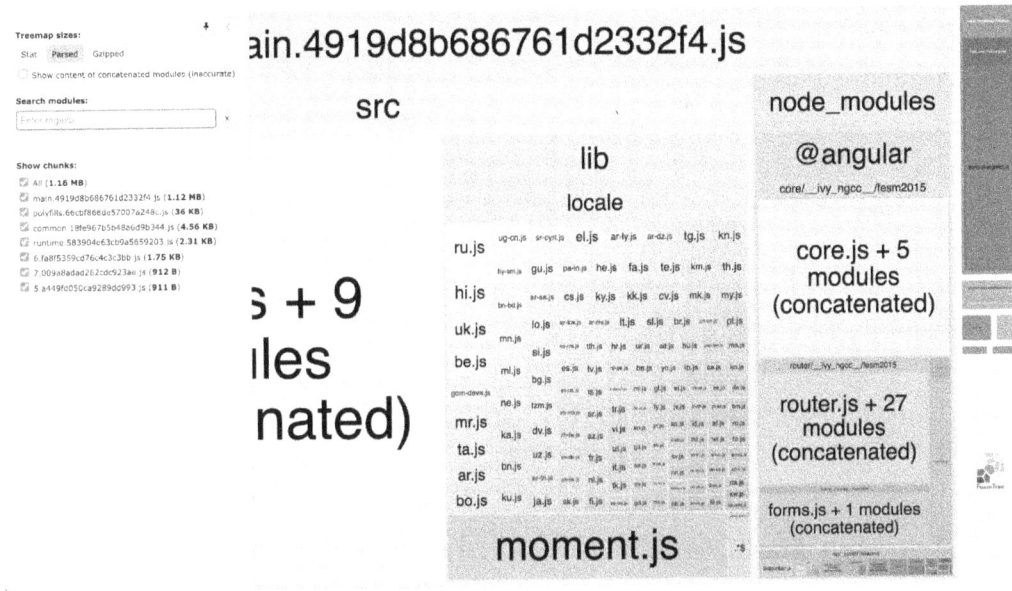

Figure 12.29 – Bundle analysis using webpack-bundle-analyzer

5. Notice that the `lib` folder takes a huge portion of the bundle size—648.29 KB, to be exact, which you can check by just doing a mouseover on the `lib` box. Let's try to optimize the bundle size. Let's install the `date-fns` package so that we can use it instead of `moment.js`. Run the following command from your project root:

```
npm install --save date-fns
```

6. Now, update the `app.component.ts` file to use the `date-fns` package's `format()` method instead of using the `moment().format()` method. We'll also just import the `Scene` class from the `Three.js` package instead of importing the whole library. The code should look like this:

```
import { Component } from '@angular/core';
import { Router } from '@angular/router';
import { AuthService } from './services/auth.service';
import { format } from 'date-fns';
import { Scene } from 'three';

@Component({...})
export class AppComponent {
    ...
    constructor(private auth: AuthService, private router:
```

```
Router) {
    const scene = new Scene();
    console.log(format(new Date(), 'LLL do yyyy'));
}
...
}
```

7. Run the `ng build --configuration production --stats-json` command, and then run `npm run analyze-bundle`.

    Once `webpack-bundle-analyzer` runs you should see the analysis, as shown in the following screenshot. Notice that we don't have the `moment.js` file or the `lib` block anymore, and the overall bundle size has reduced from 1.15 MB to 831.44 KB:

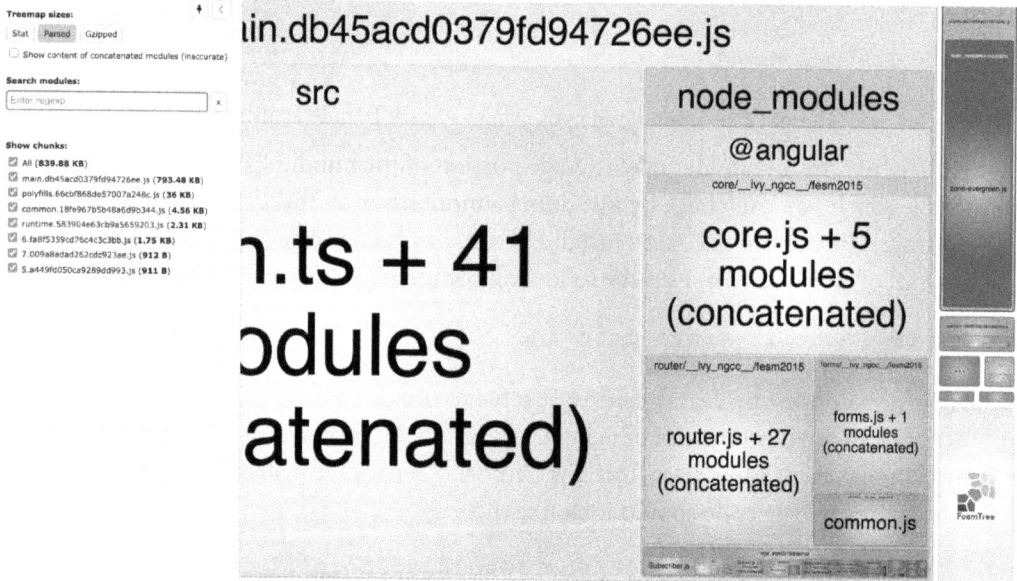

Figure 12.30 – Bundle analysis after using date-fns instead of moment.js

Woohoo!!! You now know how to use the `webpack-bundle-analyzer` package to audit bundle sizes in Angular applications. This is a great way of improving overall performance, because you can identify the chunks causing the increase in the bundle size and then optimize the bundles.

See also

- Getting started with webpack (`https://webpack.js.org/guides/getting-started/`)

- `webpack-bundle-analyzer`—GitHub repository (`https://github.com/webpack-contrib/webpack-bundle-analyzer`)

13
Building PWAs with Angular

PWAs or **Progressive Web Apps** are web applications at their core. Although they are built with enhanced features and experiences that are supported by modern browsers, if a PWA is run in a browser that doesn't support the modern features/enhancements, the user still gets the core experience of the web application. In this chapter, you're going to learn how to build Angular apps as PWAs. You'll learn some techniques to make your apps **installable, capable, fast, and reliable**. The following are the recipes we're going to cover in this chapter:

- Converting an exsisting Angular app into a PWA with the Angular CLI
- Modifying the theme color for your PWA
- Using Dark Mode in your PWA
- Providing a custom installable experience in your PWA
- Precaching requests using an Angular service worker
- Creating an App Shell for your PWA

Technical requirements

For the recipes in this chapter, make sure you have **Git** and **Node.js** installed on your machine. You also need to have the `@angular/cli` package installed, which you can do with `npm install -g @angular/cli` from your terminal. You also need to install the `http-server` package globally. You can install it by running `npm install -g http-server` in your terminal. The code for this chapter can be found at `https://github.com/PacktPublishing/Angular-Cookbook/tree/master/chapter13`.

Converting an existing Angular app into a PWA with the Angular CLI

A PWA involves a few interesting components, two of which are the service worker and the web manifest file. The service worker helps to cache the static resources and caching requests, and the web manifest file contains information about app icons, the theme color of the app, and so on. In this recipe, we'll convert an existing Angular application to a PWA. The principles apply to a fresh Angular app as well if you were to create it from scratch. For the sake of the recipe, we're going to convert an existing Angular app. We'll see what changed in our Angular web app and how the `@angular/pwa` package converts it into a PWA. Also, how it helps to cache the static resources.

Getting ready

The project that we are going to work with resides in `chapter13/start_here/angular-pwa-app` inside the cloned repository:

1. Open the project in Visual Studio Code.

2. Open the terminal and run `npm install` to install the dependencies of the project.

3. Once done, run `ng build --configuration production`.

4. Now run `http-server dist/angular-pwa-app -p 4200`.

    This should run the app at `http://localhost:4200` in production mode, and should look as follows:

Figure 13.1 – angular-pwa-app running on http://localhost:4200

Now that we have the app running locally, let's see the steps of the recipe in the next section.

How to do it

The app we're working with is a simple counter application. It has a min and max value, and some buttons that can increment, decrement, and reset the counter's value. The app saves the value of the counter in `localStorage` but it is not a PWA yet. Let's convert it into a PWA:

1. First, let's see if our application works offline at all, because that's one of the traits of PWAs. Open Chrome DevTools for the app. Go to the **Network** tab and change **Throttling** to **Offline** as follows:

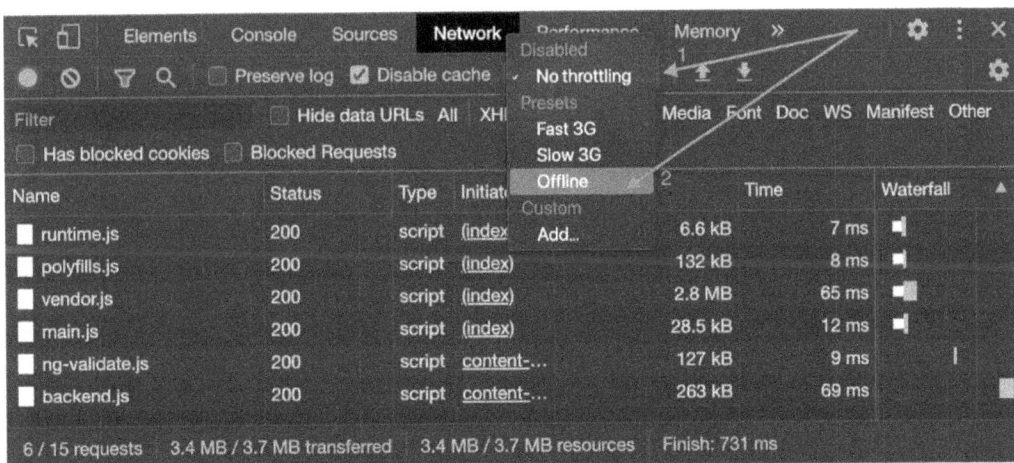

Figure 13.2 – Changing network throttling to Offline to see the offline experience

2. Now stop the `http` server by exiting the process from your terminal. Once done, refresh the app's page. You should see that the app doesn't work anymore, as shown in the following figure:

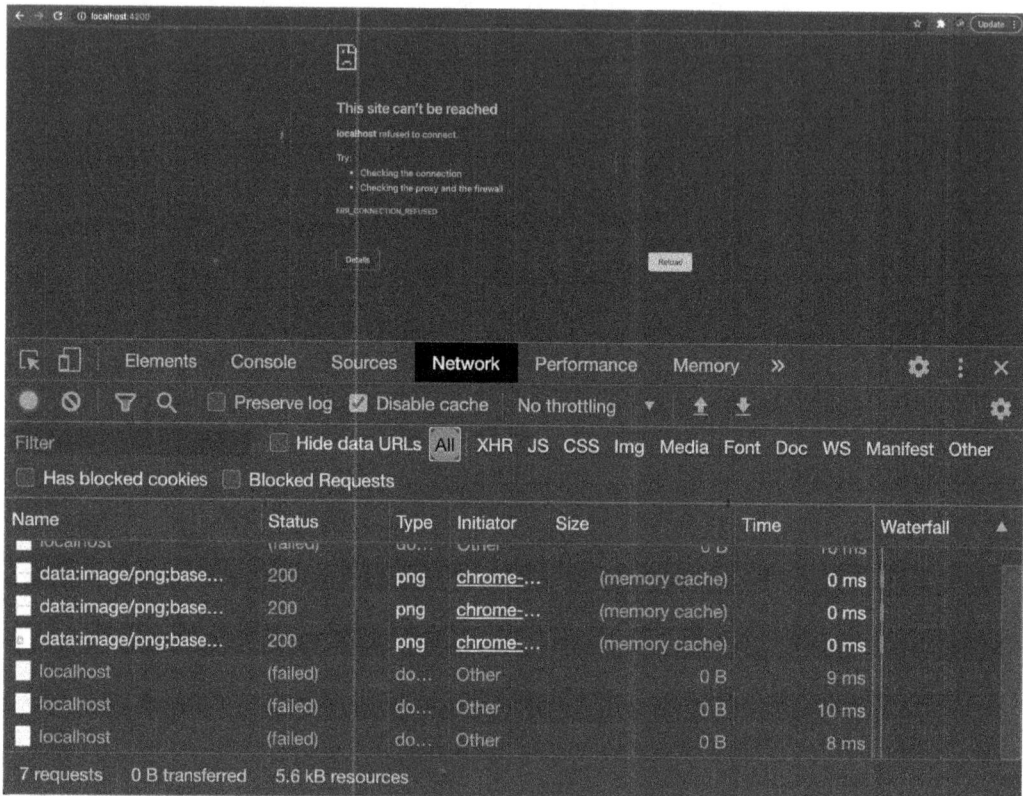

Figure 13.3 – App not working offline

3. To convert this app into a PWA, open a new terminal window/tab and make sure you're inside the `chapter13/start_here/angular-pwa-app` folder. Once inside, run the following command:

```
ng add @angular/pwa
```

You should see a bunch of files created and updated as the process from the command finishes.

4. Now build the app again by running `ng build --configuration production`. Once done, serve it using the `http-server dist/angular-pwa-app -p 4200` command.

5. Now make sure you have turned off throttling by switching to the **Network** tab and setting **No throttling** as the selection option, as shown in *Figure 13.4*. Also, notice that the **Disable cache** option is turned off:

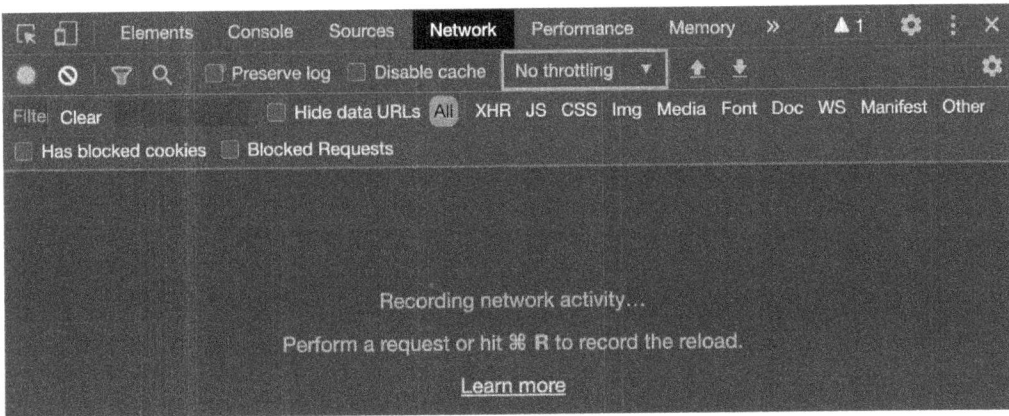

Figure 13.4 – Turning off network throttling

6. Now refresh the app once. You should see the app working and the network logs showing that assets such as JavaScript files were loaded from the server as shown in *Figure 13.5*:

Figure 13.5 – Assets downloaded from the source (Angular server)

7. Now refresh the app once again and you'll see that the same assets are now downloaded from the cache using the service worker, as shown in *Figure 13.6*:

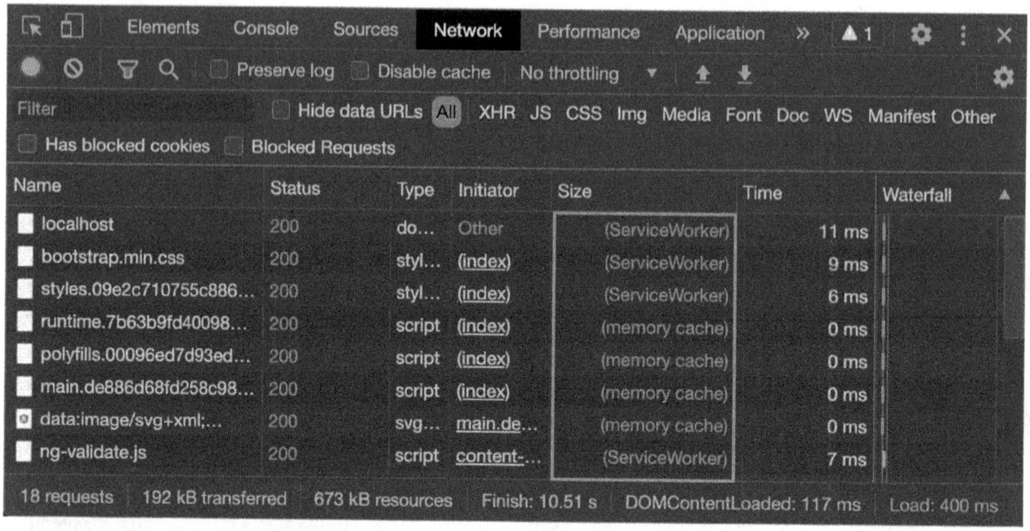

Figure 13.6 – Assets downloaded from the cache using the service worker

8. Now is the moment we've been waiting for. Change the network throttling back to **Offline** to go into the **Offline** mode and refresh the app. You should still see the app working in the **Offline** mode because of the service worker, as shown in *Figure 13.7*:

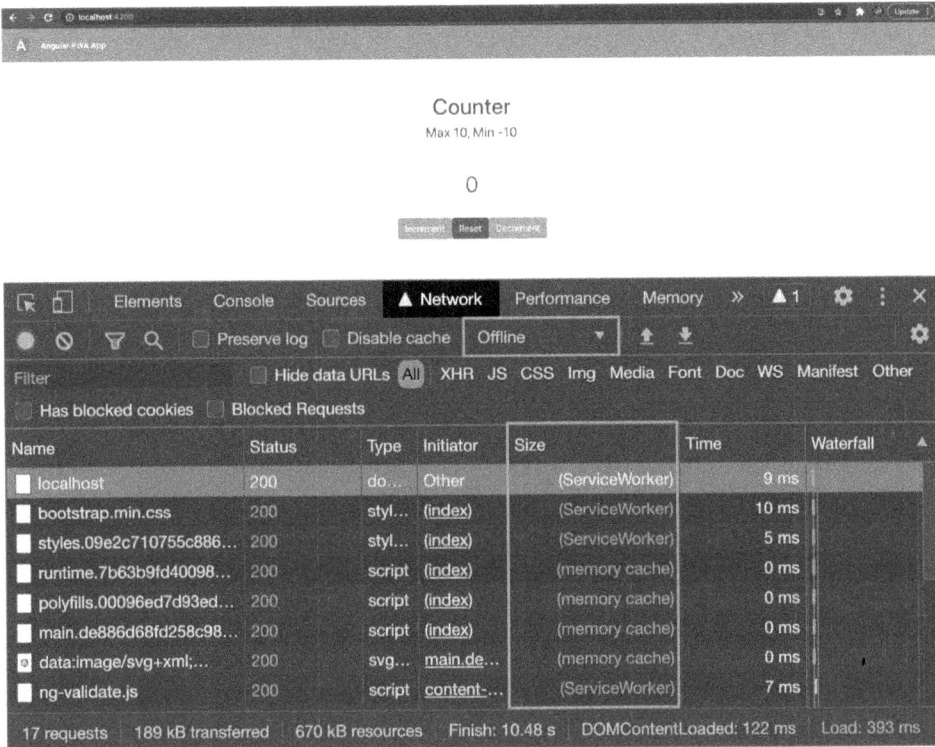

Figure 13.7 – Angular app working offline as a PWA using a service worker

9. What's more, you can actually install this PWA now on your machine. Since I'm using a MacBook, it is installed as a Mac app. If you're using Chrome, the installation option should be around the address bar, as shown in *Figure 13.8*:

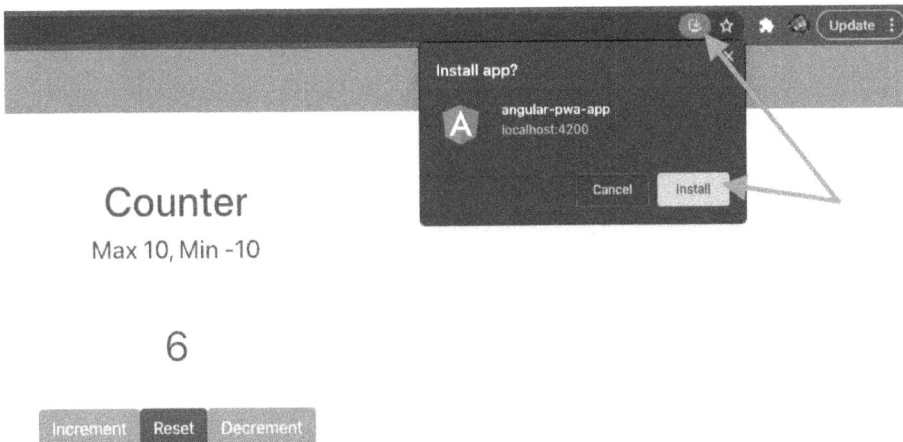

Figure 13.8 – Installing the Angular PWA from Chrome

Kaboom! Just by using the @angular/pwa package, with zero configuration done ourselves, we converted our existing Angular app into a PWA. We are now able to run our application offline, and we can install it as a PWA on our devices. See *Figure 13.9* to see how the app looks – just like a native app on macOS X:

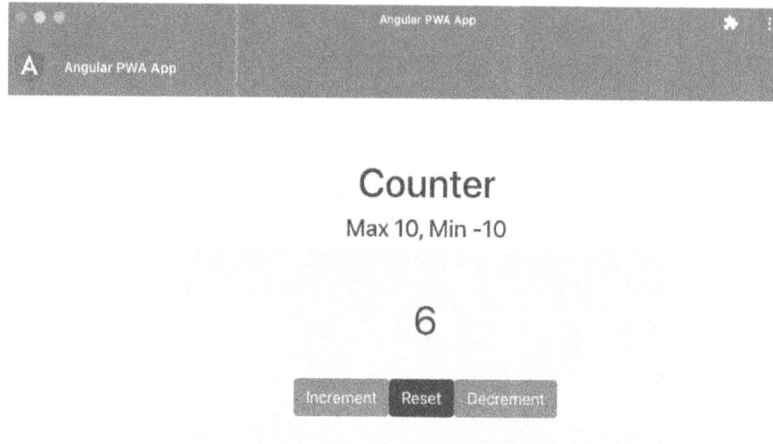

Figure 13.9 – How our Angular PWA looks as a native app on macOS X

Cool, right? Now that you know how to build a PWA with the Angular CLI, see the next section to understand how it works.

How it works

The Angular core team and the community have done an amazing job with the @angular/pwa package and, in general, with the ng add command, which allows us to add different packages to our applications using Angular schematics. In this recipe, when we run ng add @angular/pwa, it uses schematics to generate the app icons along with the web app manifest. If you look at the changed files, you can see the new files, as shown in *Figure 13.10*:

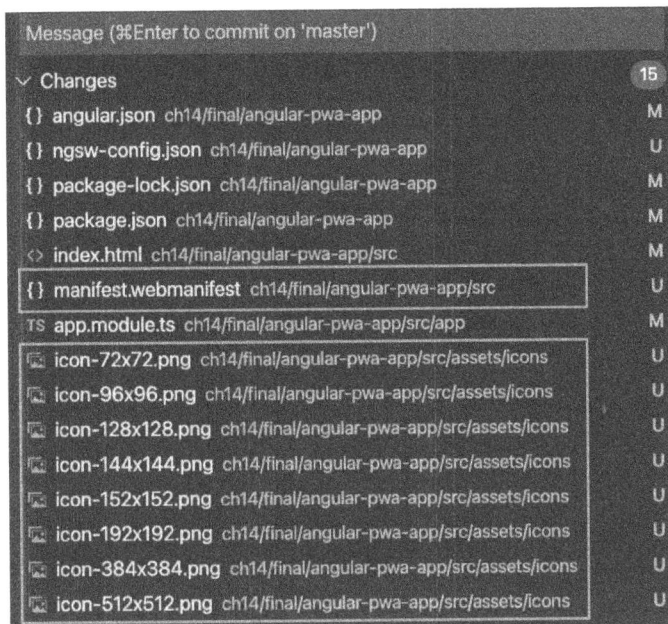

Figure 13.10 – Web manifest file and the app icon files

The manifest.webmanifest file is a file that contains a JSON object. This object defines the manifest for the PWA and contains some information. The information includes the name of the app, the short name, the theme color, and the configuration for different icons, for different devices. Imagine this PWA installed on your Android phone. You definitely need an icon in your home drawer to tap on the icon to open the app. This file holds the information regarding which icon to use based on different device sizes.

We also see the file ngsw-config.json, which contains the configuration for the service worker. Behind the scenes, while the ng add command is running the schematics, it also installs the @angular/service-worker package in our project. If you open the app.module.ts file, you'll see the code to register our service worker as follows:

```
...
import { ServiceWorkerModule } from '@angular/service-worker';
...
@NgModule({
  declarations: [AppComponent, CounterComponent],
  imports: [
    ...
    ServiceWorkerModule.register('ngsw-worker.js', {
```

```
        enabled: environment.production,
        // Register the ServiceWorker as soon as the app is
        stable
        // or after 30 seconds (whichever comes first).
        registrationStrategy: 'registerWhenStable:30000',
      }),
    ],
    ...
  })
}
export class AppModule {}
```

The code registers a new service worker file named `ngsw-worker.js`. This file uses the configuration from the `ngsw-config.json` file to decide which resource to cache and using which strategies.

Now that you know how the recipe works, see the next section for further reading.

See also

- Angular service worker intro (`https://angular.io/guide/service-worker-intro`)

- What are PWAs? (`https://web.dev/what-are-pwas/`)

Modifying the theme color for your PWA

In the previous recipe, we learned how to convert an Angular app into a PWA. And when we do so, the `@angular/pwa` package creates the web app manifest file with a default theme color, as shown in *Figure 13.9*. However, almost every web app has its own branding and style. And if you want to theme your PWA's title bar according to your branding, this is the recipe for you. We'll learn how to modify the web app manifest file to customize the PWA's theme color.

Getting ready

The project for this recipe resides in `chapter13/start_here/pwa-custom-theme-color`:

1. Open the project in Visual Studio Code.

2. Open the terminal and run `npm install` to install the dependencies of the project.

3. Once done, run `ng build --configuration production`.

4. Now run `http-server dist/pwa-custom-theme-color -p 5300` to serve it.

5. Open `localhost:5300` to view the application.

6. Finally, install the PWA as shown in *Figure 13.8*.

    If you open the PWA, it should look as follows:

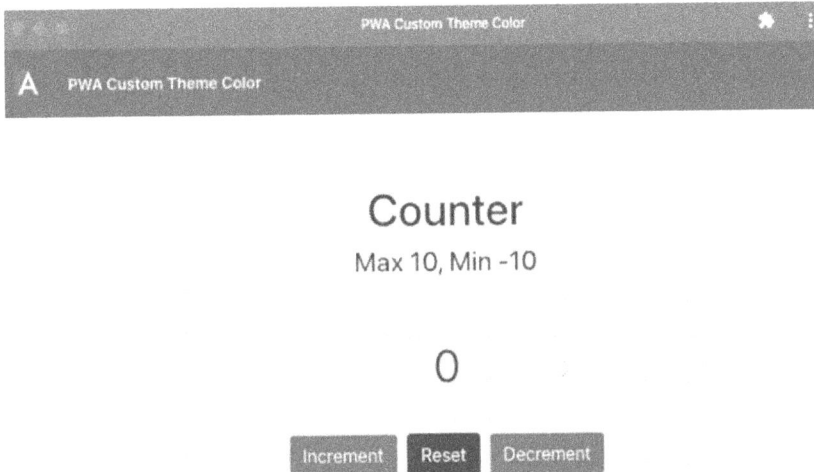

Figure 13.11 – PWA Custom Theme Color app

Now that we have the app running, let's see the steps of the recipe in the next section.

How to do it

As you can see in *Figure 13.11*, the header of the app has a bit of a different color than the app's native header (or toolbar). Due to this difference, the app looks a bit weird. We'll modify the web app manifest to update the theme color. Let's get started:

1. Open the `src/manifest.webmanifest` file in your editor and change the theme color as follows:

```
{
    "name": "pwa-custom-theme-color",
    "short_name": "pwa-custom-theme-color",
    "theme_color": "#8711fc",
```

```
    "background_color": "#fafafa",
    "display": "standalone",
    "scope": "./",
    "start_url": "./",
    "icons": [...]
}
```

2. We also have `theme-color` set in our `index.html` file. By default, that has precedence over the web app manifest file. Therefore, we need to update it. Open the `index.html` file and update it as follows:

```
<!DOCTYPE html>
<html lang="en">
  <head>
    ...
    <link rel="manifest" href="manifest.webmanifest" />
    <meta name="theme-color" content="#8711fc" />
  </head>
  <body>
    ...
  </body>
</html>
```

3. Now, build the app again using the `ng build --configuration production` command. Then serve it using `http-server` as follows:

```
http-server dist/pwa-custom-theme-color -p 5300
```

4. Open the PWA app again and uninstall it as shown in *Figure 13.12*. Make sure to check the box that says **Also clear data from Chrome (...)** when prompted:

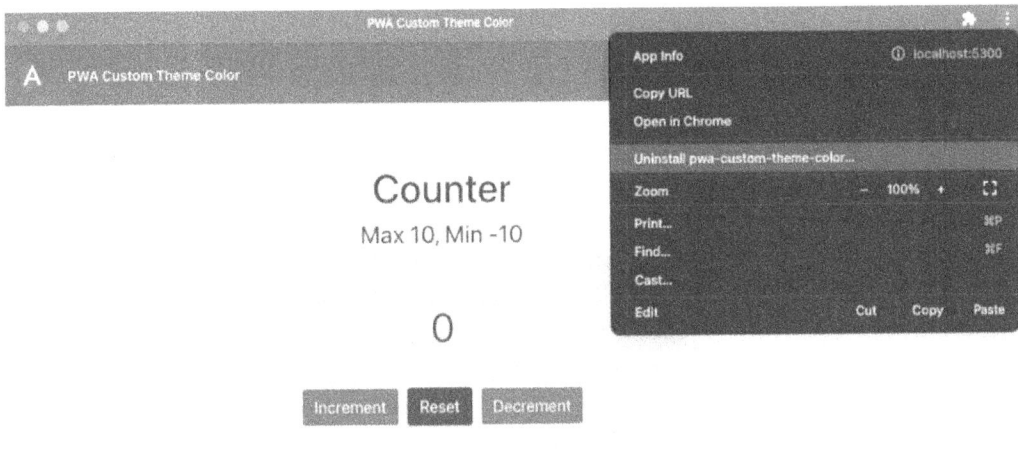

Figure 13.12 – Uninstalling the pwa-custom-theme-color app

5. Now open the Angular app in a new Chrome tab at `http://localhost:5300` and install the app again as a PWA as shown in *Figure 13.8*.

6. The PWA should already be opened. If not, open it from your applications and you should see the updated theme color as shown in *Figure 13.13*:

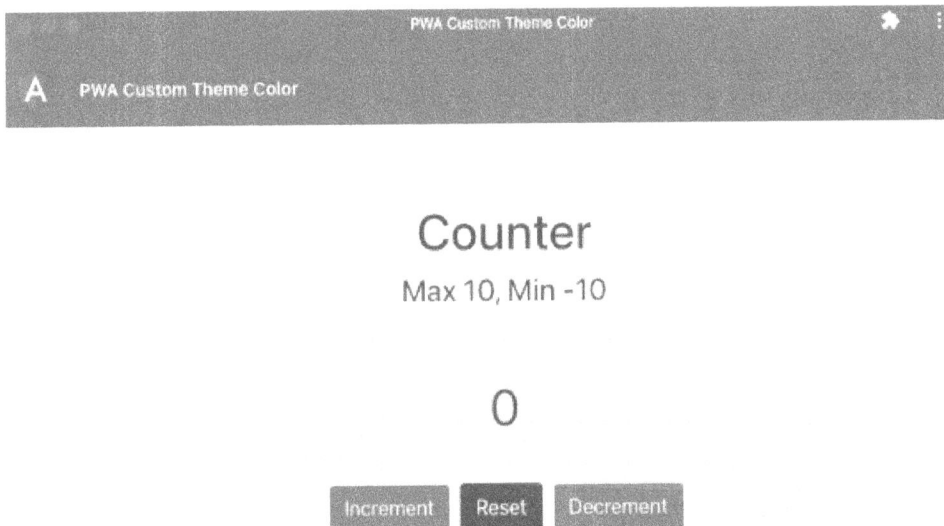

Figure 13.13 – PWA app with the updated theme color

Awesomesauce! You've just learned how to update the theme color for an Angular PWA. Now that you've finished the recipe, see the next section for further reading.

See also

- Creating a PWA with the Angular CLI (`https://web.dev/creating-pwa-with-angular-cli/`)

Using Dark Mode in your PWA

In the modern age of devices and applications, the preferences of end users have evolved a bit as well. With the increased usage of screens and devices, health is one of the major concerns. And we know that almost all screen devices now support dark mode. Considering this fact, if you're building a web app, you might want to provide dark mode support for it. And if it is a PWA that presents itself as a native app, the responsibility is much greater. In this recipe, you'll learn how to provide a dark mode for your Angular PWA.

Getting ready

The project for this recipe resides in `chapter13/start_here/pwa-dark-mode`:

1. Open the project in Visual Studio Code.

2. Open the terminal and run `npm install` to install the dependencies of the project.

3. Once done, run `ng build --configuration production`.

4. Now run `http-server dist/pwa-dark-mode -p 6100` to serve it.

5. Finally, install the PWA as shown in *Figure 13.8*

6. Now make sure you have the Dark theme enabled on your machine. If you're running macOS X, you can open **Settings | General** and select the **Dark** appearance as shown in *Figure 13.14*:

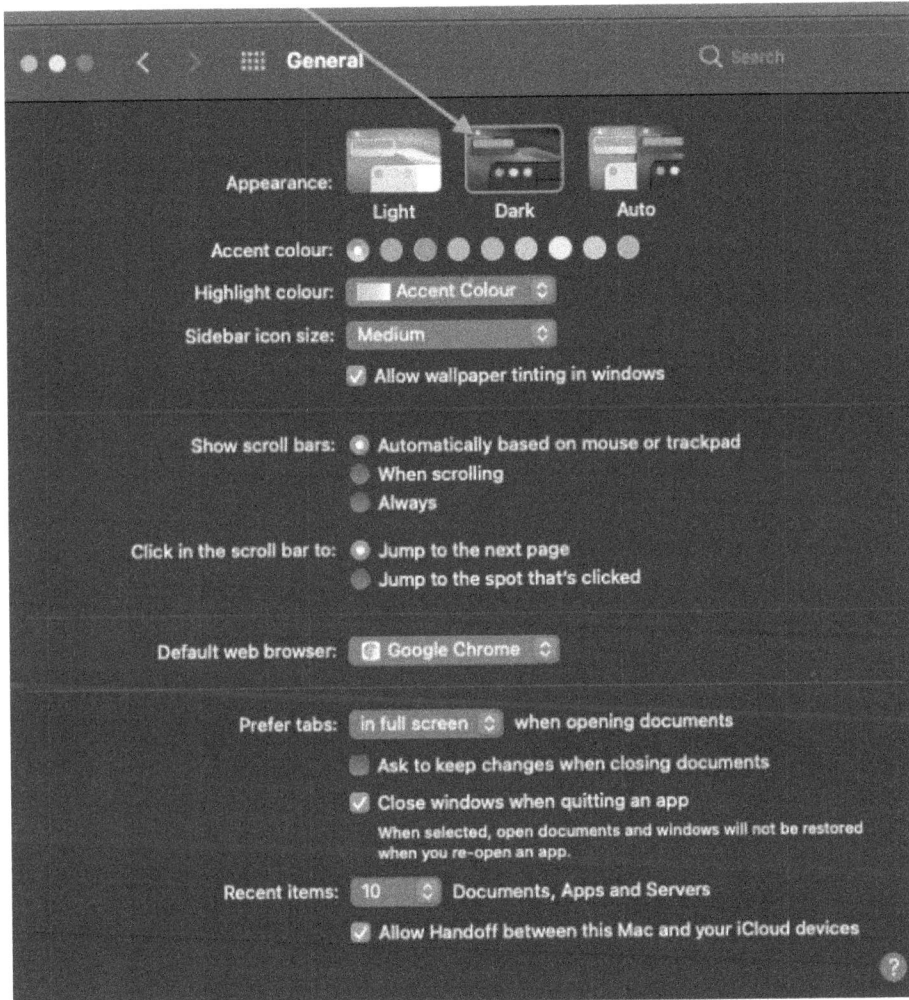

Figure 13.14 – Changing the system appearance to Dark mode in macOS X

7. Once done, open the PWA as the native app and you should see it as shown in *Figure 13.15*:

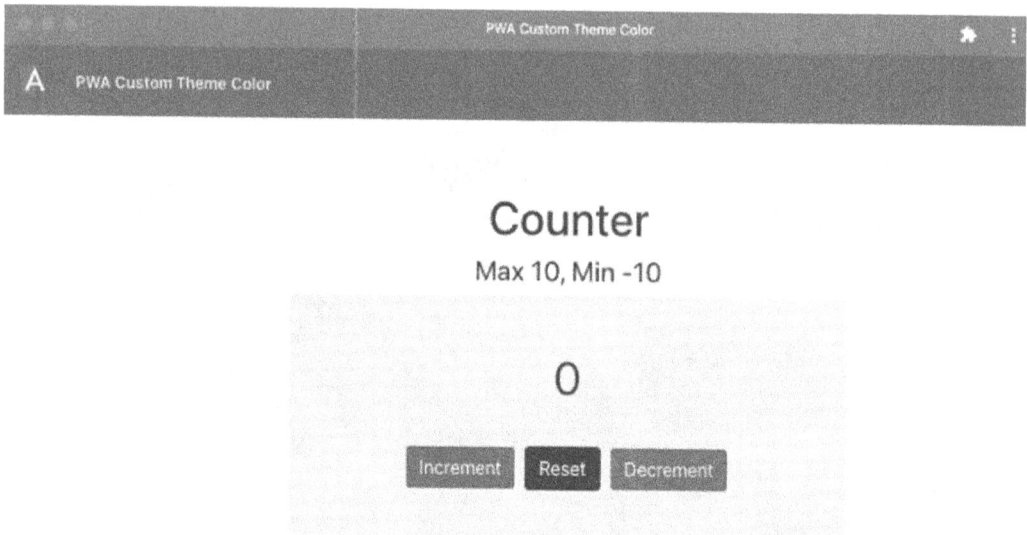

Figure 13.15 – PWA Custom Theme Color app in System Dark Mode appearance

Now that we have the PWA running as a native app, and the Dark mode applied to the system, let's see the steps of the recipe in the next section.

How to do it

As you can see, the Angular app doesn't have support for dark mode at the moment. We'll begin by serving the app in development mode and by adding different colors for dark mode. Let's get started:

1. Serve the app in development mode by running the command `ng serve -o --port 9291`.

    This should serve the app in a new browser tab at `http://localhost:4200`.

2. Now, open the `styles.scss` file to use the `prefers-color-scheme` media query. We'll use a different value for our global CSS variables to create a different view for dark mode. Update the file as follows:

```
/* You can add global styles to this file, and also
import other style files */


:root {...}
```

```
html,
body {...}

@media (prefers-color-scheme: dark) {
  :root {
    --main-bg: #333;
    --text-color: #fff;
    --card-bg: #000;
    --primary-btn-color: #fff;
    --primary-btn-text-color: #333;
  }
}
```

If you refresh the app again in the browser tab, you'll see a different dark mode view based on the `prefers-color-scheme` media query as shown in *Figure 13.16*:

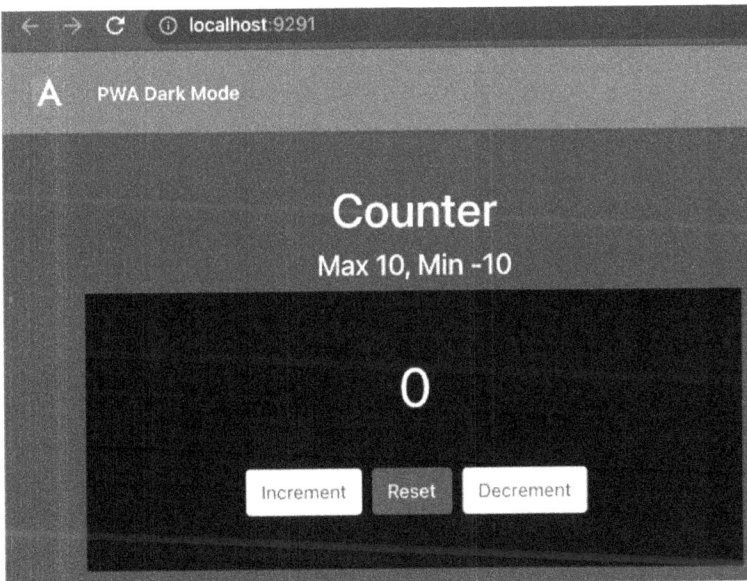

Figure 13.16 – The dark mode view using the prefers-color-scheme media query

> **Important note**
>
> It is possible that you already have run a PWA at `localhost:4200`; that is why in *step 1* we're targeting port `9291`. If even that has been used earlier, please make sure to clear the application cache and then refresh.

3. Let's simulate the dark and light modes using Chrome DevTools as it provides a really nice way to do so. Open Chrome DevTools and then open the **Command** menu. On macO SX, the keys are *Cmd + Shift + P*. On Windows, it is *Ctrl + Shift + P*. Then type `Render` and select the **Show Rendering** option as shown in *Figure 13.17*:

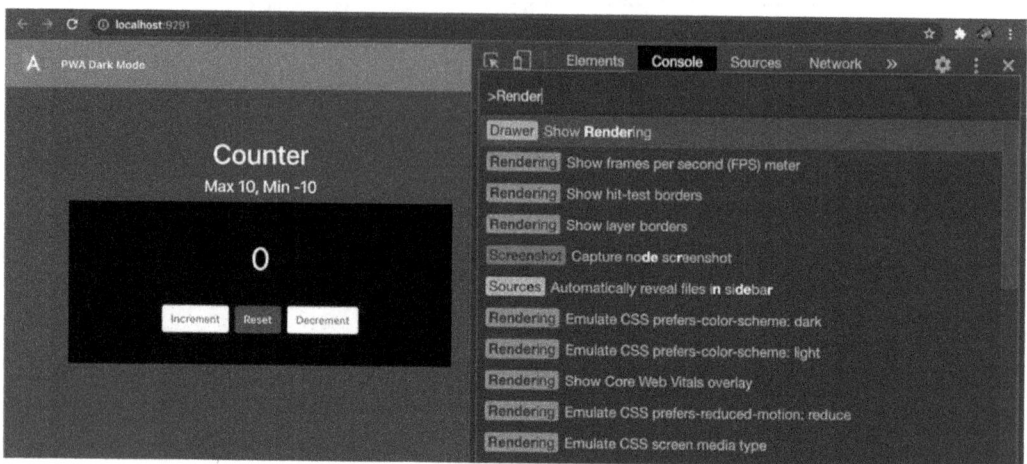

Figure 13.17 – Open the rendering view using the Show Rendering option

4. Now, in the **Rendering** tab, toggle the `prefers-color-scheme` emulation for light and dark modes as shown in *Figure 13.18*:

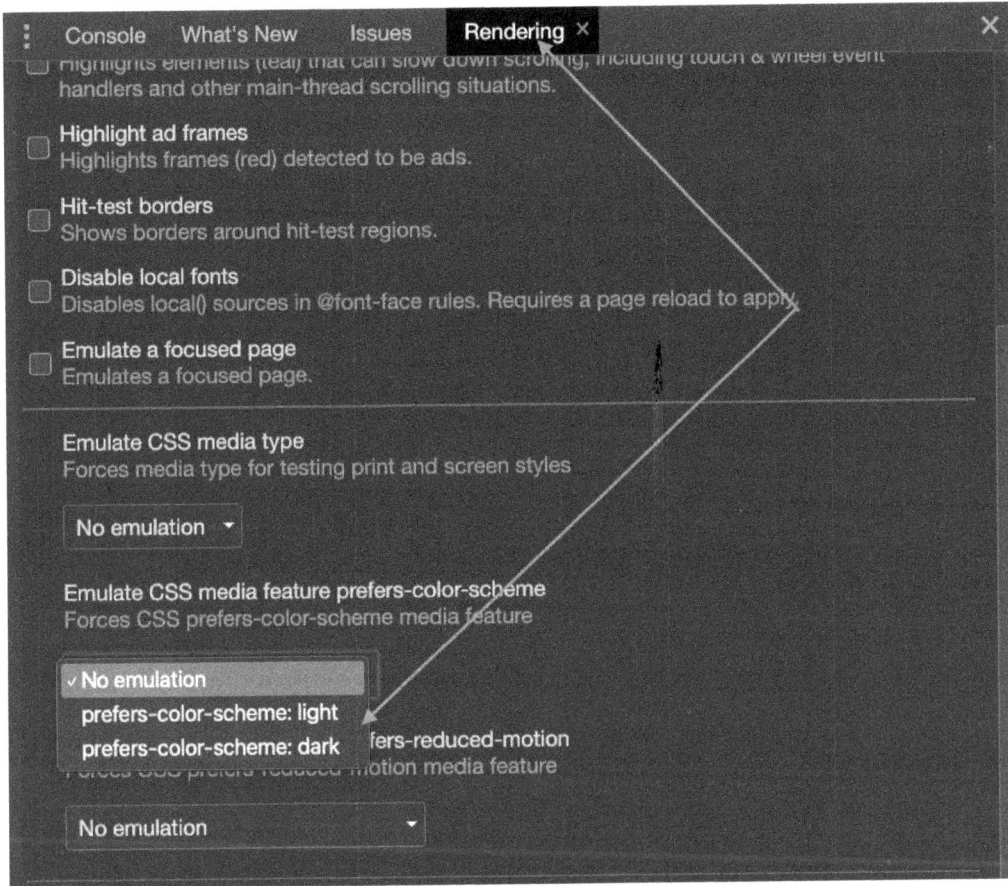

Figure 13.18 – Simulate prefers-color-scheme modes

5. Now that we've tested both modes. We can create the production build and re-install the PWA. Run the `ng build --configuration production` command to build the app in production mode.

6. Now uninstall the existing PWA by opening it and then selecting the **Uninstall** option from the **More** menu as shown in *Figure 13.12*. Make sure to check the box that says **Also clear data from Chrome (...)** when prompted.

7. Run the following command to serve the built app on the browser and then navigate to `http://localhost:6100`:

```
http-server dist/pwa-dark-mode -p 6100
```

8. Wait for a few seconds for the **Install** button to show up in the address bar. Then install the PWA similar to *Figure 13.8*.

9.   As soon as you run the PWA now, you should see the dark mode view as shown in *Figure 13.19*, if your system's appearance is set to dark mode:

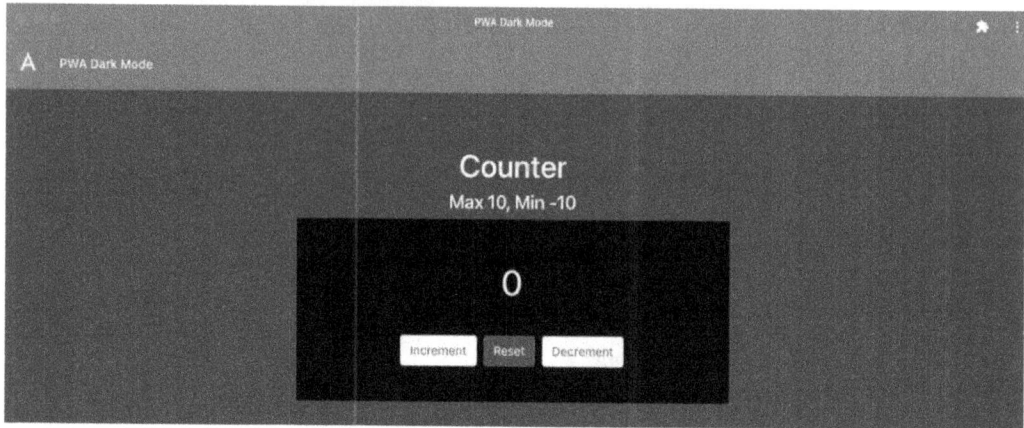

Figure 13.19 – Our PWA supporting dark mode out of the box

Awesome! If you switch your system appearance from dark mode to light mode or vice versa, you should see the PWA reflecting the appropriate colors. Now that you know how to support dark mode in your PWA, see the next section to see links for further reading.

See also

- Prefers color scheme (`https://web.dev/prefers-color-scheme/`)
- Using color scheme with prefers-color-scheme (`https://web.dev/color-scheme/`)

Providing a custom installable experience in your PWA

We know that PWAs are installable. This means they can be installed on your devices like a native application. However, when you first open the app in the browser, it totally depends on the browser how it shows the **Install** option. It varies from browser to browser. And it also might not be very prompt or clearly visible. And also, you might want to show the **Install** prompt at some point in the app instead of the app launch, which is annoying for some users. Luckily, we have a way to provide our own custom dialog/prompt for the installation option for our PWAs. And that is what we'll learn in this recipe.

Getting ready

The project for this recipe resides in `chapter13/start_here/pwa-custom-install-prompt`:

1. Open the project in Visual Studio Code.

2. Open the terminal and run `npm install` to install the dependencies of the project.

3. Once done, run `ng build --configuration production`.

4. Now run `http-server dist/pwa-custom-install-prompt -p 7200` to serve it.

5. Navigate to `http://localhost:7200`. Wait for a while and you should see the install prompt as shown in *Figure 13.20*:

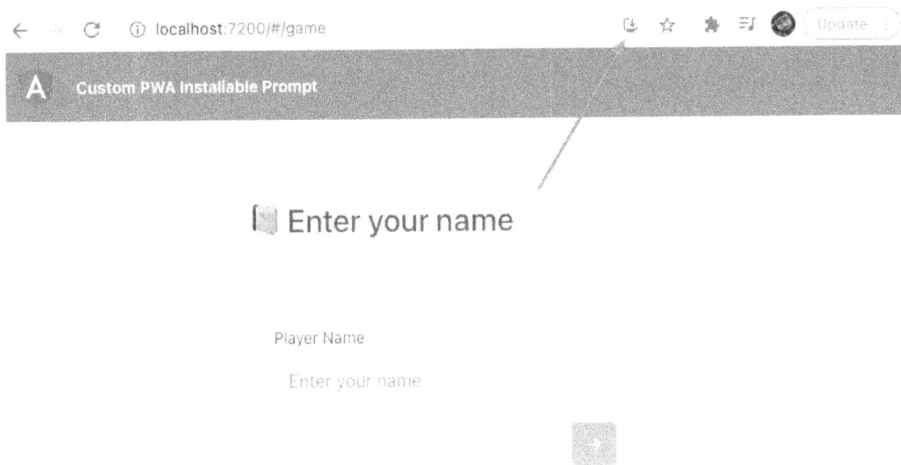

Figure 13.20 – pwa-custom-install-prompt running at http://localhost:7200

Now that we have the app running, let's see the steps of the recipe in the next section.

How to do it

We have the Dice Guesser application in which you roll the dice and guess the output. For this recipe, we'll prevent the default installation prompt and will show it only when the user has made a correct guess. Let's begin:

1. First of all, create a service that will show our custom installable prompt in the next steps. In the project root, run the following command:

```
ng g service core/services/installable-prompt
```

2. Now open the created file, `installable-prompt.service.ts`, and update the code as follows:

```
import { Injectable } from '@angular/core';

@Injectable({
  providedIn: 'root',
})
export class InstallablePromptService {
  installablePrompt;
  constructor() {
    this.init();
  }

  init() {
    window.addEventListener(
      'beforeinstallprompt',
      this.handleInstallPrompt.bind(this)
    );
  }

  handleInstallPrompt(e) {
    e.preventDefault();
    // Stash the event so it can be triggered later.
    this.installablePrompt = e;
    console.log('installable prompt event fired');
    window.removeEventListener('beforeinstallprompt',
    this.handleInstallPrompt);
  }

}
```

3. Now, let's build the custom dialog/prompt we'll show to the user. We're going to use the **Material** dialog from the `@angular/material` package that we already have installed in the project. Open the `app.module.ts` file and update it as follows:

```
...
import { MatDialogModule } from '@angular/material/
```

```
dialog';
import { MatButtonModule } from '@angular/material/
button';
@NgModule({
  declarations: [... ],
  imports: [

    ...

    BrowserAnimationsModule,
    MatDialogModule,
    MatButtonModule,

  ],
  providers: [],
  bootstrap: [AppComponent],
})
export class AppModule {}
```

4. Let's create a component for the **Material** dialog. In the project root, run the following command:

```
ng g component core/components/installable-prompt
```

5. We'll use this component in `InstallablePromptService` now. Open the `installable-prompt.service.ts` file and update the code as follows:

```
...
import { MatDialog } from '@angular/material/dialog';
import { InstallablePromptComponent } from '../
components/installable-prompt/installable-prompt.
component';
@Injectable({...})
export class InstallablePromptService {
  installablePrompt;
  constructor(private dialog: MatDialog) {...}
...

  async showPrompt() {
    if (!this.installablePrompt) {
      return;
```

```
    }
    const dialogRef = this.dialog.
    open(InstallablePromptComponent, {
      width: '300px',
    });
  }
}
```

6. We also need to show the browser's prompt based on our selection from our custom installable prompt. For example, if the user clicks the **Yes** button, it means they want to install the app as a PWA. In this case, we'll show the browser's prompt. Update the `installable-prompt.service.ts` file further as follows:

```
...
export class InstallablePromptService {

  ...

  async showPrompt() {



    ...

    const dialogRef = this.dialog.
    open(InstallablePromptComponent, {
      width: '300px',
    });
    dialogRef.afterClosed().subscribe(async (result) => {
      if (!result) {
        this.installablePrompt = null;
        return;
      }
      this.installablePrompt.prompt();
      const { outcome } = await this.installablePrompt.
      userChoice;
      console.log(`User response to the install prompt:
      ${outcome}`);
      this.installablePrompt = null;
    });
  }
}
```

7. Now that we have set up the main code for the browser's prompt. Let's work on the template of our custom installable prompt. Open the `installable-prompt.component.html` file and replace the template with the following code:

```
<h1 mat-dialog-title>Add to Home</h1>
<div mat-dialog-content>
  <p>Enjoying the game? Would you like to install the app
  on your device?</p>
</div>
<div mat-dialog-actions>
  <button mat-button [mat-dialog-close]="false">No
  Thanks</button>
  <button mat-button [mat-dialog-close]="true"
cdkFocusInitial>Sure</button>
</div>
```

8. Finally, let's show this prompt whenever the user makes a correct guess. Open the `game.component.ts` file and update it as follows:

```
. . .
import { InstallablePromptService } from '../core/
services/installable-prompt.service';
. . .
@Component({...})
export class GameComponent implements OnInit {
  . . .
  constructor(
    private leaderboardService: LeaderboardService,
    private instPrompt: InstallablePromptService
  ) {}
  . . .

  showResult(diceSide: IDiceSide) {
    . . .
    this.scores = this.leaderboardService.setScores({
      name: this.nameForm.get('name').value,
      score: 50,
    });
    this.instPrompt.showPrompt();
```

```
    }
}
```

9. Let's test the application now. Build the app in production mode and serve it using the `http-server` package on port `7200` by using the following commands:

```
ng build --configuration production
http-server dist/pwa-custom-install-prompt -p 7200
```

10. Before we test it out, you might want to clear the app's cache and unregister the service worker. You can do it by opening Chrome DevTools and navigating to the **Application** tab. Then click the **Clear site data** button as shown in *Figure 13.21*. Make sure the option **Unregister service workers** is checked:

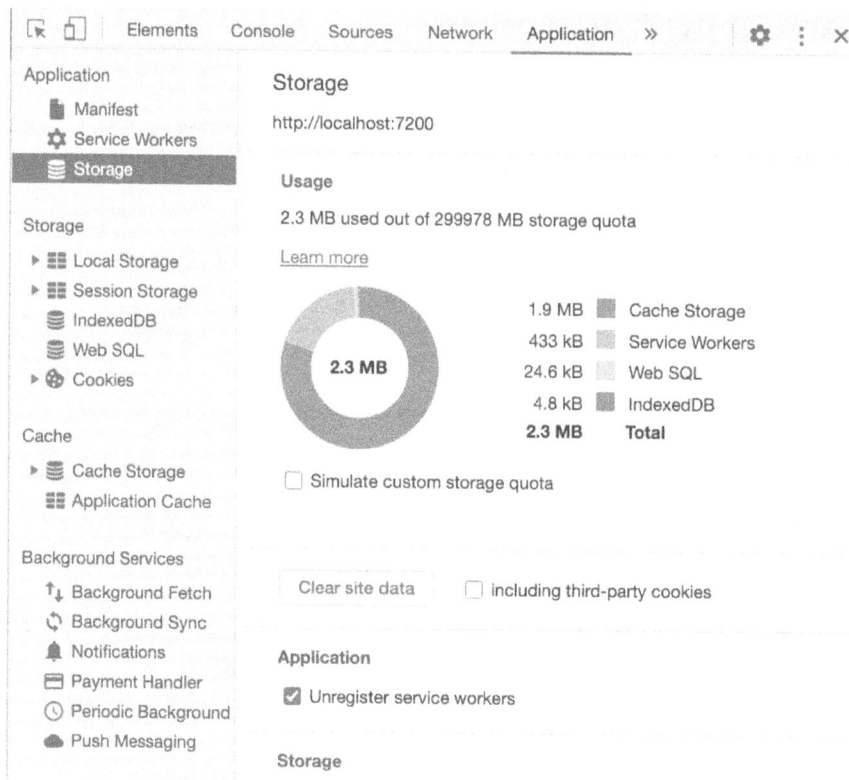

Figure 13.21 – Clearing site data including service workers

11. Now play the game until you guess one right answer. As soon as you get it, you'll see the custom installable prompt as shown in *Figure 13.22*. Click the **Sure** button and you should see the browser's prompt:

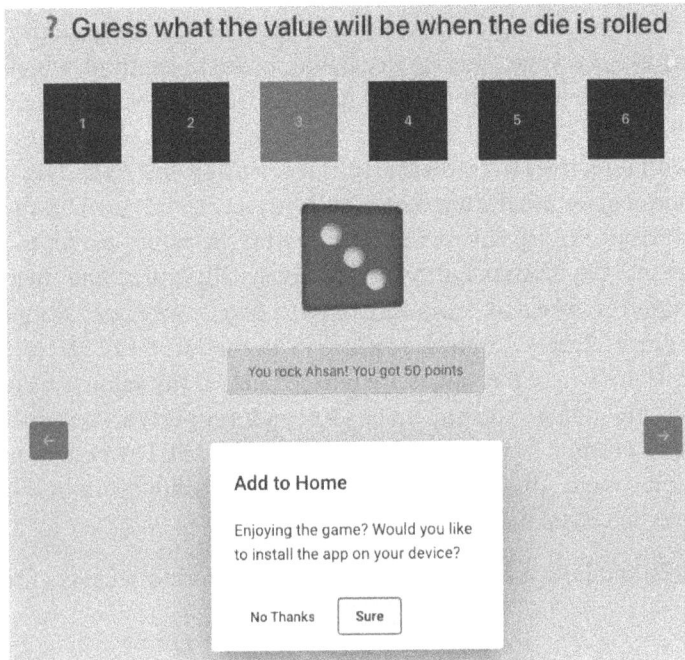

Figure 13.22 – Custom installable prompt for our PWA

Awesome! You can now play around with the app by installing and uninstalling the PWA a few times and trying out all the combinations of the user choosing to install or not to install the app. It's all fun and games. And now that you know how to implement a custom installation prompt for an Angular PWA, see the next section to understand how it works.

How it works

The heart of this recipe is the `beforeinstallprompt` event. It is a standard browser event that is supported in the latest version of Chrome, Firefox, Safari, Opera, UC Browser (Android version), and Samsung Internet, that is, almost all major browsers. The event has a `prompt()` method that shows the browser's default prompt on the device. In the recipe, we create `InstallablePromptService` and store the event in its `local` property. This is so we can use it later on-demand when the user has guessed a correct roll value. Note that as soon as we receive the `beforeinstallprompt` event, we remove the event listener from the `window` object so we only save the event once. That is when the app starts. And if the user chooses not to install the app, we don't show the prompt again within the same session. However, if the user refreshes the app, they will still get the prompt one time for the first correct guess. We could go one step further to save this state in `localStorage` to avoid showing the prompt after the page refreshes, but that's not a part of this recipe.

For the custom installation prompt, we use the `MatDialog` service from the `@angular/material` package. This service has an `open()` method, which takes two parameters: the component to show as a Material dialog and `MatDialogConfig`. In the recipe, we create the `InstallablePromptComponent`, which uses some HTML elements with directives from the `@angular/material/dialog` package. Note that on the buttons, we use the attribute `[mat-dialog-close]` in the `installable-prompt.component.html` file. And the values are set to `true` and `false` for the **Sure** and **No Thanks** buttons respectively. These attributes help us send the respective value from this modal to `InstallablePromptService`. Notice the usage of `dialogRef.afterClosed().subscribe()` in the `installable-prompt.service.ts` file. That's where the values are passed back. If the value is `true`, then we use the event, that is, the `this.installablePrompt` property's `.prompt()` method to show the browser's prompt. Note that we set the `installablePrompt` property's value to `null` after its usage. This is so we don't show the prompt again in the same session until the user refreshes the page.

Now that you understand how it all works, see the next section to see links for further reading.

See also

- Angular Material Dialog examples (`https://material.angular.io/components/dialog/examples`)

- MatDialogConfig (`https://material.angular.io/components/dialog/api#MatDialogConfig`)

- How to provide your own app-install experience (web.dev) (`https://web.dev/customize-install/`)

Precaching requests using an Angular service worker

With the addition of service workers in our previous recipes, we've seen that they already cache the assets and serve them using the service worker if we go into **Offline** mode. But what about network requests? If the user goes offline and refreshes the application right now, the network requests fail because they're not cached with the service worker. This results in a broken offline user experience. In this recipe, we'll configure the service worker to precache network requests so the app works fluently in **Offline** mode as well.

Getting ready

The project that we are going to work with resides in `chapter13/start_here/`
`precaching-requests` inside the cloned repository:

1. Open the project in Visual Studio Code.

2. Once done, run `ng build --configuration production`.

3. Now run `http-server dist/precaching-requests -p 8300` to serve it.

4. Navigate to `http://localhost:8300`. Refresh the app once. Then switch to
   Offline mode as shown in *Figure 13.2*. If you go to the **Network** tab and filter the
   requests using the query `results`, you should see that the requests fail as shown
   in *Figure 13.23*:

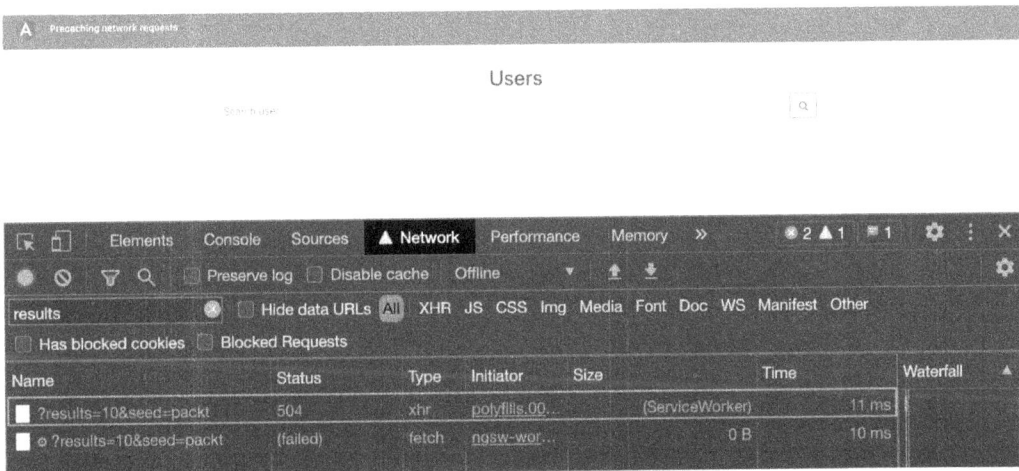

Figure 13.23 – Offline experience broken due to not caching the network request

Now that we see the network requests failing, lets see the steps of the recipe in the next
section to fix this.

How to do it

For this recipe, we have the users list and search app that fetches some users from an API endpoint. As you can see in *Figure 13.23*, if we go into **Offline** mode, the fetch call fails as well as the call for the request to the service worker. This is because the service worker isn't configured yet to cache the data request. Let's get started with the recipe to fix this:

1. In order to cache the network requests, open the ngsw-config.json file and update it as follows:

```
{
    "$schema": "./node_modules/@angular/service-worker/config/schema.json",
    "index": "/index.html",
    "assetGroups": [...],
    "dataGroups": [
        {
            "name": "api_randomuser.me",
            "urls": ["https://api.randomuser.me/?results*"],
            "cacheConfig": {
                "strategy": "freshness",
                "maxSize": 100,
                "maxAge": "2d"
            }
        }
    ]
};
```

2. Let's test the application now. Build the app in production mode and serve it using the http-server package on port 8300 by using the following commands:

```
ng build --configuration production
http-server dist/precaching-requests -p 8300
```

3. Now navigate to http://localhost:8300. Make sure you're not using **Network throttling** at this moment. That is, you are not in **Offline** mode.

4. Clear the app data using Chrome DevTools as shown in *Figure 13.21*. Once done, refresh the app page.

5. In Chrome DevTools, go to the **Network** tab and switch to **Offline** mode as shown in *Figure 13.2*. Now filter the network requests using the query `results`. You should see the results despite being offline. And the network call is served from the service worker as shown in *Figure 13.24*:

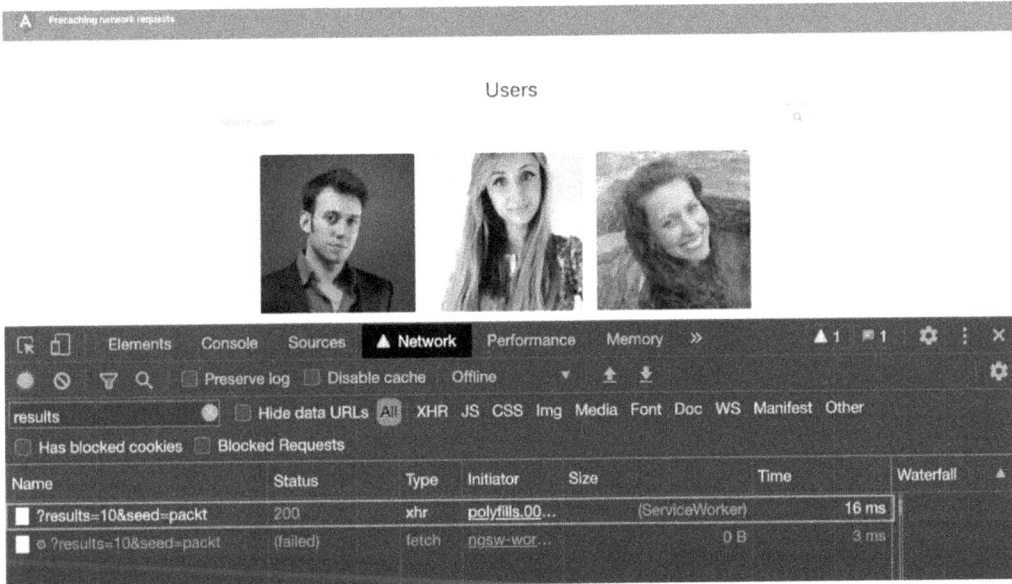

Figure 13.24 – Network call working offline using the service worker

And boom! Even if you click a card now, you should still see the app working flawlessly because all the pages use the same API call, hence, served from the service worker. And with that, you've just learned how to configure a service worker in an Angular app to cache network/data requests. And you can install the PWA and use it even if you're offline. Awesome! Right?

Now that we've finished the recipe, let's see in the next section how it all works.

How it works

The heart of this recipe is the `ngsw-config.json` file. This file is used by the `@angular/service-worker` package when generating the service worker file. The file already contains a JSON object out of the box when we use the `@angular/pwa` schematics by running `ng add @angular/pwa`. This JSON contains a property called `assetGroups`, which basically configures the caching of the assets based on the provided configuration. For this recipe, we wanted to cache network requests along with the assets. Therefore, we added the new property `dataGroups` in the JSON object. Let's have a look at the configuration:

```
"dataGroups": [
    {
        "name": "api_randomuser.me",
        "urls": ["https://api.randomuser.me/?results*"],
        "cacheConfig": {
            "strategy": "freshness",
            "maxSize": 100,
            "maxAge": "2d"
        }
    }
]
```

As you can see, `dataGroups` is an array. We can provide different configuration objects as elements to it. Each configuration has a `name`, an array of `urls`, and a `cacheConfig` that defines the caching strategy. For our configuration, we use a wildcard with the API URL, that is, we use `urls: ["https://api.randomuser.me/?results*"]`. For the `cacheConfig`, we're using the `"freshness"` strategy, which means the app will always fetch the data from its origin first. If the network is unavailable, then it will use the response from the service worker cache. An alternate strategy is `"performance"`, which first looks up the service worker for a cached response. If there's nothing in the cache for the particular URL (or URLs), then it fetches the data from the actual origin. The `maxSize` property defines how many requests can be cached for the same pattern (or set of URLs). And the `maxAge` property defines how long the cached data would live in the service worker cache.

Now that you know how the recipe works, see the next section for links for further reading.

See also

- Angular Service Worker Intro (`https://angular.io/guide/service-worker-intro`)

- Angular Service Worker Config (`https://angular.io/guide/service-worker-config`)

- Creating an offline fallback page (web.dev) (`https://web.dev/offline-fallback-page/`)

Creating an App Shell for your PWA

When it comes to building fast user experiences for web apps, one of the major challenges is minimizing the critical rendering path. This includes loading the most critical resources for the target page, parsing and executing JavaScript, and so on. With an App Shell, we have the ability to render a page, or a portion of the app, at build time rather than runtime. This means the user will see the pre-rendered content initially, until JavaScript and Angular kick in. This means the browser doesn't have to work and wait a while for the first meaningful paint. In this recipe, you'll create an App Shell for an Angular PWA.

Getting ready

The project that we are going to work with resides in `chapter13/start_here/pwa-app-shell` inside the cloned repository:

1. Open the project in Visual Studio Code.

2. Open the terminal and run `npm install` to install the dependencies of the project.

3. Once done, run `ng serve -o`.

This should open a tab and run the app at `http://localhost:4200` as shown in *Figure 13.25*:

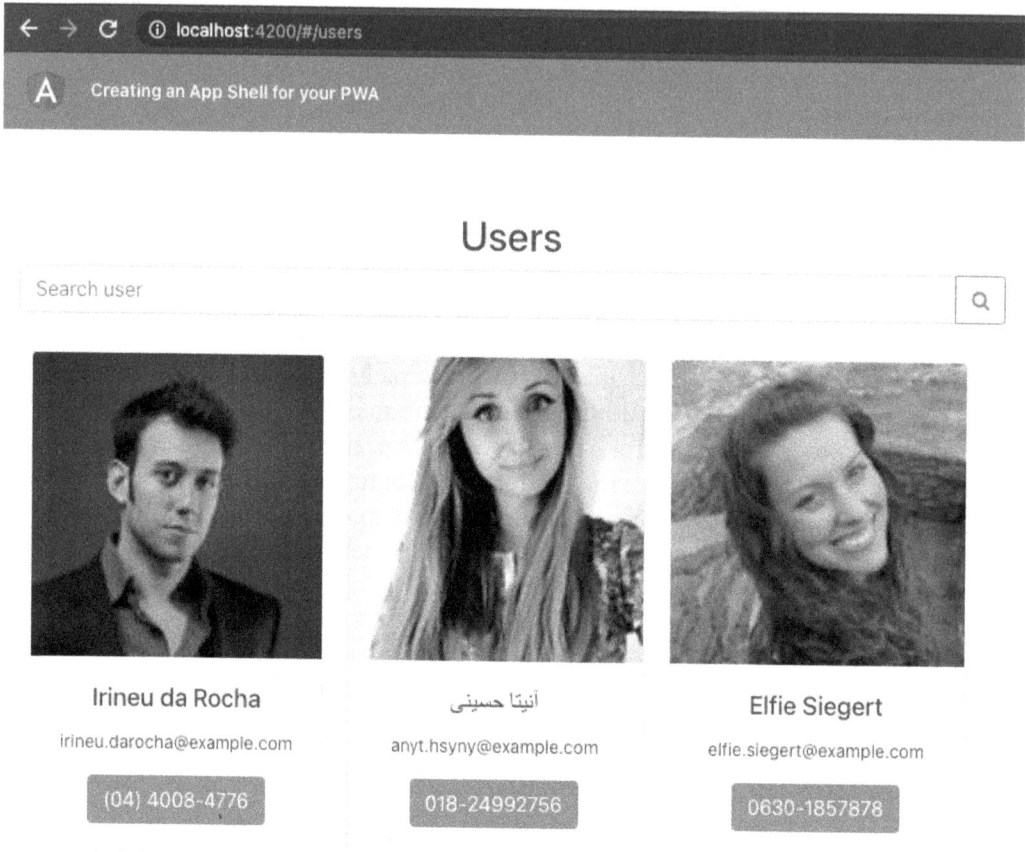

Users

Figure 13.25 – pwa-app-shell running on http://localhost:4200

Now we'll disable JavaScript to simulate taking a lot of time to parse JavaScript. Or, to simulate that there's no App Shell in place yet. Open Chrome DevTools and open the command panel. The shortcut is *Cmd + Shift + P* on macOS X and *Ctrl + Shift + P* on Windows. Type `Disable JavaScript`, select the option, and hit *Enter*. You should see the message that follows:

Please enable JavaScript to continue using this application.

Figure 13.26 – No App Shell present in the app

Now that we have checked the absence of the App Shell, let's see the steps of the recipe in the next section.

How to do it

We have an Angular application that fetches some users from an API. We will create an App shell for this app so it can provide the first meaningful paint faster as a PWA. Let's get started:

1. First, create the App Shell for the app by running the following command from the project root:

    ```
    ng generate app-shell
    ```

2. Update app.module.ts to export the components so we can use them to render the **Users** page in the App Shell. The code should look as follows:

    ```
    ...
    @NgModule({
        declarations: [...],
        imports: [... ],
        providers: [],
        exports: [
            UsersComponent,
            UserCardComponent,
            UserDetailComponent,
            AppFooterComponent,
            LoaderComponent,
        ],
        bootstrap: [AppComponent],
    })
    export class AppModule {}
    ```

3. Now open the app-shell.component.html file and use the <app-users> element so we render the whole UsersComponent in the App Shell. The code should look as follows:

    ```
    <app-users></app-users>
    ```

4. Now that we have the code written for the App Shell. Let's create it. Run the following command to generate the App Shell in development mode:

```
ng run pwa-app-shell:app-shell
```

5. Once the App Shell is generated in *Step 4*, run the following command to serve it using the `http-server` package:

```
http-server dist/pwa-app-shell/browser -p 4200
```

6. Make sure that the JavaScript is still turned off for the app. If not, open Chrome DevTools and press *Cmd + Shift + P* for macOS X to open the Command Panel (*Ctrl + Shift + P* on Windows). Then type `Disable Javascript` and hit *Enter* selecting the option as shown in *Figure 13.27*:

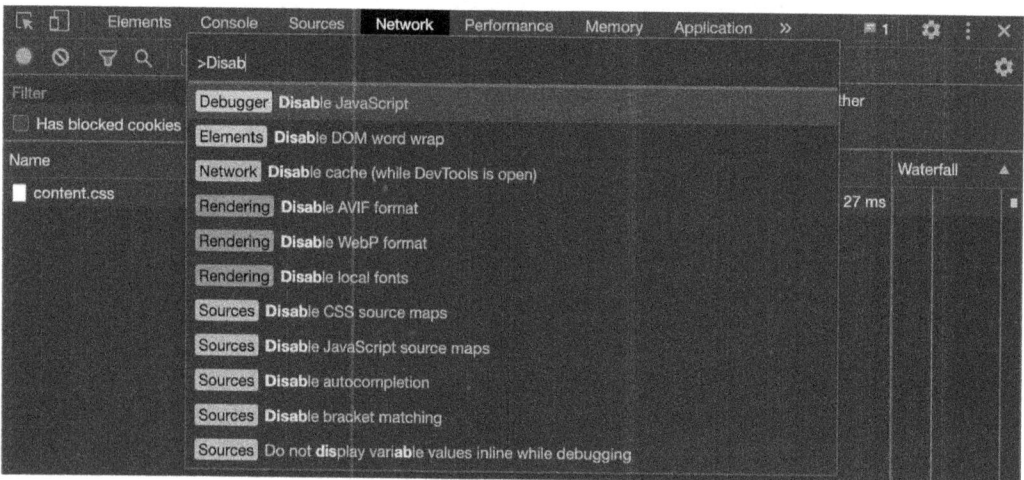

Figure 13.27 – Disable JavaScript using Chrome DevTools

7. Refresh the app while JavaScript is disabled. You should now see the app still showing the pre-rendered users page, despite JavaScript being disabled as shown in *Figure 13.28*. Woohoo!

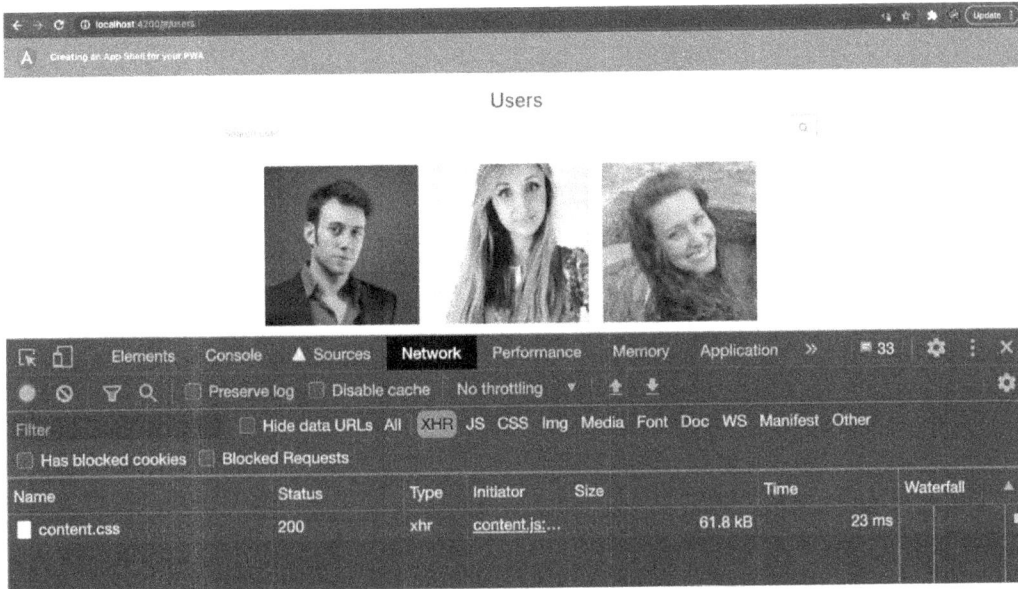

Figure 13.28 – App Shell showing the pre-rendered Users page

8. To verify that we are pre-rendering the users page at build time, inspect the generated code at `<project-root>/dist/pwa-app-shell/browser.index.html`. You should see the entire rendered page inside the `<body>` tag as shown in *Figure 13.29*:

Figure 13.29 – index.html file containing the pre-rendered Users page

9. Create the production build with the App Shell and serve it on port `1020` by running the following commands:

```
ng run pwa-app-shell:app-shell:production
http-server dist/pwa-app-shell/browser -p 1020
```

10. Navigate to `http://localhost:1020` in your browser and install the app as a PWA as shown in *Figure 13.8*. Once done, run the PWA and it should look as follows:

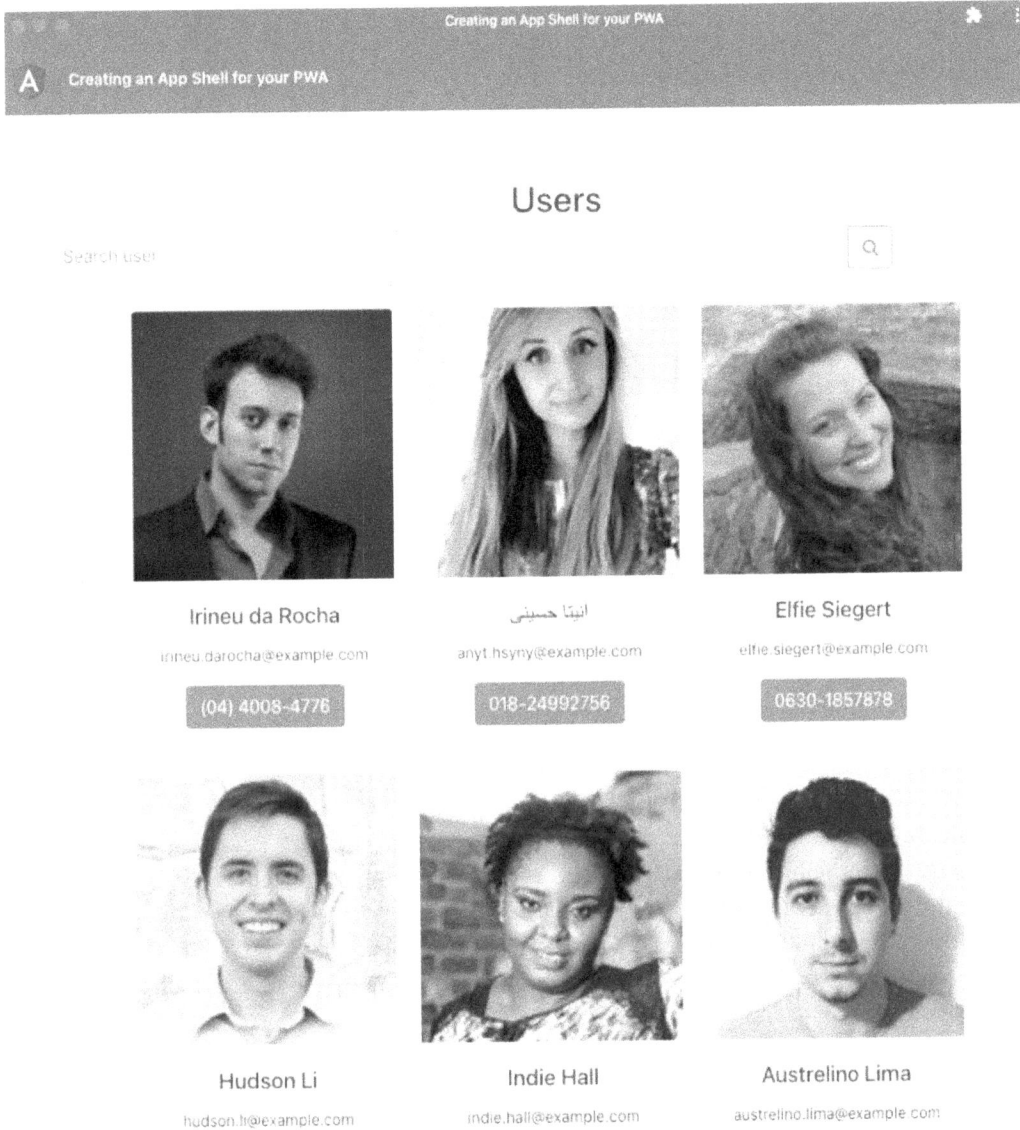

Figure 13.30 – pwa-app-shell running as a native app after installation

Great!!! You now know how to create an App Shell for your Angular PWAs. Now that you've finished the recipe, see the next section on how it works.

How it works

The recipe begins with disabling JavaScript for our application. This means when the app runs, we only show static HTML and CSS since there's no JavaScript execution. We see a message about JavaScript not being supported, as shown in *Figure 13.26*.

We then run the `ng generate app-shell` command. This Angular CLI command does the following things for us:

- Creates a new component named `AppShellComponent` and generates its relevant files.

- Installs the `@angular/platform-server` package in the project.

- Updates the `app.module.ts` file to use the `BrowserModule.withServerTransition()` method so we can provide the `appId` property for server-side rendering.

- Adds some new files, namely `main.server.ts` and `app.server.module.ts`, to enable server-side rendering (build time rendering for our App Shell, to be exact).

- Most importantly, it updates the `angular.json` file to add a bunch of schematics for server-side rendering as well as for generating the `app-shell`.

In the recipe, we export the components from `AppModule` so we can use them in the App Shell. This is because the App Shell is not part of the `AppModule`. Instead, it is part of the newly created `AppServerModule` in the `app.server.module.ts` file. As you can see, in the file, we have `AppModule` already being imported. Although, we can't use the components unless we export them from `AppModule`. After exporting the components, we update the `app-shell.component.html` (the App Shell template) to use the `<app-users>` selector, which reflects the `UsersComponent` class. That is the entire Users page.

We verify the App Shell by running the `ng run pwa-app-shell:app-shell` command. This command generates an Angular build in development mode with the App Shell (non-minified code). Note that in a usual build, we would generate the `pwa-app-shell` folder inside the `dist` folder. And inside, we would have `index.html`. However, in this case, we create two folders inside the `pwa-app-shell` folder, that is, the `browser` folder and the `server` folder. And our `index.html` resides in the `browser` folder. As shown in *Figure 13.29*, we have the code of the entire Users page inside the `<body>` tag in the `index.html` file. This code is pre-rendered at build time. This means Angular opens up the app, makes the network call, and then pre-renders the UI as the App Shell at build time. So as soon as the app opens, the content is pre-rendered.

To generate the production build with the App Shell, we run the `ng run pwa-app-shell:app-shell:production` command. This generates the production Angular build with minified code for the App Shell as well. And finally, we install the PWA to test it out.

Now that you know how the recipe works, see the next section for links for further reading.

See also

- Angular App Shell Guide (`https://angular.io/guide/app-shell`)
- The App Shell Model (Web Fundamentals by Google) (`https://developers.google.com/web/fundamentals/architecture/app-shell`)

Packt>

Other Books You May Enjoy

If you enjoyed this book, you may be interested in these other books by Packt:

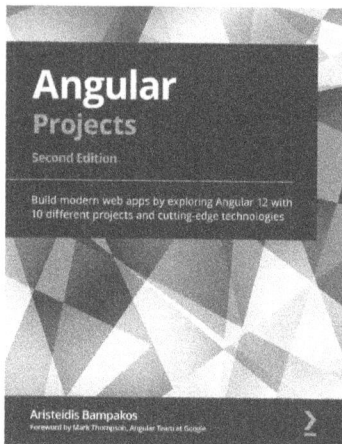

Angular Projects - Second Edition

Aristeidis Bampakos

ISBN: 978-1-80020-526-0

- Set up Angular applications using Angular CLI and Nx Console
- Create a personal blog with Jamstack and SPA techniques
- Build desktop applications with Angular and Electron
- Enhance user experience (UX) in offline mode with PWA techniques
- Make web pages SEO-friendly with server-side rendering

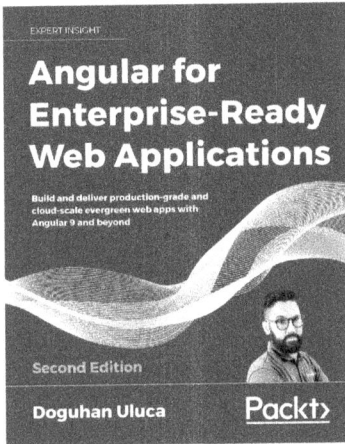

Angular for Enterprise-Ready Web Applications - Second Edition

Doguhan Uluca

ISBN: 978-1-83864-880-0

- Adopt a minimalist, value-first approach to delivering web apps
- Master Angular development fundamentals, RxJS, CLI tools, GitHub, and Docker
- Discover the flux pattern and NgRx
- Implement a RESTful APIs using Node.js, Express.js, and MongoDB
- Create secure and efficient web apps for any cloud provider or your own servers
- Deploy your app on highly available cloud infrastructure using DevOps, CircleCI, and AWS

Packt is searching for authors like you

If you're interested in becoming an author for Packt, please visit `authors.packtpub.com` and apply today. We have worked with thousands of developers and tech professionals, just like you, to help them share their insight with the global tech community. You can make a general application, apply for a specific hot topic that we are recruiting an author for, or submit your own idea.

Hi!

I am Muhammad Ahsan Ayaz, author of *Angular Cookbook*. I really hope you enjoyed reading this book and found it useful for increasing your productivity and efficiency in Angular.

It would really help me (and other potential readers!) if you could leave a review on Amazon sharing your thoughts on *Angular Cookbook*.

Go to the link below or scan the QR code to leave your review:

`https://packt.link/r/1838989439`

Your review will help me to understand what's worked well in this book, and what could be improved upon for future editions, so it really is appreciated.

Best Wishes,

Index

www.ingramcontent.com/pod-product-compliance
Lightning Source LLC
Chambersburg PA
CBHW082115210326
41600CD00001B/377